A DIFFERENT APPROACH TO COSMOLOGY

This is a different kind of book about cosmology, a field of major interest to professional astronomers, physicists, and the general public. All research in cosmology adopts one model of the universe, the hot big-bang model. But Fred Hoyle, Geoffrey Burbidge and Jayant Narlikar take a different approach. Starting with the beginnings of modern cosmology, they then conduct a wide ranging and deep review of the observations made from 1945 to the present day. Here they challenge many conventional interpretations. The latter part of the book presents the authors' own account of the present status of observations and how they should be explained. The controversial theme is that the dependency on the hot big-bang model has led to an unwarranted rejection of alternative cosmological models. Writing from the heart, with passion and punch, these three cosmologists make a powerful case for viewing the universe in a different light.

Sir Fred Hoyle held the Plumian Chair of Astronomy at the University of Cambridge 1958–72. Geoffrey Burbidge is at the Center for Astrophysics and Space Sciences, University of California, San Diego. He has held positions at the California Institute of Technology, the University of Chicago, and Harvard University and was Director of the Kitt Peak National Observatory 1978–84. Jayant Narlikar is with the Inter-University Centre for Astronomy and Astrophysics, Pune, India.

A DIFFERENT APPROACH TO COSMOLOGY

From a static universe through the big bang towards reality

F. HOYLE

G. BURBIDGE

Center for Astrophysics and Space Sciences, University of California, San Diego

AND

J. V. NARLIKAR

Inter-University Centre for Astronomy and Astrophysics, Pune

CAMBRIDGE
UNIVERSITY PRESS

PUBLISHED BY THE PRESS SYNDICATE OF THE UNIVERSITY OF CAMBRIDGE
The Pitt Building, Trumpington Street, Cambridge, United Kingdom

CAMBRIDGE UNIVERSITY PRESS
The Edinburgh Building, Cambridge CB2 2RU, UK   www.cup.cam.ac.uk
40 West 20th Street, New York, NY 10011-4211, USA   www.cup.org
10 Stamford Road, Oakleigh, Melbourne 3166, Australia
Ruiz de Alarcón 13, 28014 Madrid, Spain

First published 2000
Reprinted 2000

Printed in the United Kingdom at the University Press, Cambridge

Typeface Times 11/13pt. 3B2 [KW]

A catalogue record for this book is available from the British Library

Library of Congress Cataloging in Publication data

Hoyle, Fred, Sir.
    A different approach to cosmology: from a static universe through
the big bang towards reality / F. Hoyle, G. Burbidge and
J.V. Narlikar.
      p. cm.
    Includes bibliographical references.
    ISBN 0 521 66223 0
    1. Cosmology. I. Burbidge, Geoffrey, R. II. Narlikar, Jayant
Vishnu, 1938– . III. Title.
QB981.H754 2000
523.1–dc21   99-15821 CIP

ISBN 0 521 66223 0 hardback

Contents

Preface

This is a different kind of book about cosmology. In the past 10 or 15 years, there has been a great spate of textbooks, monographs and popular books on this subject. There is no question but that cosmology has developed into a major topic of study for astronomers, physicists and laymen alike. The reasons for this are not hard to find. We are all keenly interested in where we came from and where we are going, and cosmology, the study of the whole universe, is supposed to give us the answers to these basic questions. In this sense it has much the same attraction for many as does religion. Both have been dominated in the past by a small number of facts and large measures of belief. Some genuine progress in cosmology has come, however, mostly from advances in observational astronomy leading to new information being obtained at an increasing rate in the past 40 years.

These advances have nearly all come from advances in technology at optical wavelengths, and investigations of the universe at radio, infrared, and ultraviolet frequencies, and at X-ray and γ-ray energies. Naturally this has led to a huge increase in the number of scientists in the field, and to much larger sums of money being required. This is inevitable, but it has a downside. For example, physicists with no background in astronomy have moved into cosmology in large numbers. However, without knowledge of what has gone before, there is always a lack of perspective, and this shows up very strongly when any attempt is made to consider alternative explanations of what are often considered to be well-established facts.

Also, astrophysics and cosmology is now big science, which requires large sums of money to support it. The major sources of funds are National Government Agencies, such as the National Science Foundation and the National Aeronautics and Space Administration in the USA.

The agencies providing the money then make a considerable impact on the way we do science. For example, if you are funded by NASA, you are expected to engage in a great deal of publicity and media hype. A fact of life which is well known to the professionals, but which has probably not yet been digested by the world at large, is that the effort and propaganda required to get a major instrument funded – say the Hubble Space Telescope, or the Cosmic Background Explorer Satellite (COBE) – requires that before the instruments can be built, extravagant claims must be made about what we shall find. Not surprisingly, when these instruments eventually work, we get a succession of even stronger claims to the effect that what was expected **has indeed been found**. Some of the claims are amazing even to those who work in the field, e.g. about the microwave background radiation and its fluctuations:

> 'They have found the Holy Grail of Cosmology', Dr Michael Turner, University of Chicago.
> 'Well, if you are a religious person, it's like seeing the face of God', Dr George Smoot, University of California, Berkeley.

In an era when serious scientists can take such an approach, there is no room for the discovery of phenomena which are not already expected. Those few scientists who want to look for things that they may not be able to explain simply cannot get telescope time. The people who assign observing time on the telescopes using the so-called peer review system demand that you explain beforehand what you will find. And those who do find such phenomena by accident mostly can be relied on to be so conformist that they put them aside. And, of course, if you know little of the history of science, and particularly if you are a particle physicist, you are confident that there is nothing to learn from astronomy about fundamental physics.

For many reasons, and these are only some of them, all the books on cosmology written in the present era focus on one model of the universe – the currently popular hot big-bang model. While they may sometimes pay lip service to alternatives, anyone who reads any of these books can be left in no doubt as to the truth, as it is seen by the prestigious author.

In this monograph, we use a different approach. We start at the true beginning of the modern era – around 1914 – and give an historical account of how the subject has progressed. In the middle chapters we discuss the sequence of events from about 1945 to about 1965, and the newer observational results and the developments of the past 30 years. Much of what has gone on has not simply been due to advances in observation and theory. It has clearly been affected by the attitudes, personalities, prejudices and beliefs of the scientists who made the advances. Since we have often been involved personally in these advances (and retreats) alongside many who are, or have been, personal friends, our account of the past 40 years or more of history will reflect our own beliefs and prejudices as well as those of our colleagues.

In the latter part of the book we give a full account of our own ideas. Then we turn to a review of the present observational situation and then a summary of the current state of affairs. Since we believe that there are many fundamental questions which remain unanswered, in the last chapter we describe several of the major unsolved problems.

Acknowledgements

This book has been written by three authors who normally live on three different continents. Thus while much of the work has been done by telephone and fax, a good deal of travel has been involved.

In England we wish to thank Barbara Hoyle and Nicola Hoyle who have helped in many ways in Bournemouth. We also wish to thank Alex Boksenberg and Jasper Wall for hospitality afforded to us at the Royal Greenwich Observatory in the summers of 1996 and 1997.

In India we have been given much help by the staff at IUCAA. Also two of us have been given much hospitality at Pune when we have been able to visit on several occasions.

In the USA we have received much assistance from several people in La Jolla. In particular, Betty Travell at the Center for Astrophysics and Space Science who typed the whole of the manuscript in several drafts in 1996 and 1997. Without her the book would never have been completed. We also want to thank Kathy Steffen at CASS who has drawn or redrawn the bulk of the diagrams and the figures, Del Hewitt who made many of them, and Kate Ericson who helped us extensively in 1998. Margaret Burbidge has contributed in many ways in the latter stage and has done much editing and proofreading.

We also wish to thank many astronomers for providing us with illustrations and diagrams. In particular we wish to thank Chip Arp from the Max-Planck-Institut für Astrophysik in Munich for providing material and lots of enthusiasm. We thank John Bahcall, from the Institute for Advanced Study, John Huchra from the Harvard-Smithsonian Astrophysical Observatory, Nigel Sharp from Kitt Peak National Observatory, Allan Sandage and John Bedke from the Carnegie Observatories and Space Telescope Science Institute, Arthur Wolfe and Lisa Storrie-Lombardi from UCSD, and Charles Bennett from NASA/Goddard for providing us with many remarkable illustrations.

1

Introduction

Both Newton and Einstein thought that the universe must be static on the large scale and also homogeneous and isotropic. However, until early in the twentieth century there was no understanding of the scale of the universe nor any evidence that anything lay beyond the Milky Way. By the latter part of the nineteenth century, it was known that there were two kinds of nebulae – spiral structure had first been detected in one type in Messier 51, by Lord Rosse in 1855, and there was some speculation that such systems might be very distant Milky Way like objects. However by 1905, in her well-known popular book *The System of the Stars*, Agnes Clerke[1], basing her remarks on the results and conclusions of professional astronomers, stated with some confidence:

> The question whether nebulae are external galaxies hardly any longer needs discussion. It has been answered by the progress of research. No competent thinker, with the whole of the available evidence before him, can now, it is safe to say, maintain any single nebula to be a star system of co-ordinate rank with the Milky Way. A practical certainty has been attained that the entire contents, stellar and nebula, of the sphere belong to one mighty aggregation, and stand in ordered mutual relations within the limits of one all embracing scheme.

It would be as well, whenever cosmologists meet nowadays to discuss their certainty over the state of the universe, that they should recall these words of Agnes Clerke, who was widely regarded in her time as a gifted expositor of the views of the then-accepted trade union of competent thinkers.

Ten years after Agnes Clerke's summary, in 1915, Einstein[2] formulated his equations of gravitation and soon felt the need to apply them to the universe as a whole. Since the universe is the largest system of matter imaginable and also in reality, he considered that a theory of gravitation should say how this system would behave. Newton's attempts to solve a similar problem had failed: a model of a static universe in stable equilibrium had not been possible.

In his pioneering paper on cosmology, Einstein devoted the initial part to the Newtonian problem. His requirements on a realistic model led him to postulate that the gravitational potential of matter should tend to a constant at spatial infinity. This led him to a spherically symmetric distribution with the density tapering off at infinity faster than the inverse square of distance from the centre of symmetry. By applying Boltzmann's law of gas distribution to his matter, Einstein was then able to argue that the density had to be zero everywhere, if it was to vanish at infinity.

He then showed that if he modified the Newtonian force law with the addition of a cosmological term, adding a force of repulsion in proportion to distance, the above difficulty disappeared. Taking his cue from the Newtonian problem he then went over to his theory of relativity and found that, and we quote again:

> The conclusion I shall arrive at is that the field equations of gravitation which I have championed hitherto still need a slight modification, so that on the basis of the general theory of relativity those fundamental difficulties are avoided which have been set forth as confronting the Newtonian theory.

In other words, his basic considerations were entirely theoretical and not motivated by what the real universe might be like. When one inserts the cosmological term into the field equations of general relativity, and uses the assumptions of homogeneity and isotropy in addition to the assumption of a static universe, the following conclusions emerge unequivocally.

  (i) To get a universe of constant positive density, the cosmological constant must be positive (i.e., it must correspond to a repulsive force).
  (ii) The universe must be closed and have a constant positive curvature.

Einstein admitted in his paper that in the actual universe 'the curvature of space is variable in time and place, according to the distribution of matter, but we may roughly approximate to it by means of a spherical space'.

The idea of a finite and closed universe in which the density of matter determines the radius of curvature of the spherical space was entirely in keeping with Einstein's expectations that the matter–energy distribution should determine the spacetime geometry. Being then under the influence of Ernst Mach, he considered this a demonstration of Mach's ideas and hoped that his solution would turn out to provide a unique and hence an authentic picture of the universe.

In concluding his paper Einstein underscored his priorities for a cosmological model:

2

Early relativistic cosmology

As was just mentioned, like Newton, Einstein also thought that the universe is static on the large scale. Like Newton's attempted model of the universe, Einstein's universe was also imagined to be a homogeneous and isotropic distribution of matter. These criteria have been followed by most model makers in cosmology. We briefly discuss the implications of these assumptions.

The adjectives *homogeneous* and *isotropic* imply the following characteristics. Suppose that, viewed on the large scale, the universe looks the same from all vantage points. There is *no* preferred observing position in the universe; all positions are alike. This is the property of *homogeneity*. Furthermore, as we observe the universe from any such vantage point, should we notice any differences in the structure of the universe as we look in different directions? If we do not notice any directional differences, then we say that the universe is *isotropic*. In other words, if you are taken blindfolded from one spot to another in a homogeneous and isotropic universe, after removing your blindfold you cannot tell where you are or in what direction you are looking.

Even with these simplifying assumptions about the large-scale structure of the universe, the *quantitative* details were still lacking in Einstein's model. To determine these details, Einstein needed his theory of gravitation – the general theory of relativity.

The geometry of spacetime is different from Euclid's geometry in the neighborhood of a massive object like the Sun. It is the main feature of general relativity that any distribution of matter (and energy) should affect the geometry of spacetime around it. For example, the geometry of spacetime is different from Euclid's in the neighborhood of a massive object like the Sun. Einstein therefore expected that the distribution of matter (in the form of stars, galaxies, etc.) should determine the geometry of the large-scale structure of the universe. But here he encountered a major difficulty.

The equations of general relativity, as obtained by Einstein in 1915, permitted models of the universe that were homogeneous and isotropic but *not static*. This difficulty is in fact no different from that which had troubled Newton two centuries earlier. How can matter remain stationary in spite of its self-gravity?

To appreciate this difficulty within the framework of general relativity we need to begin with Einstein's equations

$$R_k^i - \frac{1}{2} g_k^i R = -\frac{8\pi G}{c^4} T_k^i \tag{2.1}$$

as obtained by him in his 1915 formulation[1]. The left-hand side of these equations contains tensors describing the geometry of spacetime whereas the right-hand side has the energy–momentum tensor for the physical contents of the universe[a]. These equations tell us in quantitative terms how the physical contents of the universe determine its geometrical structure.

The assumptions of homogeneity and isotropy tell us that space has constant curvature which Einstein assumed to be positive. We will discuss the motivation for this supposition later. Such a space is finite but unbounded, being the hypersurface of a sphere in four dimensions, expressed, say, by the cartesian coordinate relation

$$x_1^2 + x_2^2 + x_3^2 + x_4^2 = S^2. \tag{2.2}$$

To use coordinates intrinsic to the surface define

$$x_4 = S\cos\chi, \quad x_1 = S\sin\chi\cos\theta, \quad x_2 = S\sin\chi\sin\theta\cos\phi, \tag{2.3}$$

$$x_3 = S\sin\chi\sin\theta\sin\phi.$$

The spatial line element *on* the surface S is then given by

$$d\sigma^2 = (dx_1)^2 + (dx_2)^2 + (dx_3)^2 + (dx_4)^2 \tag{2.4}$$

$$= S^2\left[d\chi^2 + \sin^2\chi(d\theta^2 + \sin^2\theta d\phi^2)\right].$$

The ranges mathematically permitted of θ, ϕ and χ are given by

$$0 \le \chi \le \pi, \quad 0 \le \theta \le \pi, \quad 0 \le \phi < 2\pi. \tag{2.5}$$

But there are two geometrical alternatives open to us. The first is that χ takes the entire range $0 \le \chi \le \pi$, and this gives us what is commonly known

---

[a] We will assume that the reader is familiar with general relativity. For one who is not, the above discussion can be found in a number of elementary texts on the subject. In particular we follow the treatment given by Narlikar[2].

as *spherical space*. If, however, we identify antipodal points of this sphere, its connectivity is changed and the space is then known as *elliptical space*.

Another way to express $d\sigma^2$ is through coordinates r, θ, ϕ, with $r = \sin \chi \, (0 \leq r \leq 1)$. In elliptical space r runs through this range once; in spherical space it does so twice. In either case the spatial line element (2.4) takes the form

$$d\sigma^2 = S^2 \left[\frac{dr^2}{1 - r^2} + r^2 (d\theta^2 + \sin^2 \theta d\phi^2) \right]. \tag{2.6}$$

The constant S is called the 'radius' of the universe. The spacetime line element for the Einstein universe is therefore given by

$$ds^2 = c^2 dt^2 - d\sigma^2 \tag{2.7}$$

$$= c^2 dt^2 - S^2 [d\chi^2 + \sin^2 \chi (d\theta^2 + \sin^2 \theta d\phi^2)]$$

$$= c^2 dt^2 - S^2 \left[\frac{dr^2}{1 - r^2} + r^2 (d\theta^2 + \sin^2 \theta d\phi^2) \right].$$

Note that we have derived the line element entirely from the various assumptions of symmetry. The field equations have not yet been used. We will now see what happens when we substitute the above line element into the left-hand side of Einstein's equations. We get, with S independent of time for a static universe,

$$R_0^0 - \frac{1}{2} R = -\frac{3}{S^2} \tag{2.8}$$

$$R_1^1 - \frac{1}{2} R = R_2^2 - \frac{1}{2} R = R_3^3 - \frac{1}{2} R = -\frac{1}{S^2}. \tag{2.9}$$

To complete the field equations, Einstein used the energy tensor for dust at rest and of uniform density ρ_0 in the above frame of reference, which gives

$$T_0^0 = \rho_0 c^2 \tag{2.10}$$

$$T_1^1 = T_2^2 = T_3^3 = 0.$$

This leads to two independent equations:

$$-\frac{3}{S^2} = -\frac{8\pi G}{c^2} \rho_0, \qquad -\frac{1}{S^2} = 0. \tag{2.11}$$

Clearly no sensible solution is possible from these equations, thus suggesting that no static homogeneous isotropic model of the universe with $\rho_0 > 0$ is possible under the Einstein equations.

It was this inability to generate such a model that led Einstein to modify his equations to

$$R_k^i - \frac{1}{2} g_k^i R + \lambda g_k^i = -\frac{8\pi G}{c^4} T_k^i. \tag{2.12}$$

If we introduce this additional constant λ into the picture, our equations in (2.11) are modified to

$$\lambda - \frac{3}{S^2} = -\frac{8\pi G}{c^2} \rho_0 \tag{2.13}$$

and

$$\lambda - \frac{1}{S^2} = 0. \tag{2.14}$$

We now do have a sensible solution. We get

$$S = \sqrt{\frac{1}{\lambda}} = \frac{c}{2\sqrt{\pi G \rho_0}}. \tag{2.15}$$

Einstein considered this solution as justifying his conjecture that with sufficiently high density it should be possible to 'close' the universe. In (2.15) we have the radius S of the universe as given by the matter density ρ_0, with the result that the larger the value of ρ_0, the smaller is the value of S. However, if λ is a given universal constant like G, both ρ_0 and S are determined in terms of λ (as well as G and c). How big is λ?

In 1917 very little information was available about ρ_0, from which λ could be determined. The value of

$$S \approx 10^{26} - 10^{27} \text{cm}$$

quoted in those days is therefore only of historical interest. If we take ρ_0 as $\sim 10^{-31}$ g cm^{-3} as the rough estimate of mass density in the form of galaxies, we get $S \approx 10^{29}$ cm and $\lambda \approx 10^{-58}$ cm^{-2}.

The λ-term introduces a force of repulsion between two bodies that increases in proportion to the distance between them. The above value of λ is too small to make any detectable difference from standard general relativity (that is, with $\lambda = 0$) in any of the Solar System tests. Thus the Einstein universe faced no threat from the local tests of gravity. The model, however, did not survive much longer than a decade, for reasons discussed next.

Apart from finding the solution for the large-scale structure of the universe, Einstein had further expectations from his model. First, he expected the solution to be unique, given the assumptions of homogeneity and isotropy. This would have provided a reason why the universe is the only one of its kind. Further, he believed that with the λ-term there was no possible empty space solution. In this belief he was influenced by Mach's principle which required inertia to be fully determined by matter, implying that it should be impossible to determine the spacetime geometry and test particle trajectories in the absence of bulk matter.

These expectations were not realized. In fact shortly after the publication of Einstein's paper[3], W. de Sitter published another solution of Einstein's field equations[4]. This had the line element given by

$$ds^2 = \cos^2\left(\frac{\rho}{S}\right)c^2 dT^2 - d\rho^2 - S^2 \sin^2\frac{\rho}{S}(d\theta^2 + \sin^2\theta d\phi^2), \qquad (2.16)$$

with ρ used here as a radial coordinate, with the density of matter zero and $\lambda = 3/S^2$. Defining

$$R = S\,\sin\frac{\rho}{S}, \qquad (2.17)$$

then the above line element becomes

$$ds^2 = \left(1 - \frac{R^2}{S^2}\right)c^2 dT^2 - \frac{dR^2}{1 - \dfrac{R^2}{S^2}} - R^2\left[d\theta^2 + \sin^2\theta d\phi^2\right]. \qquad (2.18)$$

A further transformation of coordinates

$$R = r\exp Ht, \qquad T = t - \frac{1}{2H}\ln\left(1 - \frac{H^2 R^2}{c^2}\right) \qquad (2.19)$$

then takes the line element to

$$ds^2 = c^2 dt^2 - e^{2Ht}\left[dr^2 + r^2\left(d\theta^2 + \sin^2\theta d\phi^2\right)\right] \qquad (2.20)$$

where $H = c/S$.

Both (2.18) and (2.20) are the better-known forms of the de Sitter line element although in his 1917 paper de Sitter used only (2.16).

The de Sitter universe, being empty, offered a counter-example to Einstein's second expectation. Further, by offering an alternative model for the universe for a non-zero λ, de Sitter demonstrated that the Einstein universe is not a unique solution of the cosmological problem.

Much has been written about these theoretical implications of the de Sitter model and its role *vis-à-vis* the Einstein model. It is true that the primary issues of inertia and Mach's principle had been the motivating concepts for the genesis of the Einstein universe. Its contact with astronomical observations had been minimal except for the assumption (which later turned out to be wrong!) that the universe is static[b]. Nevertheless, de Sitter himself seems to have paid considerable attention to the then available observations in order to place limits on the cosmological parameters of his model. It is of interest to briefly give a sample of his arguments.

In units which appear strange in today's usage, de Sitter chose to express physical quantities in terms of a day for the time unit, the astronomical unit (AU) for distance and solar mass $M_\odot$ for mass. Thus $c = 173$, $G = 2.96 \times 10^{-4}$, and unit density of matter corresponds to 6×10^{-7}g cm^{-3}.

Taking the Einstein model first and noting that some of the spiral nebulae are galaxies similar to ours, de Sitter put their linear diameters as $d \approx 10^9$ and angular diameters $\simeq 5'$ (the limits of observations) to set their distance at $\leq 6 \times 10^{11}$ AU. Taking this upper limit as the value of S, the maximum extent in the elliptical space gives $\pi S/2 \leq 10^{12}$. Using the density formula (2.15) he then estimated the mass in the whole universe

$$M = \frac{2\pi^2 S}{\kappa}, \quad \kappa = \frac{8\pi G}{c^2} \tag{2.21}$$

so that $M \sim 8 \times 10^7 S$, with a density of $8 \times 10^6/S^2$ in de Sitter's units. Taking the mass of our own Galaxy as $\sim \frac{1}{3} \times 10^{10} M_\odot$ (based on Kapteyn's estimate) he found $S = 41$ which was absurdly low! Taking instead a density value of $\sim 10^{-17}$ based on the star density at the Galactic center, he found $S = 9 \times 10^{11}$ giving a total mass $M = 7 \times 10^{19} M_\odot$.

Extending arguments beyond our Galaxy, de Sitter argued that if the whole universe is filled with galaxies with their typical separation $\sim 10^{10}$, large compared to their linear dimensions, then the total mass in the form of galaxies worked out to only $\sim 2 \times 10^{16}$. He then argued that 'According to this view, only a small portion of the world-matter would be condensed into ordinary matter.'

This argument rings a bell! Today, when many cosmologists find the astronomically observed matter to be insufficient to account for the amount required by a theoretical model, they jump to the conclusion that the balance

---

[b] A parallel may be drawn with the genesis of special relativity which was *not* motivated by the Michelson–Morley experiment, but by considerations of accommodating electrodynamics within the framework of the principle of relativity of motion.

must be made up by 'non-ordinary' matter. de Sitter, however, went on to revise the observed estimate of the mean density of matter from 10^{-17} down to $\sim \frac{1}{3} \times 10^{-20}$ and managed to obtain a consistent picture with $S \leq 5 \times 10^{13}$. He concluded:

> We can thus consider the value ($S \leq 5 \times 10^{13}$) as an upper limit – subject, of course, to the uncertainty (which is considerable) of the hypothesis and of the numerical data thus derived.

So far as his own model was concerned, de Sitter referred to the cosmological redshift due to expansion: 'The lines in the spectra of very distant stars or nebulae must therefore be systematically displaced towards the red, giving rise to a spurious positive radial velocity.' Using the then available data on what he called 'helium stars', he used the predictions of his model to estimate $S = \frac{2}{3} \times 10^{10}$. This is perhaps the earliest application of the cosmological redshift hypothesis.

The astronomical data today would vitiate these estimates but their merit lies in the boldness with which the meagre observations were used to set limits on parameters of the theory. They also remind us that when one attempts to match theories with observations that have been obtained with considerable uncertainties (inevitable when the instruments have been stretched to their limits) the results may not always be correct.

The Friedmann–Robertson–Walker models

With the data on nebular redshifts still very uncertain in the early 1920s, this work of de Sitter was not taken seriously by astronomers. Nor was the work of Friedmann of 1922[5] and 1924[6] that extended the investigations of Einstein (matter without motion) and de Sitter (motion without matter) to world models that had both matter and motion. Friedmann discussed open as well as closed models and obtained a dynamical differential equation describing the change of the scale factor S with time.

Einstein made a brief reference to this work, and although by the mid-1920s there were sufficient data on nebular redshifts (cf. next chapter) neither he nor Friedmann himself felt the need to relate these non-stationary models to them. In fact Friedmann's papers were hardly known for nearly a decade after their publication[c].

---

[c] One of us, (JVN) recalls that at the 1962 Warsaw Conference on General Relativity and Gravitation, some Russian relativists were complaining that the cosmology community did not give due recognition to Friedmann when referring to the standard big-bang models. The belated recognition began to come in the late 1960s.

Lemaitre[7] and Robertson[8] in the late 1920s independently developed cosmological models similar to Friedmann's. In his 1927 paper, Lemaitre derived the two equations

$$\frac{\dot{S}^2 + c^2}{S^2} = \frac{1}{3}(\lambda c^2 + \kappa \rho c^2) \tag{2.22}$$

$$\frac{2\ddot{S}}{S} + \frac{\dot{S}^2 + c^2}{S^2} = \lambda c^2 - \kappa p. \tag{2.23}$$

Whereas Friedmann had considered 'dust' models ($p = 0$), Lemaitre considered dust as well as radiation ($p_r = \rho_r c^2/3$). He also derived the energy conservation equation

$$\frac{d}{dt}\left\{S^3(\rho c^2)\right\} + 3p_r S^2 \dot{S} = 0 \tag{2.24}$$

where ρ is the combined density of matter and radiation and p_r the radiation pressure.

Lemaitre's models included Einstein's and de Sitter's models as special cases. The Friedmann models also were special cases (with zero radiation). Lemaitre also derived the cosmological redshift formula

$$1 + z = \frac{S(t_0)}{S(t_1)} \tag{2.25}$$

for radiation that left the source at epoch t_1 and arrived at the observer at epoch t_0. An approximate Hubble law also followed. Robertson[8] found a similar result in 1928.

It is interesting how prejudices have continued to govern theorists about whether the cosmological models are open or closed. At the 1931 meeting of the British Association, the Bishop of Birmingham[9] declared: 'It is fairly certain that our space is finite, though unbounded. Infinite space is simply a scandal to human thought... the alternatives are incredible.' It was against this background where only closed models were thought to matter, that Einstein and de Sitter[10] wrote a joint paper in 1932 describing a simple open model with the line element

$$ds^2 = c^2 dt^2 - S^2(t)\left[dr^2 + r^2(d\theta^2 + \sin^2\theta d\phi^2)\right] \tag{2.26}$$

and with $S(t) \propto t^{2/3}$. Known as the Einstein–de Sitter model it is perhaps the simplest of the Friedmann models. This model had $\lambda = 0$ and it belonged to a later period when Einstein had given up the idea of a cosmological constant.

This is the model that currently goes under the name 'flat $\Omega = 1$ model', and we wonder how many cosmologists of today can trace its genesis to Einstein and de Sitter.

By 1932 the observational paper of Hubble published in 1929 (next chapter) had become well known and the idea of the expanding universe had begun to take root. The solutions obtained so far were, however, somewhat *ad hoc*, being based on the simplifying assumptions decided separately by each author. The purist may have worried at the emergence of a unique time coordinate from a theory which was generally covariant. Did such 'time-dependent' solutions have a physical meaning?

A rigorous approach to cosmological models finally emerged from the independent work of Robertson[11] and Walker[12]. Starting from two well-defined assumptions, viz. the Weyl postulate and the cosmological principle of homogeneity and isotropy, they were able to obtain the most general line element as

$$ds^2 = c^2 dt^2 - S^2(t)\left[\frac{dr^2}{1 - kr^2} + r^2(d\theta^2 + \sin^2\theta d\phi^2)\right], \qquad (2.27)$$

with $k = 1$ for closed models and $k = 0, -1$ for open models. Here (r, θ, ϕ) are the constant comoving coordinates of a typical galaxy. The fact that such coordinates can be defined rests on the assumption that the world lines of galaxies form a bundle of non-intersecting geodesics diverging from a spacetime point in the past. Thus through each spacetime point a unique member of the bundle passes. The time coordinate is that measured by a galaxy as its proper time. This is Weyl's postulate. The cosmological principle tells us that the hypersurfaces $t =$ constant are homogeneous and isotropic. What Robertson and Walker did was to give a mathematical derivation of the line element (2.27) from these postulates. Thus the Weyl postulate and the cosmological principle single out a global coordinate system. The time coordinate t, commonly called the cosmic time, arises in this way. There is no contradiction between this global symmetry and the local covariance of general relativity.

In Chapter 12 we will return to this discussion and review it in the modern context. Most of the so-called 'standard model' in cosmology today is based on the early work of Friedmann. Whereas the cosmologists of the 1930s and the 1940s were content with modest extrapolations of the present universe, their modern counterparts are more adventurous. Their extrapolations lead them to a state of the universe that was $\sim 10^{-43}$s old with a kinetic temperature $\sim 10^{30}$K! Indeed, to what extent these ideas can

3

The observational revolution

In 1914, spectra of the first spiral nebulae were obtained by V.M. Slipher[1] at the Lowell Observatory in Flagstaff, USA. This work was continued by him and then by F.G. Pease[2] at Mount Wilson in the period 1915–20, and by M.L. Humason[3] at Mount Wilson starting around 1927. By the early 1930s, spectra of some 90 nebulae had been measured.

It was this work and more on the nature of the spirals that led to the major revolution which led us to the concept of the universe that we believe in today. Thus starting less then ten years after Agnes Clerke had written with the assurance quoted earlier, a discovery had been made that was destined to change our ideas on cosmology almost completely.

From the early spectroscopic work it was clear that the spectral shifts interpreted as velocities had the following properties.

1. The vast majority of them were positive, suggesting velocities of recession.
2. They extended over a huge range compared to anything else that had been measured up to that time, varying from a few hundred km s^{-1} up to nearly 20 000 km s^{-1}.
3. Groupings of nebulae with approximately the same velocities had been detected by Humason[4]. Thus by 1932 he had already measured radial velocities in the Virgo, Perseus, Coma and the Pegasus and Leo clusters.

Going parallel with this observational work of the pioneering spectroscopists, attempts were being made to determine the nature of the spiral nebulae. In 1917, Curtis[5] used the detection of what he believed were novae to derive the distances of some spirals. He assumed an average apparent magnitude of $+5$ and an average distance of about 10^4 light years (ly) for novae in the Milky Way, to derive a difference of 13 magnitudes between the novae in the Milky Way and in some isolated spirals. This gave an average distance of about 4×10^6 ly for the spirals. The range covered was between about 0.5×10^6 ly for M31 and 10×10^6 ly for the more distant objects.

It was this result which weighed very heavily with Curtis in the famous 'debate' between him and Shapley on the distances of the spirals held in 1919[6]. Lundmark[7] gave the absolute magnitudes of the novae as -6.1. But a source of confusion in all of this was that the nova in the Andromeda Nebula (M31), S Andromedae, was far brighter than the average nova (~ -15.1) and it was not realized for more than another 15 years that S Andromedae was not a nova at all, but a member of the new class of object to be called supernovae[8].

The more accurate method of measuring the distances using the period luminosity relation for the cepheids which had been introduced earlier was applied by Hubble[9] to M31 and M33. He found 50 variables in M31, of which 40 were cepheids, and 35 cepheids in M33. In addition there were known to be nine in NGC 6822, and Shapley[10] had measured 105 in the Small Magellanic Cloud (SMC).

Shapley had found that the distance modulus $(m - M) = 17.55$ for the SMC. The corresponding modulus for M33 was $(m - M) = 22.1$ giving a distance D of about 850 000 ly, and for M31 $(m - M) = 22.2$ or $D \approx 900 000$ ly.

Of course, Hubble realized and stressed that these distances would be affected by any change in the zero point of the period–luminosity relation as it had been derived by Shapley. Several attempts were made in these early years to revise Shapley's value for the zero point, by Adams, Joy and Humason[11] and by Kipper[12]. Both of these revisions appeared to bring M33 and M31 about 40% closer.

The very large distances derived for the spirals were compatible with many attempts to determine the proper motion of the spirals, which had always led to null results (a velocity of 1000 km s^{-1} transverse to the line of sight would correspond to an annual proper motion of the order of $0''.0007$, a value far below what could be measured).

However, there was a stumbling block which led many to question these large distances. This arose from the claim by A. van Maanen[13] at Mount Wilson to have determined proper motions *in the spiral arms* of a number of nearby spiral galaxies including M33, M81, M101, NGC 2403, NGC 4051, 4736, 5055, and 5195. Van Maanen was a classical observer of the old school – an astronomer's astronomer, very conservative and priding himself on never being wrong, and his weighty pronouncement on this subject coming from the world's greatest observatory could not easily be put aside by the scientific establishment of that time. Moreover, Lundmark had also apparently found such motions. At an indicative distance of 10^6 ly an annual motion $\sim 0.''01$, which was of the order of what was being claimed, corresponds to a velocity of $\sim 15\,000$ km s^{-1}. Thus van Maanen's results correspond to periods of rotation of the spirals, if they were assumed to have sizes similar to the Milky Way, of the order 10^7 years or less, implying ejection of matter on similar time scales, so that the spirals would disintegrate in times of this order. Jeans went as far as to develop a theory of spiral evolution based on his belief in the van Maanen measurements.

How then was this problem resolved? We are not sure. Both van Maanen and Hubble were staff members at the leading observatory in the world – and van Maanen never officially retracted his results. We have been told that Hubble was finally able to persuade van Maanen to allow another astronomer at Mount Wilson, Seth Nicholson, to try to reproduce his results by remeasuring plates which he, van Maanen, already had measured – and that van Maanen's measurements could not be confirmed. While Hubble was very diplomatic and never spoke against van Maanen's results, possibly because the Director of the Mount Wilson Observatory would not let him publish a paper critical of van Maanen's work, it is clear from the literature that by the mid-1930s almost everyone believed in the island universe hypothesis, i.e. that the spirals lie at great distances, and therefore the majority of the astronomical community no longer accepted van Maanen's claims. In private conversation in the early 1950s, Walter Baade told one of us (FH) that in the stand-off situation existing at Mount Wilson Observatory in the early 1930s, Hubble persuaded Walter Adams, the Director, that an outsider be brought in to resolve the difference between himself and van Maanen. According to Baade, this was one of the reasons he was offered a post on the Mount Wilson staff. Baade further said that in his capacity as arbiter, he was permitted access to van Maanen's plates, from which he could soon see that the claimed proper motions were bogus. It was decided, however, by the Director that his finding would not be exposed unless van Maanen could be persuaded to accept it. Since this did not prove possible, nothing was published, and the world was left to make out of it what it would. Had Hubble's personality not been strong enough to carry the day, and had things been the other way round, Mount Wilson could well have lost its considerable share in the discovery of the expanding universe.

The velocity–distance relation

The idea that there might be a correlation between radial velocity and distance of the spiral nebulae was in the air in the early 1920s. Wirtz[14] showed in

1924 that for the 42 spirals for which radial velocities were known, the velocities increased with decreasing diameters of the spirals. Lundmark[15] plotted the distances of the nebulae (measured in units such that the distance to M31 was unity) against the velocities and he found a slight but systematic correlation. Dose[16] made a similar analysis. Hubble[17] made a plot of radial velocity against distance. This is the best known of these analyses. Hubble clearly showed a linear relation (Fig. 3.1). In retrospect his result is very puzzling, since a large number of galaxies in his plot are very nearby systems which we now know do not partake in the so-called Hubble flow. They include M31, M33, the Magellanic Clouds and the M81 group, etc. However, Hubble may have been influenced by the fact that, in the same issue of the *Proc. Natl. Acad. Sci.* and immediately before his paper, Humason[18] published a redshift much greater than any in Hubble's list. The galaxy is NGC 7619 which has a redshift of about $3800\,\mathrm{km\,s^{-1}}$, and Hubble's relation when extrapolated passed through the point determined by that galaxy. This redshift is clearly large enough for 'local' velocity effects to be negligible. So it appeared that the published relation might be reliable,

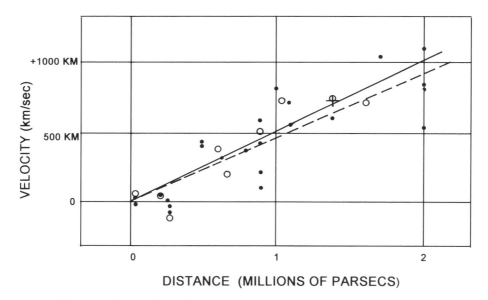

Fig. 3.1. The radial velocities, corrected for solar motion, plotted against estimated distances as plotted by Hubble[17] in 1929. The full line represents his fit using individual galaxies ('nebulae', black dots) and the dashed line represents his solution combining galaxies ('nebulae', open circles) into groups.

Table 3.1

Cluster	Diam. of cluster (degrees)	No. of redshifts	Mean cz (km s^{-1})	Av m_{pg}	D (Mpc)
Virgo	12	7	890	12.5	1.8
Pegasus	1	5	3810	15.5	7.25
Pisces	0.5	4	4630	15.4	7
Canis	1.5	—	4820	16.0	9
Perseus	2.0	4	5230	16.4	11
Coma	1.7	3	7500	17.0	13.8
Ursa Major	0.7	1	11 800	18.0	22
Leo	0.6	1	19 600	19.0	32
Group I	—	16	2350	13.8	—
Group II	—	21	630	11.6	—

but mainly because of Humason's more distant point which was not plotted in Hubble's paper.

By 1931 Hubble and Humason[19] had increased the number of redshifts significantly and extended the relation out to nearly 20 000 km s^{-1} with results from eight clusters and two groups. The data are shown in Table 3.1.

The velocities in Table 3.1 are obtained from cz, with z the measured redshift. They are derivative quantities, which however had a seriously bemusing effect on the astronomers of the time. Figure 3.2 is a plot of log cz against m_{pg}, i.e. between observed quantities, the points being well-fitted by the straight line

$$\log cz = 0.2 m_{pg} + 0.507. \tag{3.1}$$

Figure 3.3 is a plot of $v = cz$ against the distance D in megaparsecs given in the last column of Table 3.1. The distances were not observed quantities, but were derived by combining the observed m_{pg} with an extended and (by modern standards) awkward argument.

The first step consisted of obtaining the average absolute visual magnitude of nearby galaxies judged to be similar to those observed at greater distances in the clusters of Table 3.1. This was done from ten very nearby galaxies containing cepheid variables which gave an average value of $M_{vis} = -14.7$, and also from a further 40 in which it was thought that brightest stars could be resolved. Taking the apparently brightest stars to have absolute visual magnitudes of -6.1, these 40 galaxies gave $M_{vis} = -15.0$. The similarity of

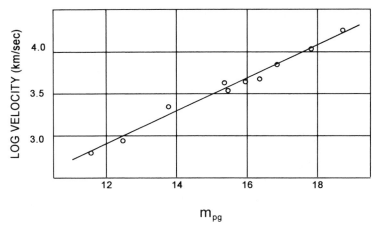

Fig. 3.2. Log (velocity) plotted against photographic magnitude m_{pg} from data by Hubble and Humason[19]. (Plot taken from Tolman[22].)

these estimates encouraged the view that $M_{vis} = -14.9$ could be adopted as a cautious compromise (the present day value is in the region of -20).

The next step was to convert from a visual magnitude to a photographic magnitude, by means of a color index (*CI*) with

$$CI = +1.10 \pm 0.02, \tag{3.2}$$

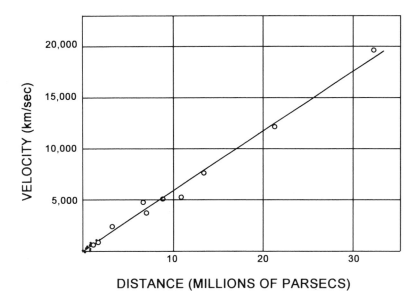

Fig. 3.3 Velocity (km s^{-1}) plotted against the derived distances. (Plot taken from Tolman[22].)

the two decimal places being again indicative of the astonishing accuracy of the claims, and they gave $M_{pg} = -13.8$. Were it not for the redshifts of the clusters in Table 3.1, the distances D in megaparsecs could then be calculated from $\log D = (m_{pg} - M_{pg}) - 5$, on the assumption that the cluster galaxies for which m_{pg} was measured were intrinsically the same as those in the survey of the nearby galaxies. However, despite the redshifts, one can still write,

$$\log D = (m_{pg} + \delta m_{pg} - M_{pg}) - 5, \tag{3.3}$$

where δm_{pg} is the change of m_{pg} due to the redshift.

In the terminology of the day, photographic apparent magnitudes of objects emitting a Planck spectrum were considered to be related to bolometric magnitudes m_b (where the subscript b stands for 'blue') by

$$m_{pg} = m_b + \Delta m_v + HI + CI, \tag{3.4}$$

where Δm_v was an empirical correction to be added in order to obtain a radiometric magnitude as measured thermally *after* absorption by the earth's atmosphere, HI was the empirical value of the so-called heat index, and CI the color index as before. Stellar measurements had given the values of these quantities, shown as a function of temperature in Table 3.2.

On the further assumption that the nearby galaxies for which M_{pg} had been obtained possessed colors corresponding to a Planck distribution of temperature 5760 K (type G3), the redshifted spectra would have a Planck distribution of temperature $5760/(1 + z)$ K. Using the observed value of z, a temperature was thus obtained and by interpolation in Table 3.2 values of Δm_v, HI and CI were obtained for $5760/(1 + z)$. Comparing these values with those for 5760 K then gave the differences $\delta(\Delta m_v)$, $\delta(HI)$, $\delta(CI)$ appearing in

Table 3.2

Spectral type	Temp K	Δm_v	HI	CI
F5	6500	0.44	0.30	0.62
dG0	6000	0.43	0.32	0.72
dG5	5600	0.41	0.39	0.83
dK0	5100	0.40	0.55	0.99
dK5	4400	0.48	1.10	1.26
dM	3400	0.53	1.40	1.76

$$\delta m_{pg} = \delta m_b + \delta(\Delta m_v) + \delta(HI) + \delta(CI), \qquad (3.5)$$

whence the value of δm_{pg} to be used in (3.3) could be obtained, subject to δm_b being also obtainable. Here

$$\delta m_b = 2.5 \log(1 + z) \qquad (3.6)$$

was used.

The effect of the redshift on m_b is actually two-fold. The energy of each photon is decreased by $(1 + z)$ and the rate of reception of photons is also decreased by $(1 + z)$, giving $\delta m_b = 5 \log(1 + z)$. This mistake was made deliberately in order to introduce a sense of caution into the investigation! Inserting (3.5) in (3.4), and then (3.4) in (3.3) with $M_{pg} = -13.8$ and m_{pg} the observed values of the photographic magnitude plotted in Fig. 3.2, gave the values of D in Table 3.1, plotted in Fig. 3.3.

The slope of the line in this lower panel gave

$$v = H_o D,$$
$$H_o = 558 \text{ km s}^{-1} \text{ Mpc}^{-1},$$

where H_o is what we now know as the Hubble constant. This value of H_o was used from the early 1930s until the IAU meeting in Rome in 1952. Because it dominated cosmology for two decades, and because of the landmark discovery of the linear relation shown in Fig. 3.3, we have discussed its derivation here at some length.

This completes our discussion of the early evidence. By 1931 Hubble and Humason had established without question a tight linear relation between apparent magnitudes and redshifts. In the next period, 1931–8, came further advances. But before describing these, it is necessary to add a few words about the nature of the redshifts.

The concept that the universe is expanding, which was based on these data, and the theoretical models of Lemaitre and of Friedmann (Chapter 2), had a tremendous impact on science and society around 1929. The result made the front page of the *New York Times*, and was one of the first of the twentieth-century results in science to receive heavy publicity. In different ways, Einstein, Eddington and Hubble all became media figures as can be seen by reading the journals and the popular press of the times (cf. biography of Hubble by Christianson[20]).

There was one snag in the argument. While the nebulae nearly all showed redshifts, there was no proof that they were due to expansion. Some well-known physicists, among them E. MacMillan and Fritz Zwicky, suggested that the redshifts might be due to scattering processes in intergalactic space –

the so-called tired light hypothesis. While at the time this appeared to some to be a genuine possibility, it was generally ignored, as is inevitably the case when the establishment has made up its mind, as it had by around 1930. There are two points to be added. The first is that we now know that the scattering of photons by intergalactic matter would lead to a smearing of the images of distant galaxies which is not seen. Thus although this topic was raised again in the 1950s by E. Findlay-Freundlich and by Max Born, who were aware of the problem of the scattering process, the mechanism cannot explain the redshifts. More important is the fact that the expansion hypothesis can be tested observationally, since the surface brightness of an extended source is a function of z, and the form of that function depends on the nature of the shift. While the possibility of making this test has been known for 60 years, it was only finally carried out successfully in 1990 by Sandage and Perelmuter[21] using the surface brightnesses of elliptical galaxies. They found that the expansion hypothesis gave much the best fit. To those who believe in the band wagon approach to cosmology, which normally means that the apparently most simplistic argument is preferred, this outcome in 1990 was presumably highly satisfactory, since the investigators had never seriously thought otherwise.

References

Chapter 3

1. Slipher, V.M. 1914, *Lowell Bull.*, **2**, 56; 1917, *Astron. Nacht.*, **213**, 47; 1917, *Proc. Amer. Phil. Soc.*, No. 5.
2. Pease, F.G. 1915, *Publ. Astron. Soc. Pacific*, **27**, 133.
3. Humason, M.L. 1931, *Ap. J.*, **74**, 35.
4. Humason, M.L. 1932, quoted by Curtis, H.D. 1933, *The Nebulae* (Berlin, Springer-Verlag), Chap. 6, Table 14.
5. Curtis, H.D. 1917, *Lick Obs. Bull.*, **9**, 108; 1919, *Proc. Natl. Acad. Sci.*, **9**, 218.
6. Shapley, H. and Curtis, H.D. 1921, *Bull. Natl. Research Council*, Vol. **2**, part 3, No. 11.
7. Lundmark, K. 1924, *M.N.R.A.S.*, **84**, 747.
8. Baade, W. and Zwicky, F. 1934, *Proc. Natl. Acad. Sci.*, **20**, 254.
9. Hubble, E. 1926, *Ap. J.*, **63**, 236.
10. Curtis, H.D. 1933, *The Nebulae*, Hdb V/2 (Springer), p. 861.
11. Adams, W.S., Joy, A.H. and Humason, M.L. 1929, *Pub. Astron. Soc. Pacific*, **41**, 195.
12. Kipper, A. 1931, *Astron. Nacht.*, **241**, 249.
13. van Maanen, A. 1916–30, *Mt Wilson Contr. Nos.*, 111, 136, 158, 182, 204, 237, 270, 290, 321, 356, 391, 405–8.
14. Wirtz, C. 1924, *Astron. Nacht.*, **222**, 21.
15. Lundmark, K. 1924, *M.N.R.A.S.*, **84**, 747.
16. Dose, A. 1927, *Astron. Nacht.*, **229**, 157.

17. Hubble, E. 1929, *Proc. Natl. Acad. Sci.*, **15**, 168.
18. Humason, M. L. 1929, *Proc. Natl. Acad. Sci.*, **15**, 167.
19. Hubble, E. and Humason, M.L. 1931, *Ap. J.*, **74**, 43.
20. Christianson, G.E. 1995, *Edwin Hubble* (New York, Farrar, Strauss and Giroux).
21. Sandage, A. and Perelmuter, J.-M., 1990, *Ap. J.*, **350**, 481.
22. Tolman, R.C. 1934, *Relativity, Thermodynamics and Cosmology* (Oxford, Clarendon Press), p. 455.

4

The observational trail 1931–56, the determination of H_0 and the age dilemma

The decade of the 1930s was the period during which Hubble and Humason at Mount Wilson, and Nicholas Mayall at the Lick Observatory, extended and consolidated our understanding of the redshift–apparent magnitude relation. On the theoretical side the view that the observations must be interpreted as due to the expansion of the universe, using as a model the Friedmann–Lemaitre solutions to Einstein's equations, became the majority position.

Nothing about the universe that was discovered in this later period could possibly match the earlier discovery of the redshift–apparent magnitude relation – the expansion of the universe as it was commonly described. This was clearly one of the few truly great discoveries of the twentieth century. Hubble's work with Humason was very widely recognized and it is a pity that they were not awarded the Nobel prize for this discovery. We have been told that this would probably have happened in 1953 had Hubble lived. He died in September 1953 and the award is not given posthumously.

Hubble remained cautious about the nature of the discovery. In many of his papers and in the Silliman Lectures at Yale (*The Realm of the Nebulae*, Yale, 1936) he always qualified remarks about 'velocity shifts' as they were called, stressing that they might only be 'apparent velocity shifts.' In fact, as was mentioned in Chapter 3, the demonstration that the redshifts of normal galaxies are genuine expansion shifts was not established until more than 60 years after the discovery of the redshift–apparent magnitude relation.

As well as extending the redshift–apparent magnitude relation to fainter galaxies, a program which was pursued by Hubble, Mayall and Humason until 1941 (when the USA became involved in the Second World War) and which was continued after Hubble's death in 1953 by Humason, Mayall and Sandage, Hubble began a program to study the large-scale geometry of the universe by using the distribution of galaxies in space. To do this he tried to use the galaxies to measure the curvature of space directly by counting them as a function of apparent magnitude, and indirectly by measuring any deceleration there might be of the redshift with time. As was mentioned in Chapter 2, the determination of the latter through the so-called parameter q_0 leads to the space curvature through the equations of general relativity.

Of course in the 1930s very little was known about galaxies. While Hubble did extensive work on their morphology and devised his famous classification system, practically nothing was known about their basic physics.

Walter Baade, after he became a colleague of Hubble at Mount Wilson, argued that this approach of Hubble's was not the right way to go. Baade felt quite strongly that it was necessary to understand more about the nature of the objects being used as markers in the universe – the galaxies – before they could be used for cosmological investigation. Specifically, it is necessary to understand how the parameters of a galaxy, such as the apparent magnitude and angular diameter, vary in time if they are to be used for cosmological purposes. As Baade's own work showed, he was correct in this. At the same time Hubble, with the bit between his teeth, was determined to push as far and as fast as he could. However, having heard Baade declaiming on this topic ourselves, we must admit that he was not simply a dispassionate by-stander – a major bone of contention was clearly that in the period up to 1941, Hubble and Humason had first call on the dark time (the time in the month when moonlight is at a minimum) on the 100-inch telescope, so that those who wanted to take the other approach, and in particular Baade and Rudolf Minkowski, came off second best.

As we shall see later, no one learns from the earlier mistakes of others. When the radio sources were first discovered, there was a headlong rush to use them to do cosmology, despite the fact that most of them remained unidentified. And confusion reigned. Close upon the heels of the discovery of radio sources, and through them, came the discovery of quasi-stellar objects. We still don't understand what they are, although many of our colleagues apparently think they do, and are using them as cosmological probes. But we leave discussion of this issue to later chapters.

The attempt to measure the curvature of space

By the mid-1930s, theoreticians had shown that the observed near homogeneity and isotropy of the universe in the distribution of faint galaxies must permit a particularly simple coordinate description to be used, provided it is assumed that any other important physical content there might be in the universe also has similar homogeneity and isotropy. Thus a time coordinate could be chosen such that the three-dimensional time-sections $t = $ constant were homogeneous and isotropic. What could not be decided from theory alone, however, was whether such time-sections possessed positive, negative or zero curvatures with respect to the four dimensions of the universe as a whole. These three cases were referred to by a parameter k which was $+1$ for positive curvature, -1 for negative curvature and 0 for zero curvature; $k = 0, \pm 1$ (cf. Chapter 2).

It is probable that Hubble heard of these theoretical developments from his friend Richard Tolman, and that in consequence he set himself to determine which of these three cases applied to the observed universe. This he proposed to decide through a counting of the number of galaxies brighter than an assigned magnitude m, a number denoted by $N(m)$.

The apparent bolometric luminosity l of a galaxy of redshift z and intrinsic bolometric luminosity L can be used to define a radial coordinate r from the equation

$$l = \frac{L}{4\pi r^2 (1 + z)^2}. \tag{4.1}$$

Assuming all galaxies to be intrinsically similar, the number distributed homogeneously in the time-sections $t = $ constant could be seen to be proportional to spatial proper volume, and the spatial proper volume to distance r was dependent on the parameter k according to

$$\int_0^r \frac{x^2}{\sqrt{1 - kx^2}}\, dx,$$

that is to say, proportional to

$$\frac{1}{2}(\sin^{-1} r - r\sqrt{1 - r^2}) \text{ for } k = 1, \quad \frac{1}{3}r^3 \text{ for } k = 0,$$
$$\frac{1}{2}(r\sqrt{1 + r^2} - \sinh^{-1} r) \text{ for } k = -1. \tag{4.2}$$

Taking all galaxies to have $M_{pg} = -13.8$ and by applying similar corrections as before involving Δm_v and the *HI* and *CI* indices, a conversion to a bolometric value of L could be obtained. m_{pg} determined by observation could also be converted to a bolometric magnitude, written now as m. Whence from m, the value of l to be used in (4.1) was considered known. Then a measurement of z would determine r for the galaxy in question. Indeed this was what had in effect been done to obtain the values of D in Table 3.1. But the last step here of determining z from observations could not be managed for a large number of galaxies, only for a few specifically selected ones. But *if* a relation could be specified between z and the coordinate r, then (4.1) could be solved for r, given L and given a measurement of l, i.e. given m. And the count $N(m)$ of all galaxies, assuming that they all have the same luminosity L, would give the number of galaxies out to the particular value of r. Thus by varying m, and obtaining the function $N(m)$, the number of galaxies as a function of r could be obtained and compared with the predicted values which depend on k. In this way the value of k would be determined. This was the program which Hubble set for himself in the early 1930s. A description of it formed the culmination to his Silliman Lectures of 1936[1].

There is much in all this to make the modern cosmologist shudder, and there was enough in it to make Hubble's colleague, Walter Baade, shudder, even in 1936. The assumption of an intrinsically constant value of L was clearly not correct. Galaxies vary in their intrinsic luminosities according to a luminosity function. However, Hubble convinced himself that he would overcome this problem provided the luminosity function did not diverge at its lower end.

As we have just remarked, Hubble was determined to push ahead as far and as fast as he could, a policy that led to many corners being cut. The corner he chose to cut on this occasion involved the relation between r and z. The radial coordinate r from (4.1) is essentially the distance D obtained previously and shown for a number of cases in Table 3.1. And this plot in Fig. 3.3 showed a linear relation between D and z for cases in which both have been measured. So Hubble decided to use a linear relation between r and z in cases where z was not measured, with an ultimately disastrous effect on the project. The correct relation has the form of a power series in z,

$$r = a_1 z + a_2 z^2 + \cdots \tag{4.3}$$

and Hubble was using only the first term in this series. If now one expands the first and third term of the expressions in (4.2) as ascending series in r, the first term is $\frac{1}{3}r^3$ in each case, implying that a count made only in first order reveals no differences between the cases $k = 0, \pm 1$. It is then only the Euclidean count. To have achieved the wanted result it was necessary to work to the second order in (4.3), i.e. the term of order z^2, which to be effective would have required measurements at larger redshifts and for much fainter galaxies.

In the case $k = 0$ the correct relation between r and z is easily given in a closed form,

$$r = \frac{2c}{H_o}\left(1 - \frac{1}{\sqrt{1+z}}\right), \tag{4.4}$$

which reduces in first order to

$$v = cz = H_o r. \tag{4.5}$$

But (4.4) has a non-zero value of a_2 in the expression (4.3). There are closed forms also in the other two cases, but these are more difficult to obtain. They are[2]

$$r = \frac{\sqrt{2q_o - 1}}{q_o^2(1 + z)}\left\{q_o z + (1 - q_o)(1 - \sqrt{1 + 2zq_o})\right\}, \quad k = 1,$$
$$r = \frac{\sqrt{1 - 2q_o}}{q_o^2(1 + z)}\left\{q_o z + (1 - q_o)(1 - \sqrt{1 + 2zq_o})\right\}, \quad k = -1. \tag{4.6}$$

The constant q_o can take any value $> \frac{1}{2}$ for $k = +1$ and any value $< \frac{1}{2}$ for $k = -1$.

These more complicated expressions were not known in the 1930s, but the series expansion (4.3) was surely known to second order for the case $k = 0$, surely known to Richard Tolman one would think. Why Hubble chose to ignore the correct forms in favor of his empirically determined plot in Fig. 3.2 is not clear. Since he was proposing to test the theory, it would have been perfectly proper to have used either (4.4) or the expanded forms of (4.6).

One of us (FH) had the pleasure of meeting Hubble on a number of social occasions in the early 1950s, occasions where, fortunately also, cosmology could be discussed. The strongest of his convictions appeared to be in the straight line of Fig. 3.2. After the first suspicion of linearity in 1929, the line had established itself and had always been maintained, however far out the observations had been taken. The belief in it was extremely deep-rooted and it continued generally for quite a while after Hubble's death. Still to this day nobody is very sure of what happens to it at large z and nobody has had the hardihood to repeat the counting process in which Hubble expressed such great faith.

Although the program led to no result of much scientific consequence, it had an immense practical consequence for cosmology. It formed the main scientific case for the construction of the 200-inch telescope on Palomar Mountain. More precisely, a repeat of the program to higher accuracy was a major part of the scientific case put forward for the telescope. Yet when the telescope was complete it was decided, over Hubble's head, not to proceed with an upgraded program of galaxy counts[a].

Martin Schwarzschild told one of the present authors (FH) that he was invited by the Observatory Director at that time, Ira Bowen, to attend the critical meeting at which the decision not to proceed in this way was taken. To begin with, Schwarzschild said, he was convinced that things would go Hubble's way – which in his view would not have been the best scientific policy. But as the meeting progressed, his own contrary position became steadily more supported, until in the end the opposite of what he had been expecting happened.

This was an example of an inversion we ourselves have observed on a few occasions: that in the beginning, when a new instrument is proposed, humans control the scientific program completely. But as the instrument is constructed, it is as if humans exercise less and less control, until in the end it is the instrument that over-rides the humans who service it and those who use it. We shall return to this issue when we discuss the use of space telescopes such as the Hubble Space Telescope and COBE.

At all events, we can report that in our experience Hubble never showed bitterness over this decision. Perhaps having suffered one heart attack he came to realize he would not have sufficient strength to see the program through, were the decision to have gone in his favour.

The redshift–apparent magnitude relation again

The program of redshift measurement went on steadily at the Mount Wilson Observatory up to 1950, and started in 1953 at Palomar. By 1936, Humason[3] had obtained redshifts for 146 galaxies, and by 1956, 620 redshifts had been obtained[4]. Starting in 1935, N.U. Mayall[5] began to obtain redshifts of the brighter galaxies using the 36-inch Crossley reflector at the Lick Observatory. The work was closely correlated with the lists of galaxies being observed by Humason and Hubble.

A total of 806 galaxies were measured and are described in detail by Humason *et al.*[4]. The vast majority of these are so-called field galaxies with $m \leq 16$. The remainder are galaxies in 18 clusters extending out to $z \approx 0.2$.

The magnitudes of the field galaxies were measured by Pettit[6] and by Stebbins and Whitford[7]. In this extensive study corrections of the raw data

---

[a] Modern work on galaxy surveys will be discussed in Chapter 19.

were made both for the redshifts and the apparent magnitudes. In the case of the redshifts it had been accepted that they must be corrected for solar motion with respect to the centroid of the Local Group, since it had been realized since 1936 that the systematic redshift does not operate within the Local Group[1,8].

The apparent magnitudes required several corrections Δm specifically for obscuration in our own Galaxy which depends on galactic latitude, and corrections into bolometric magnitudes from photographic or photovisual magnitudes, by using the K term which accounts for the effects of z.

In the linear approximation

$$cz = H_o r$$

where r in parsecs is given by

$$\log r = 0.2[m - \Delta m - K - M] + 1$$

so that

$$\begin{aligned}
m_{bol} &= m - \Delta m - K \qquad\qquad\qquad\qquad\qquad (4.7)\\
&= 5\log cz + (M - 5 - 5\log H_0)\\
&= 5\log cz + \text{constant.}
\end{aligned}$$

Humason *et al.*[4] showed that the redshift–apparent magnitude plot for the field galaxies gave a value of the observed slope of 5.028 ± 0.116. The relation was then taken to larger redshifts using the clusters. In this case, we must take into account the effect of curvature through the term $q_{\{o}$. Then

$$m_{bol} = 5\log cz + 1.086(1 - q_o)z + \text{constant.} \qquad\qquad (4.8)$$

Details of the method of measurement and corrections applied to obtain the apparent magnitudes m_{bol} for the faint galaxies in the cluster were given by Humason *et al.*[4].

It was also realized by these authors that yet another correction would have to be applied. In using the normal redshift–apparent magnitude relation it is assumed that we are dealing with standard galaxies which have a constant bolometric magnitude. However, the more distant the galaxies, the earlier is the epoch in their lives at which they are being observed. Thus we must try to take into account the dependence of the magnitude of a galaxy on its age defined as the time which has elapsed since stars were first formed, and the epoch at which it emitted the radiation that is presently being observed. Humason, Mayall and Sandage added a correction to take account of this by writing

$$m_{bol} = A \log cz + Bz + D$$

where the cosmological correction and the correction for galactic evolution are contained together in B. Using the cluster data for m_{bol} and z they made several fits assuming first that A, B and D were all unknown, second that $A = 5$, and B and D were unknown, third, that A and D were unknown and $B = 0$. As might be expected the best fit was obtained when all of the parameters were unknown, the solution in that case giving $A = 5.73$, $B = -5.62$ and $D = -8.55$ with σ, the dispersion in the magnitude residuals, equal to 0.282. However, the theory requires that $A = 5$, and in this case the best fit was given by $B = -1.180 \pm 0.875$ and $D = -5.81$, with $\sigma = 0.315$, which is not much different from 0.282. There is considerable scatter in the observational data for the clusters, and Humason *et al.* attributed this to a number of causes including a real dispersion in the absolute magnitudes of the galaxies within each cluster, and peculiar motions of the galaxies in clusters as well as the effect of measuring errors.

The numerical value of the B term was only about twice its probable error. Humason *et al.* suggested that two further corrections would explain this, the aperture effect associated with the photometry of faint galaxies which had not been completely corrected for, and possible inter-nebular obscuration. Both would tend to make the negative value of B absolutely larger.

The final conclusions from this study according to Humason *et al.*[4] were as follows.

1. That the slope of the $(\log cz, m)$ correlation (for small z) is as close to 5 as the probable errors allow.
2. That the value of B is negative enough so that unless the effect of evolution of the galaxies can overcome the term involving q_o, the expansion is decelerating. Their attempt to evaluate the galactic evolution gave a value still suggesting that the universe is decelerating. We shall return to this whole question of galactic evolution later in this chapter.
3. The expansion appears to be isotropic – this conclusion is simply based on the fact that the 12 clusters in the north and 6 in the south give points on the $\log cz - m$ plot which are indistinguishable.
4. The absolute magnitudes for the brightest galaxies in the clusters is nearly the same as for galaxies in the field.
5. If there is intergalactic absorption, its departure from uniformity must be small, giving only magnitude residuals between zero and 0.3.

To summarize: some 30 years after the first pioneering discoveries, there appeared to be strong evidence (but not yet strict proof) that the universe is expanding. The first attempts to measure the curvature of space by Hubble and Humason had failed, and it was also becoming clear that it would not be easy to measure the curvature through the higher order terms in the redshift–apparent magnitude relation. There then began several optical investigations which partly clarified, but in some aspects complicated, the picture on the observational front.

Before describing these, we turn to a discussion of the determination of the Hubble constant H_o and the successive corrections that had to be applied from the 1930s onward. Here we shall be covering the period of around 60 years up to the present.

The Hubble constant

Hubble and Humason's calibrations of the distance scale were based on the brightest stars in a sample of nearby galaxies. They were thought to be supergiants, and their absolute luminosities were in turn calibrated by using similar stars in M31 and M33, whose distances in turn were determined by the cepheid variables. The zero point of the period–luminosity law for the cepheids went back to the statistical parallax calibration by Shapley[9] and Wilson[10].

By the 1940s, it had become clear that there were two types of pulsating stars, the classical cepheid variables of Population I and the RR Lyrae stars of Population II which have two distinct period–luminosity relations. The concept of two stellar populations had been introduced by Baade in 1944[11]. Mineur[12] and Baade[13] and later Blaauw and Morgan[14] showed that this led to a correction of the zero point amounting to $-1\overset{m}{.}4 \pm 0.3$ for M31 and M33. This *increased* the distances of M31 and M33, and hence the absolute magnitudes of their brightest stars. A second correction was concerned with the objects which Hubble had originally believed were the brightest stars in nearby galaxies. Sandage[4] began a program with the 200-inch Hale telescope to test this hypothesis. By 1958 Sandage[15] had shown conclusively that Hubble had mistakenly identified HII regions with brightest stars, thus requiring a further correction, again in the sense that the nearby galaxies in which the brightest star method is used have distance moduli greater than those estimated by Hubble.

Thus the value $H_o = 558\,\mathrm{km\,s^{-1}\,Mpc^{-1}}$ obtained by Hubble and Humason was reduced successively to $180\,\mathrm{km\,s^{-1}\,Mpc^{-1}}$ and then by Sandage[15] to $75\,\mathrm{km\,s^{-1}\,Mpc^{-1}}$. By the 1960s there was general acceptance that the value of H_o was much smaller then the original value obtained by Hubble and Humason. In 1962 Sandage[16], attempting to summarize all of the estimates which had been made by him and others, gave $H_o = 98 \pm 15\,\mathrm{km\,s^{-1}\,Mpc^{-1}}$.

Table 4.1 *Estimates of the distance to the Virgo cluster (in Mpc)*

Method	Distance		
	Sandage (ref. 19)	van den Bergh (ref. 21)	Jacoby *et al.* (ref. 22)
Globular clusters	21.1 ± 2	19.7 ± 2.3	18.8 ± 3.8
Novae	20.6 ± 4	18.2 ± 2.5	21.1 ± 3.9
Supernovae	21.2 ± 2.2	19.1 ± 6	19.4 ± 5
		22.9 ± 5	
$D_n - \sigma$	23.4 ± 2	—	16.8 ± 2.4
21 cm line widths	20.9 ± 1.4	15.0 ± 1.4	15.8 ± 1.5
Size of the Galaxy	20.0 ± 1.8	—	—
Size of M31	—	17.0 ± 4	—
Size of M33	—	10.5 ± 2.5	—
Size of LMC	—	12.0 ± 2.5	—
Surface Brightness Fluctuations	—	14.9 ± 0.9	15.9 ± 0.9
Planetary Nebulae	—	14.1 ± 0.3	15.5 ± 1.1
Red Supergiants in NGC 4571	—	13.8	—
Red Supergiants in NGC 4523	—	13.2	—

In the 1970s an extensive program to measure H_o more accurately was started by Sandage and Tammann[17]. They calibrated the linear sizes of HII regions as a function of spiral galaxy luminosities, and went on to determine the distances to 39 spiral galaxies of types Sc, Sd, Sm and Ir using these values. The adopted absolute magnitudes combined with the apparent magnitudes of the Virgo spirals gave a distance to the Virgo cluster of 19.5 ± 0.8 Mpc. The Hubble diagram for first-ranked ellipticals in the Coma cluster scaled back from magnitudes in Coma to magnitudes in Virgo using $5 \log cz$ gave a redshift for the Virgo cluster of $1111 \pm 75 \, \mathrm{km \, s^{-1}}$. These values combined to give what Sandage and Tammann[18] called a first hint at the value of $H_o = 57 \pm 6 \, \mathrm{km \, sec^{-1} \, Mpc^{-1}}$.

Since that time, these authors have used a variety of methods to determine H_o and have concluded that the value has converged and must lie in the range $50 < H_o < 60 \, \mathrm{km \, s^{-1} \, Mpc^{-1}}$. One major key to the issue is the distance to the center of mass of the Virgo cluster, which they believe that they have shown by a variety of methods to lie close to 20 Mpc. In Table 4.1 we show how different methods used by them have led to this result. If the true redshift of

Table 4.2 *Values of H_0 from Sandage (ref. 19)*

Method	H_0 [km s^{-1} Mpc^{-1}]
Virgo Distance	55 ± 2
ScI Hubble Diag.	49 ± 15
M101 Diameters	43 ± 11
M31 Diameters	45 ± 12
Tully–Fisher	48 ± 5
Supernovae (B)	52 ± 8
Supernovae (V)	55 ± 8
Unweighted mean	50 ± 2
Weighted mean	53 ± 2

the Virgo cluster is $\sim 1100\,\mathrm{km\,s^{-1}}$, a value of $H_0 \sim 55\,\mathrm{km\,s^{-1}\,Mpc^{-1}}$ is obtained. Many other methods give a similar result. We show in Table 4.2 a summary of these results of the determinations of H_0 from Sandage[19].

The two experts contemporary over the past 20 years with Sandage and Tammann who have been most outspoken in their disagreement with the value of H_0 obtained by Sandage and Tammann have been the late G. de Vaucouleurs and S. van den Bergh. Working actively in the 1970s and 1980s on this problem, de Vaucouleurs[20] concluded that $90 \leq H_0 \leq 100\,\mathrm{km\,s^{-1}Mpc^{-1}}$. In a comparatively recent review, van den Bergh[21] concluded that $H_0 = 76 \pm 9\,\mathrm{km\,s^{-1}\,Mpc^{-1}}$. In the past 10 to 15 years, a younger generation of investigators bent on getting the 'true' value of H_0 have appeared, and have obtained a great deal of publicity.

To give some idea of the different values obtained by different groups using many of the same methods, we show in Table 4.1 van den Bergh's compilation of distances to the Virgo cluster using some of the same methods used by Sandage and Tammann and also by others.

A recent discussion of the different methods of distance determination has also been given by Jacoby *et al.*[22]. This paper is interesting in that it contains analyses of many methods, including comparatively new ones recently proposed by some of those authors themselves. Jacoby *et al.* have given a table of distance determinations of the Virgo cluster which we also show in Table 4.1. Jacoby *et al.*[22] concluded in 1992 that H_0 is 80 ± 11 or $73 \pm 11\,\mathrm{km\,s^{-1}\,Mpc^{-1}}$, depending on different weighting factors.

It can be seen by comparison of the values of the distance to Virgo in Table 4.1, that different methods sometimes give very different results, and the same methods in the hands of different investigators can also give different results. In general this is because there is intrinsic dispersion in the properties of the 'standard' objects chosen.

There has also been a sociological problem which has developed between the various groups, especially between Sandage and Tammann on the one hand, and van den Bergh and the younger investigators on the other. Thus despite the fact that the differences are not large, the current view that has been propagated is that we are witnessing a competition between Sandage and Tammann who have consistently claimed for 20 years that a 'low' value of H_0 is correct, and the rest who favor a 'higher value' of H_0 of $70 - 80 \, \mathrm{km \, s^{-1} Mpc^{-1}}$. A major issue is the structure and distance to the Virgo cluster.

Observational astronomers have long been aware that not only is the Virgo cluster widely extended, but it appears to involve two sub-clusters, one predominantly involving spirals which is more extended than the other sub-cluster of ellipticals[23]. In fact, de Vaucouleurs originally argued that there were two physically different clusters involved. Thus it is clear that the determination of the distance of any single galaxy in Virgo will not give a good measure of the distance to the center of mass of this complex cluster. This somewhat obvious limitation may explain the dispersion between some of the values in Table 4.1.

Another question is: what is the true redshift of the Virgo cluster? It is now generally believed that the Local Group galaxies, involving predominantly our Galaxy, M31 and M33, are perturbed by the large mass of the local supercluster, and there is a component of velocity due to this effect. Attempts to correct for this velocity term have led to various values being used for the 'true' redshift of the Virgo cluster.

A good example of the way that these uncertainties can give rise to very different values of H_0 is given by the much quoted and publicized work of Freedman *et al.*[24], when that work is compared with the results of Sandage and Tammann. Freedman *et al.*, having derived the distance of *one* galaxy in the Virgo cluster from Hubble Space Telescope observations of cepheids, use the distance 17.1 ± 1.8 Mpc of that galaxy, M100, combined with the redshift of the Virgo cluster which they take to be $+1404 \, \mathrm{km \, s^{-1}}$, to obtain a value of H_0 of $80 \pm 17 \, \mathrm{km \, s^{-1} Mpc^{-1}}$. On the other hand, if we take the distance of the center of mass of the Virgo cluster according to Sandage and Tammann to be 21 Mpc (Table 4.1) and a redshift of $1179 \, \mathrm{km \, s^{-1}}$, which Sandage and Tammann have obtained by plotting a Hubble diagram for more distant clusters[25] and extrapolating inward, $H_0 = 56 \, \mathrm{km \, s^{-1} \, Mpc^{-1}}$.

Most recently, Kennicutt, Freedman and Mould[26] in a review unfortunately still only involving authors from one side of the controversy, took a more moderate approach than that of Freedman *et al.*[24] and concluded that H_0 is currently constrained to $80 \pm 17 \, \mathrm{km \, s^{-1} \, Mpc^{-1}}$, and that when distances have been obtained in ~ 25 galaxies from observations of cepheids using the Hubble Space Telescope and combined with several secondary distance indicators, H_0 will be determined to $\pm 10\%$.

We have studied all of the literature, have taken account of what we believe are genuine uncertainties, and also have tried to allow for the prejudices of our friends, colleagues and ourselves, in order to estimate what we believe will be the final value of H_0.

The best standard 'candle' appears to be the absolute magnitude at maximum of supernovae of Type Ia. Thus a key program is to find cepheids in galaxies in which those supernovae have been reported and derive their absolute magnitudes $\overline{M}$ at maximum light. The data have shown that plotting apparent magnitudes of SN Ia against log cz gives a tight relation with a slope[27-30] of 5, the intrinsic dispersion being $\leq 0^{m}\!.3$. Thus the determination of $\overline{M}$ (*max*) should give a good value for H_0, since supernovae can be detected in galaxies far beyond the region of the supercluster where perturbations are present, effectively removing most errors in the redshift values.

Galaxies in which supernovae of the so-called 'Branch' type Ia have been identified and have had distances derived now include IC 4182 (SN 1937C), NGC 5253 (SN 1895B and SN 1972E), NGC 4536 (SN 1981B), NGC 4496 (SN 1960F), NGC 4639 (SN1990N), and NGC 3627 (SN 1989B)[31]. When the corrections are included they are small and $\overline{M}_B(max) = -19.47 \pm 0.07$ and $\overline{M}_V(max) = -19.48 \pm 0.07$. This gives $H_0 = 56 \pm 4 \, \mathrm{km \, s^{-1} \, Mpc^{-1}}$ from $\overline{M}_B(max)$ and $58 \pm 4 \, \mathrm{km \, s^{-1} \, Mpc^{-1}}$ from $\overline{M}_V(max)$.

Thus we have concluded with Sandage and Tammann that the *supernova method is the best currently available.* Having put everything together this leads us to the conclusion that the final value for H_0 lies in the range $58^{+10}_{-5} \, \mathrm{km \, s^{-1} \, Mpc^{-1}}$. We believe therefore that Sandage and Tammann were correct in their original determination some 20 years ago.

Thus one of the most fundamental parameters of cosmology which determines the scale of the universe has been shown to be smaller by a factor of approximately 10 than the value originally derived by Hubble and Humason some 70 years ago. While many people have contributed to this fundamental work, there is no question but that the real pioneers and heroes have been Edwin Hubble, Milton Humason and Allan Sandage using the Mount Wilson 60-inch and 100-inch telescopes, the Hale 200-inch telescope and more recently the Hubble Space Telescope. In the past two or three years there has been extensive publicity associated with the attempts to measure H_0 using the Hubble Space Telescope by many people. The results are now coming in and it is clear that they are confirming the value that was obtained from the ground by Sandage and Tammann more than 20 years ago.

The age dilemma

The view that we live in an expanding universe, which had become widely accepted by the 1930s, implied, using the simplest logic, that there must have been a *beginning* and that the universe has a finite age. As is shown from the elementary theory (Chapter 2), this *age* must be of order H_0^{-1} years. For the value of H_0 of 588 km s^{-1} Mpc^{-1} obtained by Hubble and Humason in 1931, $H_0^{-1} = 1.75 \times 10^9$ years.

Also in the 1930s, there was no certain understanding of the mechanism of energy production in the stars and the only ages for the stars were those estimated much earlier by Eddington and Jeans who obtained values as large as 10^{13} years based on the idea that a significant fraction of the mass of the stars was used up in generating the luminosity. However, since there was no physical theory associated with this estimate, it could more reasonably be thought of as an upper limit to the ages of the known stars.

On the other hand, the work on spontaneous radioactive decay had led to an age of the rocks from the Pb–U ratio of $\sim 1.3 \times 10^9$ years[32], and Rutherford[33] and Aston[34] from the observed ratio U^{235}/U^{238} had obtained an age of 3×10^9 years for the earth. An apocryphal story from that period is that Rutherford met Eddington one day in Cambridge and asked him 'How old is the Universe?' Eddington answered that it was not more than 2000 million years old, whereupon Rutherford took a small rock out of his pocket and simply stated 'This rock is at least 3000 million years old!'.

Whether or not the story is true, it pointed up a dilemma which has persisted ever since. The discrepancy associated with the energy generation in stars remained until Bethe's work in the late 1930s, when it became clear that energy generation in main sequence stars occurs by what were called the proton–proton cycle and the CNO-cycle, both of which convert hydrogen into helium yielding rather more than 6×10^{18} erg for each gram of hydrogen transformed, much less than a total conversion of matter to energy would have given.

A realistic time scale for the evolution of the stars only came after the energy sources had been established and the basic ideas of stellar evolution had been worked out[35]. The realization that age determinations for globular clusters and galactic clusters could be obtained from observations of the color–magnitude diagram led in the 1950s to fairly accurate age determinations of order 12×10^9 years for the oldest cluster[36].

Similarly the age of the elements made in nucleosynthesis was worked out[37,38] using Rutherford's method involving Th^{232}, U^{235}, and U^{238} giving an age for the elements in the solar system $\sim 10 \times 10^9$ years.

As we have just shown, in the same period, the value of H_o was corrected downward from $558 \, \text{km s}^{-1} \, \text{Mpc}^{-1}$, to $180 \, \text{km s}^{-1} \, \text{Mpc}^{-1}$, and then to $75 \, \text{km s}^{-1} \, \text{Mpc}^{-1}$ so that H_o^{-1} was steadily increased. To see more precisely whether or not it can be argued today that the age from cosmology is greater than the ages of the oldest stars and/or the age of the elements, the models derived from the standard theory (Chapter 2) have to be looked at a little more closely. We consider two models.

(i) The standard hot big-bang cosmology, with or without inflation,
(ii) The modified big-bang cosmology including the cosmological constant.

(i) The classical hot big-bang model follows from Einstein's equations of general relativity. In the dust approximation and choosing units such that $c = 1$, we get for the matter density ρ

$$\frac{\dot{S}^2 + k}{S^2} = \frac{8\pi G \rho}{3}, \tag{4.9}$$

$$\rho \propto S^{-3}. \tag{4.10}$$

For the inflationary version, $k = 0$ and $S \propto t^{2/3}$, and the density for $k = 0$ is given by

$$\rho = \frac{3H^2}{8\pi G} \equiv \rho_c \tag{4.11}$$

where $H \equiv \dot{S}/S$ is the Hubble constant of the epoch, and ρ_c is called the *closure* or *critical density*.

For $k \neq 0$, we write

$$\rho = \Omega \rho_c, \tag{4.12}$$

where Ω is a density parameter. For closed models ($k = 1$), $\Omega > 1$, while for open models ($k = -1$), $0 \leq \Omega \leq 1$. The age of the universe, i.e. the time elapsed from $S = 0$ to the present epoch is given in the $k = 0$ case by

$$t_0 = \frac{2}{3H_o}, \tag{4.13}$$

H_o being the present value of Hubble's constant. For the $k = 1$ models the age is less than this value whereas for $k = -1$ it lies between $\frac{2}{3} H_o^{-1}$ and H_o^{-1}.

(ii) The λ cosmologies: here the Einstein equations include the λ-term and equation (4.9) is replaced by

$$\frac{\dot{S}^2 + k}{S^2} - \frac{1}{3}\lambda = \frac{8\pi G\rho}{3}, \tag{4.14}$$

again with $c = 1$. By adjusting λ we may get cosmological models with arbitrarily large ages. The limiting model in this series is the static Einstein universe (Chapter 2), which has $S = $ constant $= S_0$ (say) and $k = 1$ with λ related to ρ by

$$\rho = \frac{\lambda}{4\pi G} \; , \; \lambda = \frac{1}{S_0^2}. \tag{4.15}$$

How far do these models go in resolving the age dilemma?

We have shown that there is now strong evidence that H_o lies in the range 50–$60\,\mathrm{km\,s^{-1}\,Mpc^{-1}}$ so that H_o^{-1} lies in the range 16–19×10^9 years. For model (i), $t_0 = 2/3H_o^{-1} = 13$–11×10^9 years. The best data suggest that the age of the elements lies in the range $12 - 15 \times 10^9$ years[39,40] while the ages of the oldest stars in our Galaxy are in the range $13 - 17 \times 10^9$ years. On this basis it appears that for either $k = 0$ or $k = 1$ the classical cosmological model (i) is close to being ruled out.

For $k = -1$ the largest age for the universe occurs in a model with $\Omega \to 0$ and is $H_0^{-1} \simeq 16$–19×10^9 years, sufficient to at least equal the ages of the oldest globular clusters. Some have preferred to adopt this model, rather than the so-called standard model with $k = 0$. But then there are fundamental theoretical difficulties in explaining the formation of galaxies and of structure generally.

It is this problem, taken seriously in modern times by Peebles and others, that has led to a revival of model (ii), where by adjusting λ it is possible to get an arbitrarily large age which will accommodate the other observationally determined quantities. However, models with $\lambda \neq 0$ raise difficulties concerning the origin of λ to which we shall return later. Of course, none of these arguments is clear cut because there still remain small uncertainties in all of these determinations. For a detailed discussion of the issues involved, see the work of Bagla, Padmanabhen and Narlikar[41].

At the same time, we are confident that for each determination the data are improving and the level of uncertainty is being lowered. Thus *we* tend to believe that the data are already good enough to conclude that (i) is ruled out by current observations. However others, driven by different devils, are still reaching different conclusions. Some are obviously so sure that (i) *must* be the correct model that they are convinced that ultimately it will all fall in line.

In one paper recently in circulation, some of the most dedicated big bangers have put forward the thesis that H_o must be no greater than $30 \, \text{km s}^{-1} \, \text{Mpc}^{-1}$. Needless to say, some would consider them to be emotionally driven theorists. Others, still convinced of the reality of a big bang, have moved on to model (ii).

As will be shown later, we are convinced that there was no hot big bang, and that the observed value of H_o is not a constraint. According to our theory, which will be discussed in the later chapters, stars and galaxies much older than H_o^{-1} must be present.

Independent of all beliefs, it appears that the measurements of H_o locally, and the ages of the oldest stars and the chemical elements, are becoming well determined, and in each case the results are converging to values in the range $\sim 10 \times 10^9$ years up to $\sim 15 \times 10^9$ years.

Recent results from the Hipparcos satellite confirm in a rather spectacular way that the ground-based observations of the galactic cepheid zero points are correct, so that the distance scale calibrations made from the ground are correct[42]. At the same time, it is possible that some of the Hipparcos results may change again our estimates of the ages of the stars.

But we have now reached a stage in which it is clear that all three age determinations are rather closely in agreement, and this in itself is a fundamental result.

References

Chapter 4

1. Hubble, E. 1936, *The Realm of the Nebulae* (Oxford University Press).
2. Mattig, W. 1958, *Astron. Nacht.*, **284**, 109.
3. Humason, M.L. 1936, *Ap. J.*, **83**, 10.
4. Humason, M.L., Mayall, N.U. & Sandage, A.R. 1956, *A. J.*, **61**, 97.
5. Mayall, N.U. 1935, *Pub. Astron. Soc. Pacific*, **47**, 319.
6. Pettit, E. 1954, *Ap. J.*, **120**, 413.
7. Stebbins, J. & Whitford, A.E. 1952, *Ap. J.*, **115**, 284.
8. Humason, M.L. & Wahlquist, H.D. 1955, *A. J.*, **60**, 254.
9. Shapley, H. 1918, *Ap. J.*, **48**, 89.
10. Wilson, R.E. 1923, *A. J.*, **35**, 35; *Ap. J.*, **89**, 218.
11. Baade, W. 1944, *Ap. J.*, **100**, 137.
12. Mineur, H. 1952, *C. R. Acad. Paris*, **220**, 495.
13. Baade, W. 1953, *Trans. IAU*, **8** (Cambridge U. Press), p. 387.
14. Blaauw, A. & Morgan, H.-R. 1954, *Bull. A. Ned.*, **12**, 95.
15. Sandage, A. 1958, *Ap. J.*, **127**, 513.
16. Sandage, A. 1962, in *Problems of Extragalactic Research*, Proc. IAU Symp. 15 (Kluwer, Dordrecht), p. 359.
17. Sandage, A. & Tammann 1974a, *Ap. J.*, **190**, 525.

18. Sandage, A. & Tammann 1974b, *Ap. J.*, **194**, 559.
19. Sandage, A. 1995, in *The Deep Universe* (Saas-Fee Advanced Course 23, eds. A.R. Sandage, R.G. Kron, M.S. Longair, Berlin, Heidelberg, Springer), p. 77, Fig. 4.2.
20. de Vaucouleurs, G. 1982, *Lectures, Austr. Natl. Univ. (Canberra)*.
21. van den Bergh, S. 1992, *Pub. Astron. Soc. Pacific*, **104**, 861.
22. Jacoby, G. H. *et al.* 1992, *Pub. Astron. Soc. Pacific*, **104**, 599.
23. de Vaucouleurs, G. see Abell, G. 1975, in *Galaxies and the Universe* (eds. A. and M. Sandage, Chicago, University Chicago Press), Ch. 15.
24. Freedman, W. L. *et al.* 1994, *Nature*, **371**, 757.
25. Sandage, A. & Tammann, G. 1990, *Ap. J.*, **365**, 1.
26. Kennicutt, R.C., Freedman, W.L. & Mould, J.R. 1995, *A. J.*, **110**, 1476.
27. Branch, D. & Tammann, G.A. 1992, *ARA&A*, **30**, 359.
28. Sandage, A. & Tammann, G.A. 1993, *Ap. J.*, **415**, 1.
29. Hamuy, M., Phillips, M.M., Maza, J., Suntzeff, N.B., Schommer, R.A. & Avilez, R. 1995, *A. J.*, **109**, 1.
30. Riess, A.G., Press, W.H. & Kirshner, R.P. 1995, *Ap. J.*, **438**, L17.
31. Sandage, A., Saha, A., Tammann, G.A., Labhardt, L., Panagia, N. & Macchetto, F.D. 1996, *Ap. J.*, **460**, L15; *Ap. J.*, **466**, 55; *Ap. J. S.*, **107**, 693.
32. Holmes, A. & Lawson, R. 1927, *Amer. J. Sci.*, **13**, 327.
33. Rutherford, E. 1929, *Nature*, **123**, 314.
34. Aston, F. 1929, *Nature*, **123**, 313.
35. Hoyle, F. & Schwarzschild, M. 1995, *Ap. J. S.*, **2**, 1; Schwarzschild, M. 1958, *Structure and Evolution of the Stars* (Princeton University Press).
36. Hoyle, F. 1959, *M.N.R.A.S.*, **119**, 124.
37. Burbidge, E.M., Burbidge, G., Fowler, W.A. & Hoyle, F. 1957, *Rev. Mod. Phys.*, **29**, 547.
38. Hoyle, F. & Fowler, W.A. 1960, *Ann. Phys.*, **10**, 280.
39. Fowler, W.A. 1987, *Q. J. R. Astron. Soc.*, **28**, 87.
40. Cowan, J.J., Thielemann, F.-K. & Truran, J.W. 1991, *ARA&A*, **29**, 447.
41. Bagla, J.S., Padmanabhan, T. and Narlikar, J.V. 1996, *Com. Astrophys.*, **18**, 275.
42. Sandage, A. & Tammann, G.A. 1998, *M.N.R.A.S.*, **293**, L23.

5

Changing times, 1945–65: new techniques and new people

So far we have largely given a historical view of the development of observational cosmology up to around 1950, although in the last chapter we have gone on to bring the reader up to date concerning the current discussion of H_0 and the age of the universe, in an evolving cosmology.

The period 1945–50 was a kind of watershed in the study of observational cosmology. Before it, all of the observations were made using optical telescopes and very few people were involved. The observational work was carried out predominantly by Edwin Hubble, Milton Humason, Nicholas Mayall, Joel Stebbins, and Albert Whitford, and the work was nearly all done with the Mount Wilson and the Lick telescopes.

But then after the Second World War many things changed and new factors entered.

1. In 1948 the 200-inch Palomar telescope (the Hale telescope) came into operation. It had been advertised as the instrument with which the cosmological problem would be solved. In the 1950s, following the death of Hubble in 1953, Sandage together with Humason started a new program to extend the redshift–apparent magnitude relation to much larger redshifts using that telescope.

2. Radio astronomy opened a new window on the universe, and while initially the bulk of the discrete radio sources were thought to lie in our own Galaxy, by the early 1950s it was known that many were extragalactic, and even while most of them remained unidentified they started to be used for cosmological investigation.

3. A number of physicists, of whom the most visible was George Gamow, with his students, Robert Herman and Ralph Alpher, but also some of the most distinguished, including Enrico Fermi, Maria Mayer and Edward Teller, began to tackle the problem of the origin of the chemical elements in an early universe while one of us (FH) looked at the same problem from

another viewpoint. This followed the work of Hans Bethe who had finally solved the problem of the energy generation in the stars. In particular George Gamow and his colleagues in the period 1948–58 tried very hard to understand how the elements could be built up by rapid neutron capture in the early phases of an evolving universe. This work and its impact on big-bang cosmology will be discussed in Chapter 9. In 1946 one of us (FH)[1] had shown that iron peak elements could be built in stellar interiors. Following that it was then shown that the CNO group could also be built in stars[2].

4. The first direct challenge to the approximately 20 year old view that we live in an evolving universe had been made by the development in 1948 of the steady-state cosmology by Hoyle and separately by Bondi and Gold.

In the following sections and chapters, we shall describe what happened in these areas. Since each of these developments had many repercussions in several areas of cosmology, it is difficult to know which to describe first. However since radio astronomy not only led to measurements of large red-shifts, but also to a major debate about the correct cosmology, and to the new problem of understanding the physics of the sources themselves, we shall start with it. We shall then go back to the attempts to extend the redshift–apparent magnitude diagram.

An outline of the early impact of radio astronomy on cosmology

We shall not retell the very early pre-Second World War history of radio astronomy here. Suffice it to say that by 1950 radio astronomers had detected radio emission, not only from the Sun and from the Milky Way, but also from what they believed were sources in our Galaxy. Theoretical ideas were rudimentary, but in the west Alfvén and Herlofson[3] had proposed that the mechanism of radiation was the incoherent synchrotron process, and Unsöld and Kiepenheuer had suggested that the sources were largely intrinsically faint dwarf flare stars. Although it was not generally known in the west, in the former USSR Ginzburg, Pikelner and Shklovsky had independently also carried out fundamental theoretical studies of the nature of the radio sources and had correctly identified the emission processes[4].

The first post-Second World War observations by Hey *et al.*[5] in 1946 showed, as had been shown earlier by Reber, that much of the emission came from the Galaxy, but that there was a very powerful source in Cygnus. Bolton and Stanley[6] showed that the Cygnus source had a diameter $< 8'$, Ryle and Smith[7] found the powerful source in Cassiopeia, and Bolton[8] had discovered several discrete sources. The first generally accepted optical identification was with the Crab Nebula, followed by other objects which also turned out to be galactic supernova remnants.

However the majority of the sources were unidentified, and their roughly isotropic distribution led most people to believe that they were genuine dwarf flare stars quite close by.

Two of us (GB and FH, not known to each other at the time) attended a meeting at University College London in 1950[9] at which in the discussion, T. Gold pointed out that the isotropy meant that either the sources were close by, or they were extragalactic and very far away. Gold obviously favored the latter hypothesis, but was immediately subjected to severe criticism from Martin Ryle and George McVittie, who both rather bluntly told him that he didn't know what he was talking about. This was the first introduction for one of us (GB), a graduate student at the time, to the real antagonisms that already existed in this field.

In 1951, F.G. Smith[10] obtained the first accurate position for the brightest radio source in Cygnus, which was called Cygnus A. This was sent to Walter Baade and Rudolph Minkowski in Pasadena, and with the 200-inch telescope they identified the source with a 17^m galaxy. Spectra obtained by Minkowski at Palomar showed strong emission lines at a redshift z of 0.057. In 1952, Baade attended the General Assembly of the IAU in Rome, and showed the spectra to many people privately, including Jan Oort, Martin Ryle and one of us (FH). Ryle was visibly dismayed. Baade later described the result at a joint discussion meeting where the protagonists were all present, and Gold, feeling vindicated, publicly made his views of Ryle's earlier criticism very clear. Thus at this early stage, before the cosmological debate had started, the battle lines were drawn.

Actually at this stage only a very few sources had been identified, and Baade and Minkowski did not publish their identification of Cygnus A until 1954[11]. In addition to the Crab Nebula, and Cas A, both supernova remnants in our Galaxy, and Cygnus A, Virgo A had been identified with M87, a bright elliptical galaxy in the Virgo cluster[12] , and in the southern sky Centaurus A had been identified with NGC 5128, a peculiar elliptical galaxy very near to us, and Fornax A with NGC 1316[13]. Thus it was known that there were both galactic and extragalactic sources, and only the isotropic distribution of the unidentified sources together with the spectacular property of Cygnus A suggested that most of the sources were very distant.

Even at this stage it was clear that much could be learned by the study of these sources if the proper procedures were followed. First it would be necessary to identify the sources and try to understand the physics. If we knew *what* they were and *where* they were they might be used as cosmological probes. The one identification of Cygnus A with a galaxy at a redshift $z = 0.06$ meant that sources as intrinsically powerful as it could be detected at very large redshifts. So, it was felt the way was open to making cosmological investigations using objects at large values of z.

The three tasks, identifying the sources with optical galaxies and obtaining their redshifts, understanding their physics, and using them to do cosmology, should ideally have been carried out in that sequence. Following Baade's dictum (Chapter 3) it can reasonably be argued in hindsight that more rapid progress would have been made had we known more about the sources before using them for cosmology. But this was not the way it went. At the best of times research is an untidy business. And in this case the subject was very hot and topical, and several first-class observational groups in radio astronomy were competing with each other. The major groups were Martin Ryle's in Cambridge, England, Bernard Lovell and Robert Hanbury-Brown's at Jodrell Bank, England, John Bolton and his group first in Australia, then at Caltech, USA, and then back at Parkes in Australia, and Bernard Mills and his group at Sydney University in Australia. They were all superb on the technical side, and they were highly competitive. Clearly much prestige, honor and much research funding were at stake. And all of the scientists had the same background – they were electrical engineers and physicists who had learned their techniques in wartime radar work.

Bernard Lovell[14], Robert Hanbury-Brown[15] and Taffy Bowen[16] have each described their work during the Second World War and the developments in radio astronomy afterwards. Lovell's major contribution to radio astronomy was the construction of the 250-foot telescope, which for many years was unique in the world.

In this crucial period in England the major decisions as to what radio telescopes got built and how they were used, and consequently what could be carried out in this new research field, were made by Martin Ryle and Bernard Lovell who divided the available funds between them. If you were a bright young scientist trained in Cambridge or Manchester and wanted to continue to do observational research in radio astronomy, there was nowhere else in the UK you could go. You either managed to stay with one of these groups, which meant staying on the right side of one of these two powerful individuals, or moved to the other, which was rarely done, or you left radio astronomy, or you emigrated to groups abroad.

In our discussion of the observational approach to cosmology using radio sources we shall be very much concerned with the Cambridge group, who made the strongest claims though not always the greatest advances. Thus we start with a few words about the way that they operated. One of us (GB) spent the period 1953–5 working with the group (nominally as a theoretician trying to understand the physics of the sources) and thus had first-hand experience of how things were done. Another of us (FH) was very soon engaged in extensive debates and disputes with them. The third of us (JVN), then a young graduate student, was drawn into the fray in the early 1960s.

As well as great talent, the Cambridge group had great cohesive strength, which was one of the reasons why it achieved so much. The members of the group treated Ryle as a demi-god, and decisions ultimately coming from him were regarded as binding on all members of the group. On every topic, and every issue, there was only one opinion. Differences were ironed out in private sessions so that which member transmitted the information to the outside world was immaterial. And often nothing was transmitted for long periods. The secretive nature of the group became legendary. To give one or two examples: at the time when the analysis of the first counts of radio sources were being carried out leading, after much correction, to the 2C catalog and then the 3C catalog, one of us (GB) was sharing an office in the Cavendish Laboratory with some of the people involved. He recalls:

> In those days I had no interest in, and little knowledge of cosmology. More to the point, I knew nothing of the coming battle with Hoyle nor did I know him. Since I was supposed to be working on theory with the group, one might have supposed that I might have been told something of what was being done and invited to take part. But since I was on the periphery – not an observer, and probably I was not to be trusted since I could never give out the party line without really believing in it, all discussion of that work and what it might lead to was carried out in strict secrecy – I was told nothing. This went on for a year or more. About a week before Ryle was to give the Halley Lecture in Oxford in 1955 at which he would announce that the work showed that the steady-state cosmology was wrong, I was told that I should come to Oxford to hear Ryle make the announcement. Needless to say, I was very angry at this treatment since I could only conclude that I had been deliberately excluded, and I refused to go.
>
> Then there was the behavior at teatime. We used to have tea in the Austin wing of the Cavendish Laboratory every afternoon. I soon found that on the occasions when Hoyle came in to tea – not to sit with us, or even talk to us, but as soon as he was seen, someone from the group was immediately dispatched to go back over to the old Cavendish wing where all of the data were, to see that they were secure and could not be seen by strangers!

This was the atmosphere in the group which achieved so much as seen by their contemporaries. The lesson clearly is that the way to immediate success is often *not* to encourage open minds and freedom of ideas.

But to return to the way that the field advanced. The surveys made by all of the groups showed that hundreds of radio sources down to the 10 Jansky level (10 flux units) at meter wavelengths existed, and that the distribution was largely isotropic. At fainter flux levels (~2 flux units) the counts in the fourth Cambridge catalog (4C), the Parkes catalog and the Molongolo catalog showed that there were many thousands of sources distributed fairly uniformly over the sky.

Optical identifications

To make optical identifications, very accurate positions of the radio sources were required. After the first identifications of powerful extragalactic sources had been made, including Cygnus A, Virgo A (M87), Perseus A (NGC 1275), Fornax A (NGC 1316), Centaurus A (NGC 5128) etc., it was clear that the most luminous sources were associated with intrinsically bright and sometimes disturbed elliptical galaxies, while many of the sources were identified with 'normal' galaxies comparatively nearby and similar to our own galaxy. The majority of the early identifications were made at Palomar. The radio positions were sent to Pasadena, where the optical astronomers made identifications and then obtained spectra of the optical sources. Pioneering work in this area was reported by Baade and Minkowski[11], Matthews, Morgan and Schmidt[17] and Schmidt[18]. Often there was ambiguity in the possible identification, in that there were several faint galaxies in the error box obtained by radio methods. Since it was also known early on that the radio sources tended to be double, with the optical source lying roughly symmetrically between the two radio lobes, this assumption was used in making the identification, i.e. if there was ambiguity, the brightest galaxy lying closest to the centroid of radio emission was taken to be the correct optical identification.

As the number of identifications and redshifts increased, a pattern in the optical properties became apparent. Apart from the intrinsically weak sources like our own galaxy, the identifications out to redshifts $z \simeq 0.2$ turned out mostly to be with giant elliptical galaxies with spectra which show the typical features associated with stars that make up such galaxies. At larger redshifts, however, the spectra showed a much higher frequency of strong broad emission lines, characteristic of highly excited gas, and a component due to starlight was much harder to detect and often impossible to find. Strong emission-line spectra had already been found in the sources Cygnus A and Perseus A, although they have $z \ll 0.2$.

A milestone was reached in 1960 when following the determination of an accurate position of 3C 295 by Elsmore, Ryle and Leslie[19], Minkowski[20] measured its redshift to be $z = 0.461$. This redshift was based on one emission line identified as [OII]λ3727. While Humason had tried in the years before he retired to measure the redshifts of normal galaxies which were faint enough for him to suppose that they had values of z significantly greater than 0.2, he had failed, even though he made exposures of many nights on single objects[a].

This result of Minkowski clearly spurred on efforts to identify faint galaxies in the radio source catalogs to reach higher and higher redshifts. However, at that time Minkowski[21] considered that his result on 3C 295 with $m_v = 21$ was close to the limit of what could be reached with the 200-inch telescope and the existing spectrographs. He believed that it might only be possible to go about one magnitude fainter than he had gone.

It was the next advance in technology that led to the detection of fainter galaxies and larger redshifts. The advent of the image-tube spectrographs in the 1960s, followed by the development of charged coupled detectors (CCDs), which could be used for both spectroscopy and direct imaging, enabled astronomers with large telescopes with mirrors in the 2–4 m class and greater to go much fainter.

Also, the continuing technical advances in radio astronomy meant that source positions could be obtained with greater and greater precision. It was really this latter development that led to the next major discovery in the field. The Owens Valley Radio Observatory of Caltech, set up by John Bolton, contained two 90-foot antennas working as an interferometer and their program was to determine very precise radio positions. Some compact sources with positions accurate to $\pm 5''$ then became available. In 1960 Matthews and Sandage[22] identified 3C 48, a compact radio source with high surface brightness with what appeared to be a faint blue star with a magnitude of about 16.7. Sandage took spectra of the object. He found broad emission lines but could not identify them. The object was clearly not a normal star since it had the colors of a white dwarf and a wisp of nebulosity. In the following two years many spectra of 3C 48 were obtained by many astronomers, but no one could identify the emission lines. And no one published any spectra.

In 1962 Matthews and Sandage[23] identified 3C 196 and 3C 286 with faint stellar objects. In early 1963 Hazard, Mackey and Shimmins[24] identified a

---

[a] Rumors persisted that Humason's lack of success was caused by the fact that another astronomer had damaged the off-set guider (the image of the galaxy is kept on the spectrograph slit by the observer guiding on a star whose position is measured relative to the faint galaxy on direct plates obtained in earlier observing runs).

much brighter stellar object (12^m8) with 3C 273 using the lunar occultation method, and Schmidt and Oke[25] identified the lines in its spectrum as red-shifted Balmer lines at $z = 0.158$. This result immediately led to the identification of the features in 3C 48 as emission lines with $z = 0.367$ by Greenstein and Matthews[26].

Thus it was clear that besides radio galaxies there was another kind of extragalactic radio source – the so-called quasi-stellar radio sources (QSRSs), which often but not always have compact radio structure. Some were classical two-lobe sources (e.g. 3C 147). The typical structure for an extragalactic radio source by now had the shape of two lobes flanking a central optical source. All of the new QSRS objects appear on photographic plates to look like stars, but they have large redshifts.

The identification process went on apace, and with the advent of image tube spectrographs attached to the 120-inch telescope at the Lick Observatory, and the 84-inch telescope at Kitt Peak, and later at other observatories, the monopoly of Palomar in this field was broken, and many more QSRSs were found to have large redshifts. Within a year of the identification and measurement of the redshift of 3C 273 being obtained, Schmidt[27] had found a QSRS, 3C 9, with a redshift of 2.012. Even with the very small sample of objects known by then, it was clear from the values of the redshifts, which ranged from 0.16 to 2, that if these objects were at their redshift distances they were very far from being distributed uniformly in space.

Sandage realized that a similar class of objects which are not detectable as radio sources might very well exist, and he set himself to look for them using color criteria. He had reached this conclusion because when making two exposures to identify radio sources, one through a blue filter, and one through an ultraviolet filter, he sometimes found ultraviolet objects not at the radio positions. Roger Lynds at Kitt Peak had found the same thing. Sandage realized that these were the same kinds of objects as those which had earlier been found by Humason, Zwicky, Luyten, Iriarte, Chavira, Haro, Feige and others, and these earlier surveys showed that there were about four objects per square degree down to $m_B \simeq 19$, far more frequent than QSRSs. Sandage looked at a small sample of such objects fainter than 15^m. Those brighter than this were clearly galactic stars. About half of his sample turned out to be extragalactic.

Sandage caused a sensation when he published a paper in 1965 announcing that he had found a major new constituent of the universe which he called the quasi-stellar galaxies[28]. This was one of the rare occasions when the editor of a journal, in this case S. Chandrasekhar, felt that the paper was so important that he published it immediately without external refereeing – so rapidly in fact that the date of publication printed on the journal was before the date of receipt of the paper. In fact, Sandage had overestimated the frequency of these objects as was shown by Kinman[29] and others almost immediately. Sandage had clearly bruised the egos of some of the people who felt that they had discovered the same phenomenon earlier. However, when the dust settled it was clear that there *is* a population of radio-quiet quasi-stellar objects (QSOs), that far outnumbers the radio-emitting QSOs (QSRSs). The estimates of the density on the sky made in the 1960s of about 10–20 per square degree down to $m_B \simeq 20$ is still correct.

Thus in about ten years the radio astronomers had made observations which led to the discovery first of radio galaxies, and then quasi-stellar objects (QSOs). In the future we shall use the term QSOs for all of the objects which others may call QSOs, QSRSs, or quasars. While we understood little about the physics of these objects, they had much larger redshifts than anything found before, and so it was thought that they would be very valuable cosmological tools. We shall discuss the way the radio galaxies have been used to extend the Hubble law in the next chapter.

References

Chapter 5

1. Hoyle, F. 1946, *M.N.R.A.S.*, **106**, 343.
2. Hoyle, F. 1954, *Ap. J. S.*, **1**, 121.
3. Alfvén, H. & Herlofson, N. 1950, *Phys.* Rev., **78**, 616.
4. Ginzburg, V.L. 1951, *Dokl. Akad. Nauk SSSR*, **76**, 377; see also 1984 *In the Early Years of Radio Astronomy* (ed. W. Sullivan, Cambridge University Press), p. 289.
5. Hey, J.S., Parsons, S.J. & Phillips, J.W. 1946, *Nature*, **158**, 234.
6. Bolton, J.G. & Stanley, G. 1948, *Nature*, **161**, 312.
7. Ryle, M. & Smith, F.G. 1948, *Nature*, **162**, 462.
8. Bolton, J.G. 1948, *Nature*, **162**, 141.
9. Sullivan, W.T. 1988, in *Modern Cosmology in Retrospect* (eds. R. Balbinot *et al.*, Cambridge U. Press).
10. Smith, F.G. 1951, *Nature*, **168**, 555.
11. Baade, W. & Minkowski, R. 1954, *Ap. J.*, **119**, 206.
12. Mills, B.Y. 1953, *Austr. J. Phys.*, **6**, 452.
13. Shain, C.A. 1959, *Proc. IAU Symposium No. 9* (ed. R.N. Bracewell, Stanford Univ. Press), p. 451.
14. Lovell, B. 1991, *Echoes of War, the Story of H_2S Radar* (Bristol, Institute of Physics Publishing Co. Ltd, Adam Hilger).
15. Hanbury-Brown, R. 1991, *Boffin* (Bristol, Institute of Physics Publishing Co. Ltd, Adam Hilger)

16. Bowen, E.G. 1987, *Radar Days* (Bristol, Institute of Physics Publishing Co. Ltd, Adam Hilger).
17. Matthews, T.A., Morgan, W. W. & Schmidt, M. 1964, *Ap. J.*, **140**, 35.
18. Schmidt, M. 1965, *Ap. J.*, **141**, 1.
19. Elsmore, B., Ryle, M. & Leslie, P. 1959, *Mem. R.A.S.*, **68**, 62.
20. Minkowski, R. 1960, *Ap. J.*, **132**, 908; *Pub. Astron. Soc. Pacific*, **72**, 354.
21. Minkowski, R. 1962, *I.A.U. Symp. No. 15* (New York, MacMillan), p. 379.
22. Sandage, A. 1961, *Sky & Telescope*, **21**, 148.
23. Matthews, T.A. & Sandage, A.R. 1963, *Ap. J.*, **138**, 30.
24. Hazard, C., Mackey, M.B. & Shimmins, A.J. 1963, *Nature*, **197**, 1037.
25. Schmidt, M. 1963, *Nature*, **197**, 1040; Oke, J.B. 1963, *Nature*, **197**, 1040.
26. Greenstein, J.L. & Matthews, T.A. 1963, *Nature*, **197**, 1037.
27. Schmidt, M. 1965, *Ap. J.*, **141**, 1295.
28. Sandage, A.R. 1965, *Ap. J.*, **141**, 1560.
29. Kinman, T. 1965, *Ap. J.*, **142**, 1241.

6

The extension of the redshift–apparent magnitude diagram to faint galaxies 1956–95

It was generally believed that when the 200-inch telescope came into operation, it would be possible to measure the redshifts and energy distribution of galaxies with large enough redshifts so that it would be possible to distinguish between cosmological models using the redshift–apparent magnitude relation alone.

In his last discussion of the observations, Hubble in the George Darwin lecture at the Royal Astronomical Society in 1953, a few months before he died, gave the first results obtained using the 200-inch telescope. Humason had re-measured the redshifts for eleven clusters in the redshift–apparent magnitude diagram which extended out to the Hydra cluster ($cz = 60\,000\,\mathrm{km\,s^{-1}}$), and the photometry had been started using modern photometric methods by Sandage who was Hubble's young assistant. Sandage has pointed out that by using the notation 'no recession factor'[a], Hubble was still doubtful if the expansion was real. By this time the 48-inch Schmidt telescope was being used to survey the whole of the northern sky and Abell[1] had begun to systematically catalog clusters of galaxies.

By 1956 Humason had measured the redshifts of 18 clusters at Palomar with the new ones coming from the 48-inch Schmidt Palomar survey. Photometric data were measured for these clusters, using the new photo-electric techniques, and the redshift–apparent magnitude plot out to $cz = 60\,000\,\mathrm{km\,s^{-1}}$ from the data by Humason, Mayall and Sandage (Reference 4 in Chapter 4) is shown Fig. 6.1. An important advance was that the photometric data had been tied to a proper standard and a well-defined zero point and proper procedures were used to measure 'total' or 'fractional' magnitudes relative to the galaxy diameters.

---

[a]Meaning no correction for the number effect.

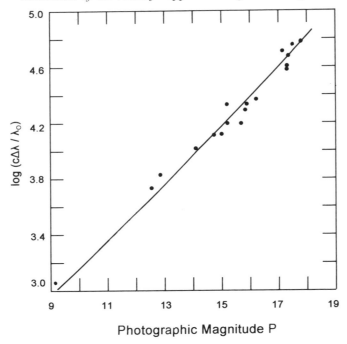

Fig. 6.1. Plot of $\log c\, \triangle\, \lambda/\lambda_0$ against photographic magnitude P (Humason, Mayall & Sandage, Reference 4 in Chapter 4, where corrections to P and to the fitted line are described).

By 1972 Sandage with his associates had measured the redshifts of many more clusters and the Hubble diagram extended out to $z \simeq 0.25$. The largest redshift was that for the radio galaxy 3C 295 with $z = 0.461$, which Minkowski had measured in 1960. The relation, which relied on the brightest elliptical galaxies in clusters, was still strictly linear (Fig. 6.2).

To extend the relation further it was necessary to find fainter clusters. This was done by photographing random fields with the 200-inch telescope[2] and by identifying clusters containing 3C radio sources where it was known that no *bright* galaxy could be the source (the so-called empty fields)[3]. Both methods led to the discovery of fainter galaxies and hence to larger redshifts.

However, while it is possible in principle to use the Hubble relation to distinguish between cosmological models, this can only be done if we take into account the fact that since the fainter galaxies are much further away, we are observing them as they were when they were much younger than they are now. This requires us to make corrections for the evolution of galaxies as a function of epoch, using models which, except at z close to zero, cannot be

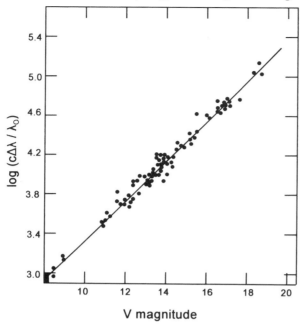

Fig. 6.2. Hubble diagram ($\log c \, \triangle \lambda / \lambda_0$ plotted against corrected V magnitude) for the brightest galaxy in 84 clusters, from Sandage[16].

checked in any detail. Thus this evolutionary correction causes problems if we are trying to determine the deceleration parameter q_0 which measures the departure of the m–$\log cz$ relation from linearity.

The general relation is of the form:

$$m_{bol} = 5\log \frac{c}{q_0^2}\left\{q_0 z + (q_0 - 1)\left[(1 + 2q_0 z)^{1/2} - 1\right]\right\} + \text{constant} \qquad (6.1)$$

which reduces to

$$m_{bol} = 5\log cz + \text{constant} \qquad (6.2)$$

for small z.

Bolometric magnitudes are never measured. However, changing to a magnitude in a particular wavelength range only requires a change in the constant in equation (6.2). This bolometric correction term can be obtained from the spectral energy distribution for galaxies at small z. To convert to objects at high z we must put in a K term, which consists of a term to take account of the fact that the galaxy is no longer being observed in a particular wavelength, but at what for small z would be a shorter wavelength, and also with a bandwidth term $2.5\log(1 + z)$ to account for the stretching of the spectrum.

Moreover, a galaxy at a high redshift is seen earlier in its life than a nearby galaxy and it has a different absolute luminosity due to the fact that the stars have not evolved as far as those in nearby galaxies. Thus there will be a change in the luminosity which we call $E(z)$, leading to an apparent magnitude given by

$$m_1 = M_1 - K_1(z) - E_1(z)$$
$$+ 5 \log c q_0^{-2} \{ q_0 z + (q_0 - 1)[(1 + 2 q_0 z)^{1/2} - 1] \} + \text{constant.}$$
$$(6.3)$$

For specific cosmological models:

$$q_0 = +1,$$
$$m = M - K(z) - E(z) + 5 \log cz + \text{constant} \qquad (6.4)$$

$$q_0 = \frac{1}{2}, \text{ and to second order in } z,$$
$$m = M - K(z) - E(z) + 5 \log cz(1 + z/4) + \text{constant} \qquad (6.5)$$

$$q_0 = 0,$$
$$m = M - K(z) - E(z) + 5 \log cz(1 + z/2) + \text{constant} \qquad (6.6)$$

$$q_0 = -1 (\text{corresponding to the steady-state model}),$$
$$m = M - K(z) - 5 \log z(1 + z) + \text{constant.} \qquad (6.7)$$

Thus, except for the steady-state model, to obtain q_0 from the observed curve, we must be able to determine $E(z)$. This problem was actively worked on starting in the 1970s by Beatrice Tinsley, and by others in the period 1968–80 (cf. Tinsley and Gunn[4]). Beatrice Tinsley, who died prematurely, became a leader of this endeavor.

A simple approach to determining $E(z)$ is to calculate the change in the luminosity of the main sequence termination point in the Hertzsprung–Russell diagram for the stars in a standard elliptical galaxy as a function of time. Here we are following the method of Sandage and Tammann[5]. The look-back time can be calculated separately, and the two factors can be combined to obtain $E(z)$.

For example, Sandage has shown that if L_N is the luminosity at $z \simeq 0$ and L_T is the intrinsic luminosity of a galaxy at redshift z, then for a model in which $q_0 = 0$,

$$E(z) = 2.5 \log \left(\frac{L_N}{L_T} \right) \approx \ln \left(\frac{1}{1+z} \right). \qquad (6.8)$$

For this cosmological model and a Hubble constant of $H_o = 50$ km s^{-1} Mpc^{-1}, $E(z, t) = 0.07^m$ per 10^9 years.

A more precise calculation would take into account the fact that the stellar evolutionary effects come both from the main sequence stars and the luminosity function along the red giant branch. According to Sandage the more detailed models do not change the result very much. We show in Figs 6.3 and 6.4 how the luminosity evolution changes the Hubble diagram[6] for z out to ~ 0.5.

A greater uncertainty comes from the possibility that the stellar contents of the galaxies used as standard candles are *not* the same from galaxy to galaxy. Many faint clusters with redshifts out to $z \sim 0.5$ have by now been investigated[3,7,8] and it has become increasingly clear that the complications are increased, so that it is now obvious that the scatter in the data and the evolutionary corrections make it difficult to determine q_0 by this method with any precision. Crudely, it has been concluded that using this method $q_0 = +1 \pm 2$.

Moreover, Butcher and Oemler[9] discovered in 1978 that in at least one faint high-redshift cluster there are many faint blue galaxies. This discovery

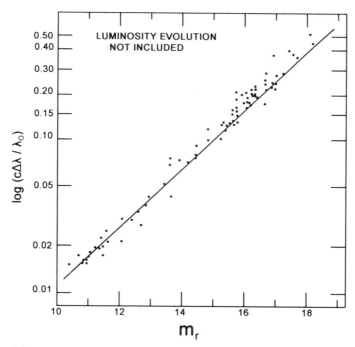

Fig. 6.3. Hubble diagram, without luminosity evolution, using corrected *R*-based magnitudes, from Sandage[6].

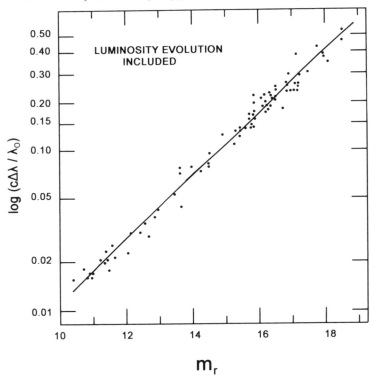

Fig. 6.4. Hubble diagram, with luminosity evolution, using corrected *R*-based magnitudes, from Sandage[6].

clearly showed that it could not be assumed that all of the faint clusters simply consisted of galaxies identical to local ($z \simeq 0$) ellipticals seen at earlier epochs. Thus the attempt to determine q_0 by measuring departures from linearity in the Hubble diagram of normal galaxies was abandoned until recently, when it was revised by using Type I supernovae in distant galaxies to measure their distances. See Chapter 7.

To extend the redshift–apparent magnitude relation to much larger red-shifts, it has appeared to some that the solution is to turn to the radio galaxies, whose first identifications were described in Chapter 5. However, while the low-redshift radio galaxies are predominantly bright ellipticals, the evidence that this is the case for the higher-redshift identifications is much weaker. Thus while much larger redshifts for radio galaxies, up to $z \simeq 3$ are now being obtained, the question of whether or not the Hubble diagram can be used safely for cosmological investigations is open to question.

High *z* radio galaxies are rare, but they are comparatively easy to find from radio surveys. This is because they have strong emission lines, and are luminous compared with ordinary galaxies at optical and infrared wavelengths as well as at radio wavelengths. These characteristics are what give them their 'Achilles heel'. Since they have these characteristics, how typical are they of galaxies in general and how reasonable is it to extrapolate any property we can deduce about their evolution to the general population of galaxies?

In 1987, the alignment effect was discovered[10,11] in radio galaxies with high *z* (HZRG), i.e. $z \gtrsim 0.6$. Here the radio, optical and infrared continua, and the emission line structure, are all extended and aligned along the main radio axis leading to the outer structures. There is no clear understanding of this effect. However, since it is generally agreed that the source of the activity lies deep in the nucleus of the galaxy, and since the explosion of matter which took place to give rise to the extended radio lobes determined the main axis, it is clear that this same event must have been responsible in some basic way for the alignment effect. Thus what caused the underlying galaxy to become a powerful radio source clearly changed its optical and infrared luminosity and its line spectrum. Thus in no sense can it be treated as a standard candle.

However, some have tried to interpret these systems as mature slowly evolving galaxies. Up to $z \simeq 2$, the infrared Hubble relation, *K* of infrared luminosity versus *z* for the 3C radio galaxies, shows small dispersion. This could be interpreted in terms of the normal stellar evolution of elliptical galaxies if a large burst of star formation took place at $z \gtrsim 5$, and the galaxy declined gradually in brightness. Lilly and Longair[12] showed that the amount of evolution present in the *K*–*z* relation is compatible with such a model.

It is then necessary to argue that the alignment is due to stars recently formed in the wake of the ejection of the radio source.

The alternative possibility that the HZRGs are very young systems has been put forward by Chambers and Charlot[13]. Here it is argued that in essence the whole galaxy, as far as star formation and luminosity is concerned, is being formed in the wake of the outburst that gives rise to the extended radio source. This leads us back to the idea that, as is the case in Cygnus A, the underlying nucleus is buried very deeply and its spectrum is not easily detected.

More than 100 HZRGs are now known[14,15] up to a redshift of about 5. Beyond the 3C sources many HZRG systems have been found by identifying the sources with the steepest radio spectra. This method works for double sources because the spectral index itself is correlated with either radio luminosity or redshift.

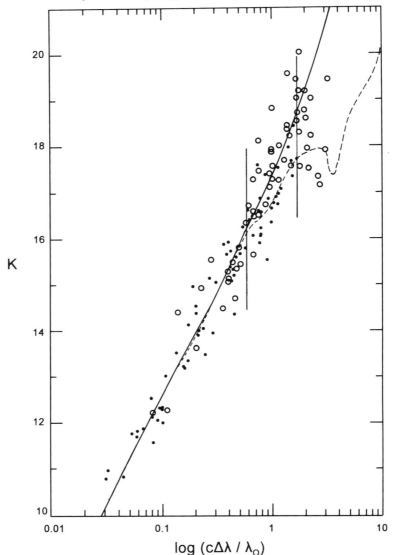

Fig. 6.5. *K*-band magnitude plotted against redshift *z* for radio galaxies, adpated from the plot by Eales and Rawlings[15]. The filled circles are galaxies in the 3C sample. The open circles are galaxies in faint samples chosen to find large redshifts. The continuous curve shows the relation expected if there is no evolution. The other (broken) curve shows the relation expected if all the stars form at a continuous rate for 10^9 years starting at $z = 10$. The vertical lines show the redshift above which the alignment effect starts to be seen ($z \simeq 0.6$), and the maximum redshift of 3C radio galaxies ($z = 1.8$).

The Hubble diagram using the K band luminosities for the radio galaxies is shown in Fig. 6.5, which is taken from Eales and Rawlings[15]. It is clear from this plot that while in a very crude sense the no-evolution curve fits the data, the large scatter is due to some of the effects described here. In particular much scatter must come from the contribution to the K luminosity from the extended emission lines and continuum. Also at large redshifts, we have a highly inhomogeneous sample of galaxies, with a wide range of K luminosities. Clearly it is not possible to determine q_0 by using these objects.

We shall discuss the attempts to detect normal galaxies at large redshifts in the later chapters.

References

Chapter 6

1. Abell, G.O. 1958, *Ap. J. S.*, **3**, 211.
2. Gunn, J.E. & Oke, J.B. 1975, *Ap. J.*, **195**, 255.
3. Kristian, J., Sandage, A. & Westphal, J.A. 1978, *Ap. J.*, **221**, 383.
4. Tinsley, B.M. & Gunn, J. E. 1976, *Ap. J.*, **203**, 52.
5. Sandage, A. & Tammann, G.A. 1983, in *Large Scale Structure of the Universe, Cosmology, and Fundamental Physics* (eds. S. Setti & L. Van Hove, Geneva, ESO/CERN), p. 127.
6. Sandage, A. 1995, in *The Deep Universe* (Saas-Fee Advanced Course 23, eds. A.R. Sandage, R.G. Kron & M.S. Longair, Berlin, Heidelberg, Springer), p. 117.
7. Schneider, D. P., Gunn, J.E. & Hoessel, J. G. 1983, *Ap. J.*, **264**, 337.
8. Dressler, A. & Gunn, J.E. 1992, *Ap. J. S.*, **78**, 1.
9. Butcher, H. & Oemler, A. 1978, *Ap. J.*, **219**, 18.
10. Chambers, K.C., Miley, G. & van Breugel, W. 1987, *Nature*, **329**, 604.
11. McCarthy, P. J., Spinrad, H. & van Breugel, W. 1995, *Ap. J. S.*, **99**, 27.
12. Lilly, S.J. & Longair, M.S. 1984, *M.N.R.A.S.*, **211**, 833.
13. Chambers, K.C. & Charlot, S. 1990, *Ap. J.*, **348**, L1.
14. McCarthy, P. 1993, *Ann. Rev. A.& A.*, **31**, 639.
15. Eales, S.A. & Rawlings, S. 1996, *Ap. J.*, **460**, 68.
16. Sandage, A. 1972, *Ap. J.*, **178**, 1.

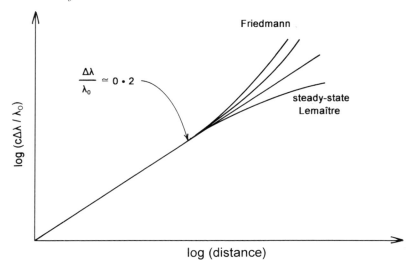

Fig 7.1. Rate of expansion of the universe ($\log c \, \Delta \lambda/\lambda_0$ plotted against log (distance)) for various cosmological methods, showing schematically how these depart from a linear relation, from Hoyle[8].

positive. As was shown in Chapter 2 for the steady-state model, $q_0 = -1$. However, as we just described in Chapter 6, we now know that the uncertainties associated with determining an accurate value of q_0 are very great, so that the argument was not very strong at all [a]. Thus it is not surprising that as soon as a new objection came along, this line of argument was heard less often.

The war of the source counts

As soon as it became clear that the radio sources are extragalactic, it was realized they could be used for cosmological investigation. By counting them as a function of flux density, Ryle and others attempted to carry out investigations similar to those made by Hubble who had tried to count optical galaxies (cf. Chapter 4).

If we suppose that all radio sources have the same luminosity L, and they are distributed through Euclidean space with uniform density ρ, the intensity of a source at distance r is $S = L/4\pi r^2$ and the number per unit solid angle

---

[a] Indeed, recent determinations of distances of high-redshift galaxies using Type I supernovae suggest that q_0 *is* negative. See Chapter 16.

within a distance r is $N = \frac{1}{3}\rho r^3$. The number having an intensity greater than S is then given by

$$N = \frac{1}{3}\left(\frac{L}{4\pi}\right)^{3/2} S^{-3/2}\rho, \tag{7.1}$$

and if we plot log N against log S we should find for sources in Euclidean space a straight line with a slope of $-3/2$. The effect of a spread in intrinsic luminosity L will not seriously affect this result provided that the range of luminosities is not too great.

Thus the method used was to construct a log$N$–log$S$ curve from the radio catalogs. When the log$N$–log$S$ method is applied to sources distributed over a large range of distances, it can be shown that the effect of the curvature of space is to flatten the log$N$–log$S$ relation so that the slope falls (numerically) below -1.5 at the faint end, the amount of flattening being determined by the cosmological model. Thus it was claimed that without knowing very much about the sources, the counts would be a powerful cosmological tool.

Ryle[9] unveiled the first result using this method applied to his early catalog of sources following a period of extreme secrecy when he delivered the Halley Lecture at Oxford on May 6, 1955. We show in Fig. 7.2 Ryle's plot of log$N$ against log$S$.

Now the steady-state cosmology predicts that, taken on a large enough scale, the sources should not change in their statistical properties. This means that the log$N$–log$S$ curve should flatten to a value that is different from the evolutionary models. Ryle concluded that since the slope of the log$N$–log$S$ curve was far steeper than -1.5 (he gave a slope of -3.0) the steady-state cosmology was ruled out. He believed that the steep slope meant that there must have been many more sources in the past, suggesting strong evolution.

The alternative possibility was that the counts were in error. In fact, by 1958 the same group[10] had concluded that there were problems with the confusion of sources in the earlier work, and from new observations at 159 MHz they found that the log$N$–log$S$ slope was close to -2.2. However, with the same messianic certainty as before, the Cambridge group claimed this corrected slope ruled out the steady-state theory. What did it matter that the slope had become -2 instead of the previous value of -3 when -2 was sufficient for their purpose? It mattered of course because, if errors could close more than half the gap between the observations and the theory, further errors could close the entire gap.

The catalog on which Ryle's original claims were based – the second Cambridge catalog 2C – was published by Shakeshaft *et al.*[11]. It contained

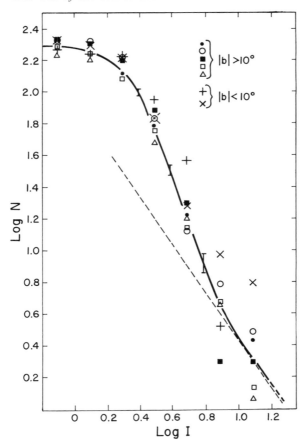

Fig 7.2. The original $\log N - \log I$ curve announced and published by Ryle[9] when he gave his Halley Lecture. It was this (completely erroneous) slope that led to the general belief that the steady-state cosmological model was wrong. (Ryle used I = intensity rather than S = flux density.)

1936 sources and very few optical identifications (< 20) had been obtained. However, in Australia B.Y. Mills was also carrying out a survey of sources using the Mills Cross, a 1500 foot long cross-shaped array of dipoles. Soon after Ryle's announcement and claim concerning the results, it was possible to compare Ryle's results with those obtained by Mills in the southern hemisphere. J. Pawsey, the leading Australian radio astronomer, came to England in 1955 for the Jodrell Bank Symposium on Radio Astronomy. He showed Mills' results in which for 500–600 sources, Mills had found that the slope of the $\log N$–$\log S$ curve was not much different from -1.5, in strong disagreement with Ryle. Ryle refused to listen to any criticism of the Cambridge results. The consequence was that Mills and Slee[12] published a paper in 1957 in which they concluded:

> We have shown that in the sample area which is included in the recent Cambridge catalog of radio sources (an area which had been surveyed by both groups) there is a striking disagreement between the two catalogs. Reasons are advanced for supposing that the Cambridge survey is very seriously affected by instrumental effects which have a trivial influence on the Sydney results. We therefore conclude that discrepancies, in the main, reflect errors in the Cambridge catalog, and accordingly deductions of cosmological interest derived from its analysis are without foundation. An analysis of our results shows that **there is no clear evidence for any effect of cosmological importance in the source counts** [our emphasis].

But the damage to the steady-state theory had been done. By making his claim so strongly and so publicly, Ryle had carried out a pre-emptive strike against the steady-state model. No matter that his data were suspect and perhaps wrong. His claims were believed by very influential scientists. For example, Jan Oort wrote to Ryle 'I have just read your and Scheuer's article on the radio stars, and do not want to wait another hour to congratulate you with the splendid discovery you have made. For me, it is the most fascinating discovery I have ever seen.'[13]

Two of us (FH and GB) knew Jan Oort very well, and it is clear that from that time on he firmly believed that the radio-source counts disproved the steady-state cosmology. This was for many years his strongest argument for an evolutionary universe. But like many others he was not qualified to make judgements as to who was right, Ryle or Mills. But many in the scientific establishment believed the above claim because Oort said so. And this is often how science progresses – or regresses.

Ryle continued to press the attack. Late in 1960 one of us (FH) had a telephone call from the headquarters of the Mullard Company in London. The Mullard Company had made extensive donations to research in radio astronomy at the Cavendish Laboratory, so much so that the radio telescopes of Ryle's group outside Cambridge at Lord's Bridge were known as the Mullard Observatories. Hoyle was told that during the coming week Ryle would be announcing new results that he might find of interest and he and his wife were invited to be present at the announcement. He accepted the invitation, and went to London. He was seated on the platform with various media representatives present. Ryle was introduced and launched into a lecture and Hoyle immediately realized that he had been set up. The results involved a larger catalog of fainter sources, the so-called 4C survey, the earlier results having been from the 3C survey.

approach (perhaps more restrained than Gold's would have been) was characteristically cynical when he congratulated Ryle for reducing the discrepancy between the data and the steady state theory once more. He further argued that he had seen the steepness of the slope come down from 3, to 2.2., to 1.8, and that he would be willing to wait longer for it to come down even further! Naturally these remarks did not go down well with Ryle and his coworkers.

In the early 1960s, inhomogeneities on $\geq$ 50 Mpc scales were indeed unimaginable. Now they are generally accepted to be real[b]. The universe just happens to be more inhomogeneous than theoreticians would have liked it to be. The writing of the detailed paper[15] on this work was carried out by the two of us (FH and JVN) and presented at the RAS in the April 14 meeting. In a second paper[16] published in the following year, a Monte Carlo computer simulation of log$N$–log$S$ curves for a variety of randomly placed observers in inhomogeneous steady-state models were made to demonstrate that the local slope can easily vary between ~ -1.3 to -1.9. This may well have been the first Monte Carlo simulation in cosmology made possible by the arrival of the second generation computers. Partly as a result of this work, and partly because other arguments against the steady-state theory were soon to take the spotlight off the radio-counting controversy, Ryle and Hoyle had little further debate on this issue. However, the general view of the steady-state theory remained that it was in conflict with the radio observations and therefore must be wrong.

The controversy had been started and, sooner or later, it had to be resolved. In 1968, one of us (FH) reviewed the matter as it then stood[8]. Of the 250 or so extragalactic sources of the 3C survey, about 160 were radio galaxies, and the rest were either unidentified or were QSOs. For the QSOs plus unidentified sources the slope was about -2, while that for the radio galaxies was about -1.3.

Already at the Paris Symposium in 1958 Robert Hanbury-Brown had pointed out that Ryle was not counting a uniform set of objects. If the QSOs are local objects (cf. Chapter 11) then the slope of their log$N$–log$S$ curve is irrelevant for cosmology. If they are not local objects, their properties are so variable from one to another that they do not satisfy the necessary criterion of being intrinsically similar objects. Thus any unequivocal conclusion to be drawn from the log$N$–log$S$ slope had to be made from the radio galaxies which, it is now known, constitute more than two-thirds of

---

[b]We discuss large-scale structure in Chapter 20. We now know that large-scale structure with scales $\sim$50–100 Mpc do exist.

the 3C sample. Already in 1968 with a slope of -1.3, it was clear that the distribution of radio galaxies was not inconsistent with the steady-state theory. Ryle still argued otherwise, always maintaining that, although he had lost every battle from the time of the conference in London in 1950, he had nevertheless won the war. And of course the scientific community believed that he had and rewarded him accordingly.

There was no means of settling this final issue until the distances of the bulk of radio galaxies in the sample became known, a situation that in 1968 seemed impossibly far away. However, the optical observers now came back into the picture. Starting in the 1970s a long observing program was begun by Hyron Spinrad and his colleagues using the Lick 3-meter telescope and the Kitt Peak 4-meter telescope. He and his group began a systematic program to obtain redshifts of all of the extragalactic objects in the 3CR catalog of radio sources. There are now known to be just 50 quasi-stellar objects (QSOs) and the remainder, nearly 200, are so-called radio galaxies. By 1985, this group, with a few measurements from others, and the early work previously described, were able to publish a catalog in which more than 95% of all of the objects have measured redshifts[17]. This work was a real tour-de-force, since when it began it appeared likely that the very faintest galaxies identified would require observations made by larger-aperture telescopes or the Hubble Space Telescope.

Following this work, a test of the steady-state theory using first-class data for a complete sample could be made, since for the radio galaxies, assuming that the redshifts give distances, we know all of the luminosities directly. The analysis made by Das Gupta together with JVN and GB was published in 1988[18].

It was first shown that within the framework of the steady-state theory, the luminosity function could be determined uniquely. Then the counts to be expected were worked out and compared with the old counts which were made when most of the sources were unidentified. Das Gupta carried out this final test by using the Monte Carlo technique. However, we describe it here in terms of old-fashioned statistics. When the difference between expected theoretical values and actual values lies within 1 standard deviation, the theoretical prediction is judged to be well-satisfied. When the difference is 2 standard deviations, it is usual to raise eyebrows. When the difference is 3 standard deviations it is considered reasonable to question the correctness of the theory and, at 4 standard deviations or more, the theory is considered wrong, unless there are strong counter-arguments in its favor. The final result obtained by Das Gupta was good to within about half a standard deviation, a typical situation for a correct theory.

So about 30 years after the original argument began, the truth emerged. It demonstrates that far-reaching conclusions should not be based on shadowy data, especially if the conclusions are being used in the sense of disproving a theory. Because in strict logic theories are never completely proved, poor data used to support a theory do less damage than poor data used to disprove a theory.

The fact remains that Ryle and his colleagues were believed by most of the opinion makers and were given many medals, prizes and financial support. Some good theorists were turned off the steady state not because they thought it was an uninteresting theory. As a Nobel laureate who knew nothing about radio astronomy told one of us, 'It has been disproved by the observations and therefore does not merit further study'.

It is of interest in concluding this discussion to see how what was thought in the 1960s to be a matter of over-riding importance has ended. Thus in the *Proceedings of IAU Symposium 124*, K.I. Kellerman and J.V. Wall[19] dismiss the matter in a single paragraph as follows:

> Between source densities of about 20 to 50 sources per steradian, the observed number of sources appears to increase rapidly with a slope which is in excess of the Euclidean value. But over the whole sky there are only about 500 sources that contribute to this part of the count, and so the statistics are limited. It is this rapid rise in the source density which occurs only over a limited range of flux density, that led to earlier claims of a 'huge excess of weak sources'. The best-fitting slope in this region of the source count is significantly steeper than the Euclidean value, but only over a limited range of perhaps three to one in flux density.

So what the immense fuss created by Ryle and his secretive group amounts to is that it involved 'only about 500 sources' and that it amounts to an upward kick in the source count curve only over a factor of about 3 in the flux density. In Chapter 16 we shall see how this upward kick is well-explained by the recent quasi-steady-state theory, a development of the old steady-state theory (cf. Fig. 16.4).

But in the 1960s, for those who had always believed that the steady-state theory must be wrong, there was a new rod at hand to beat it with. This was the existence of the microwave background, to which we turn next.

The microwave background radiation

In the next chapter we discuss the history of the discovery of the microwave background radiation. Here we only comment on its impact on the debate concerning the steady-state theory. For some years before the first modern detection of the microwave background radiation by Penzias and Wilson[20] several powerful groups of cosmologists, and particularly the Russian school of Zeldovich and his colleagues in Moscow, and R.H. Dicke and his students in Princeton, were convinced that the hot big-bang model was correct, and were promoting the view that the microwave background radiation arising from such a big bang would be found. They set the tone for the view that the Penzias–Wilson discovery not only showed that their view was correct, but that it brought the death knell to the steady-state theory.

How was this certainty displayed? For example, the Russians always called the microwave background the 'relict' radiation, implying that they already *knew* where it came from. Also when Penzias and Wilson detected the radiation at one frequency there was, of course, no evidence that the radiation had a black-body energy distribution as it must have if, according to Gamow and his colleagues, it originated in an early universe. It took some years for this to be established, but a very heavy bias was involved in all future measurements. All of the observers believed that the curve *should* have a black-body form, so that when some of the early experiments gave a flux lying above the black-body curve they were immediately suspect. Of course, ultimately the COBE data showed that those measurements were wrong and the curve does have an almost exact black-body form out to centimeter wavelengths. However, as we shall see in the next chapter, the development of ideas following the earliest work of McKellar might very well have led down a very different track.

The view in the 1960s was that to get such a black-body curve from many discrete sources, as is required in a steady-state theory, was highly contrived, and could in effect be ruled out. Thus this discovery was considered to be the strongest piece of evidence against the steady-state theory, stronger than the Stebbins–Whitford effect, the observed value of q_0, or the counts of radio sources! However, in Chapter 16 we shall consider what we believe is an excellent model that will explain the microwave background based on the quasi-steady-state model which gives rise to a near black-body curve fitting all of the COBE data.

We have shown in this chapter that personal beliefs and personal prejudice do enter strongly into cosmology. Lest the reader thinks that we are exaggerating, we conclude by quoting the comments of Professor H. Dingle in his Presidential Address to the Royal Astronomical Society in 1953[21]. His theme was that the steady-state cosmological model was not even science!

Directing his criticism to the 'perfect cosmological principle' which was a major part of the formulation of the steady-state cosmology in the work of Bondi and Gold[1], he stated:

> No great scientific work, it is true, has been done without the free and bold use of imagination, but let its products be properly assessed before they are announced as discoveries of the order of nature. Even idle speculation may not be quite valueless if it is recognized for what it is. If the new cosmologists would observe this proviso, calling a spade a spade and not a perfect agricultural principle, one's only cause for regret would be that such great talents were spent for so little profit.
>
> But I am not yet convinced that facility in performing mathematical operations must inevitably deprive its possessor of the power of elementary reasoning, though the evidence against me is strong. Let our younger cosmologists forget cosmology for the space of three years – the universe is patient – it can wait, and instead read the history of science – I mean, the work of the great scientists themselves. After asking themselves what meaning it has for the work of today, let them return to cosmology and give their attention again to the great problems into which they have prematurely rushed.
>
> I do not enjoy the task of arraigning those whose mathematical facility greatly exceeds their judgement of scientific authenticity, and who have in consequence exercised this facility on any premises that will give it scope. But one who, however unworthy, accepts the honor of presiding over one of the foremost scientific societies of the world, accepts a responsibility. The ideas to which we give publicity are accepted as genuine scientific pronouncements and as such influence the thinking of philosophers, theologians, and all who realize that in no intellectual problem, however fundamental, can scientific research now be ignored. And so when it happens we have published, in the name of science, so-called 'principles' that in origin and character are identical with the 'principles' that all celestial movements are circular and all celestial bodies immutable, it becomes my duty to point out that this is precisely the kind of celebration that science was created to displace.

We now turn to a historical discussion of the way that the microwave background radiation had its impact on cosmological ideas.

References

Chapter 7

1. Bondi, H. & Gold, T. 1948, *M.N.R.A.S.*, **108**, 252.
2. Hoyle, F. 1948, *M.N.R.A.S.*, **108**, 372.
3. Stebbins, J. & Whitford, A.E. 1948, *Ap. J.*, **108**, 413.
4. de Vaucouleurs, G. 1948, *Comptes Rendus*, **237**, 466.

5. Code, A.D. 1959, *Pub. Astr. Soc. Pacific*, **71**, 118.
6. Whitford, A.E. 1952, *Ap. J.*, **120**, 599.
7. Oke, J.B. & Sandage, A.R. 1968, *Ap. J.*, **154**, 21.
8. Hoyle, F. 1969, *Proc. R. Soc. Lond.* A, **308**, 1.
9. Ryle, M. 1955, *Observatory*, **75**, 137.
10. Archer, S., Baldwin, J., Edge, D., Elsmore, B., Scheuer, P. & Shakeshaft, J. 1959, *Proc. Paris Symposium on Radio Astronomy* (Stanford Press), p. 487.
11. Shakeshaft, J., Ryle, M., Baldwin, J., Elsmore, B. & Thomson, D. 1955, *Mem. R.A.S.*, **67**, 97.
12. Mills, B.Y. & Slee, O.B. 1957, *Austr. J. Phys.*, **10**, 162.
13. Letter quoted by Woodruff T. Sullivan, III, 1988, in *Modern Cosmology in Retrospect* (eds. R. Balbinot *et al.*, Cambridge U. Press).
14. Gold, T. & Hoyle, F. 1959, *Proc. Paris Symposium No. 9 on Radio Astronomy* (Stanford Univ. Press), p. 583.
15. Hoyle, F. & Narlikar, J.V. 1962, *M.N.R.A.S.*, **123**, 133.
16. Hoyle, F. & Narlikar, J.V. 1963, *M.N.R.A.S.*, **125**, 13.
17. Spinrad, H., Djorgovski, S., Marr, J. & Aguilar, L. 1985, *Pub. Astron. Soc. Pacific*, **97**, 932.
18. Das Gupta, P., Narlikar, J.V. & Burbidge, G.R. 1988, *A. J.*, **95**, 5. See also *Non-evolving Luminosity Functions for Radio Galaxies*, P. Das Gupta 1988, Ph.D. Thesis, Bombay University.
19. Kellermann, K. & Wall, J.V. 1987, *IAU Symposium No. 124* (eds. A. Hewitt, G. Burbidge, G. & L.-Z. Fang, Dordrecht, D. Reidel), p. 545.
20. Penzias, A.A. & Wilson, R.W. 1965, *Ap. J.*, **142**, 419.
21. Dingle, H. 1953, *Observatory*, **73**, 42.

8

The cosmic microwave background – an historical account

History up to 1965

It is generally supposed that the time order in which discoveries are made should have no eventual influence on our beliefs. But it is very evident that they do, and perhaps nowhere more importantly than with the cosmic microwave background.

The cosmic microwave background first showed itself observationally to astronomers in the late 1930s and early 1940s, when Adams and Dunham observed a number of interstellar absorption lines in the blue, identified by Swings and Rosenfeld[1], by McKellar[2], and by Douglas and Herzberg[3], as arising from the diatomic molecules CH, CH^+ and CN. Of these, those lines from CN were the most immediately important. One was from the $J = 1$ level of the ground state. The first of these was to the $J = 0$ rotational state of the higher $B^2\Sigma$ multiplet, and the others were to the $J = 1$ and $J = 2$ rotational levels of the same $B^2\Sigma$ multiplet. In 1941, McKellar[4] interpreted the required population of the $J = 1$ rotational level of the ground state as being caused by radiative excitation from the $J = 0$ level. The exciting radiation was taken to be *black body*, and the temperature required for it, in order to explain the relative intensities of the observed lines, was 2.3 K.

This detection of the microwave background was so very explicit that its discovery could quite properly be dated from 1941, if it suited astronomers to do so. The sociological problem of course in 1941 was that Europe was deeply into a disastrous war. And so, within a few more months, would be the United States. There was simply no opportunity for McKellar's work to be adequately discussed. It might have been discussed after the Second World War, but the world recovered only very slowly, and it was really not until the meeting of the IAU in Rome in 1952 that an opportunity became available. By then McKellar's paper had, unfortunately, lain too long in observatory files and had become forgotten. A further problem is that McKellar published in an Observatory publication which almost no physicist had heard of. The extent to which this still remains true is shown by the reference list of the otherwise excellent article by Boesgaard and Steigman[5] with the title 'Big Bang Nucleosynthesis: Theories and Observations'. This article contains of the order of 500 references on the cosmic microwave background. But it does not contain any reference to McKellar's determination of its temperature. The attitude of the post-1965 astronomical world is apparently that the work of the 1940 period was all a peculiar aberration best now forgotten. The hope of the present authors, however, is that by 2041, when the centenary of McKellar's discovery comes round, it will at last be remembered, and that by then it will be the strident cosmological pronouncements of the last third of the twentieth century on this matter that have become forgotten.

In 1946, George Gamow published a short but far-reaching letter[6]. Up to then the effect of radiation on the universal expansion had been thought to have no important observable consequences in the modern universe, with the result that the early cosmologists of the 1920s and 1930s had tended to omit it. What Gamow argued in 1946 was that the chemical elements as a whole were synthesized in the early universe by neutron addition, a view that was directly opposite to that of one of us (FH) who published his first paper on the theory of stellar nucleosynthesis also in 1946[7].

At that time there was little interest in the synthesis of helium, because by then it was widely accepted that helium was synthesized from hydrogen in ordinary stars like the Sun. And observational developments had not yet elevated D and ^{3}He into nuclides of importance. Nor was ^{7}Li seen as being of importance. This had to wait until the work of Wagoner, Fowler and Hoyle[8] in 1967. Thus the issue as it presented itself to astronomers gradually was essentially concerned with the synthesis of the elements from carbon upwards in atomic mass.

Nevertheless two remarkable short letters involving helium were published in 1948 by Gamow[9] and by Alpher and Herman[10]. They were remarkable in pointing out that if helium is considered to have been synthesized in the early universe, then at the present day there should be a thermodynamic radiation field with a temperature of about 5 K. In the 1950s, in private conversation with one of us (FH), Gamow raised this estimate to about 10 K. By something of a freak of circumstance, one of us (FH) argued in discussions with George Gamow that none of this could be true, because if it were, then McKellar in 1941 would have obtained 5 K or more for the excitation temperature of the ground-level of the CN molecule.

Gamow's method used to obtain this estimate was strikingly modern. It made use of the baryon–photon ratio $\eta \cdot 10^{-10}$ remaining constant between helium synthesis in an early universe and the present day. The number of quanta present in a black-body radiation field of temperature T K is $20.3T^3 \mathrm{cm}^{-2}$. Writing η_b for the baryon number density we therefore have

$$10^{-10}\eta = \frac{\eta_b}{20.3T^3}. \tag{8.1}$$

Gamow and his colleagues used an early calculation of the kind described in Chapters 9 and 10 to obtain $\eta \simeq 4$ in order that the resulting value of the mass fraction Y of helium should be 0.25, the value commonly quoted at the time for the mass fraction originally present in stars. It was then argued that the present-day mass density of baryonic material in galaxies is about $10^{-30} \mathrm{g\ cm}^{-3}$, giving $\eta_b \simeq 10^{-6} \mathrm{cm}^{-3}$, and (8.1) leads immediately to $T \simeq 5$ K. The relation (8.1) is used differently today, in the form

$$\eta_b = 2.03 \times 10^{-9}\eta T^3. \tag{8.2}$$

Thus with T considered measured at 2.73 K and with $\eta \simeq 6$ as the best estimate from the arguments of Chapters 9 and 10 gives

$$\eta_b = 2.51 \times 10^{-7} \mathrm{cm}^{-3},$$

corresponding to a present-day mass density for baryons of $4.2 \times 10^{-31} \mathrm{g\ cm}^{-3}$, a fraction 0.062 of the total present-day critical mass density ρ_c given by $3H_o^2/8\pi G$, the latter being calculated using $H_o \simeq 60$ km s^{-1} Mpc. This is the ground for arguing in the standard big-bang model that the 'missing mass' required to make up ρ_c cannot be baryonic. Thus those who argue in favor of the missing mass being brown dwarfs or large planets are in effect denying that the light elements were synthesized in this manner. And if the missing mass could indeed be shown to be brown dwarfs/large planets, or any other non-visible form of baryonic material, then it would not be correct to conclude that the light elements were synthesized in the manner claimed by Gamow and his colleagues. There are three possibilities:

    (a) η has a value in the range from 1 to 10 in an early universe of the kind considered in the standard model;
    (b) η has a value in the range 1 to 10 but in local objects where the physical conditions are similar to those of the early universe;
    (c) the light elements were synthesized elsewhere.

Those who believe in brown dwarfs/large planets as the source of *all* missing mass must also believe in either (b) or (c).

Returning to history, Gamow and his colleagues continued to work at the ideas of generating the chemical elements about 100 s after the origin of the universe. Further support for this idea was gained in the wake of a number of important papers, of which an article by Alpher and Herman[11] in *Reviews of Modern Physics* was especially outstanding. But on this occasion the universe saw fit to challenge the developing orthodoxy. By 1953–4 it had become clear that the elements from carbon upwards were produced in the stars, not in the early universe. Making the best of a bad job for the orthodox, J. Robert Oppenheimer announced that the concept of the formation of the elements in stars was a case of a correct idea emerging from a wrong one: the wrong idea being the steady-state theory, which did not have a singular initial high-density state and so it would not incorporate Gamow's theory. Oppenheimer had a mellifluent style, and by the early 1950s he had become so accustomed to being able to persuade university audiences of anything he said, that he gave the impression that he no longer troubled himself to check the facts. Otherwise, half an hour spent in the library at Princeton would have shown him that the steady-state theory dated from 1948, whereas the first extended paper on the formation of elements in stars dated from 1946. Indeed, Eddington had spoken in a qualitative way in favor of the idea from *c.*1920, even before there were any dynamic cosmological theories (the steady-state theory has the cosmological redshift effect and so it has dynamics, despite its name).

The loss of the argument over the formation of elements from carbon upwards had a depressing effect on Gamow and his colleagues, their last major paper on the subject appearing in 1953[12]. While the previous situation could have been maintained for ^{4}He, with the values of Y in stars still arising from nucleosynthesis in an early universe, helium was not an element over which there was much excitement in the early 1950s. As remarked above, helium was known to be synthesized from hydrogen in ordinary stars, and that seemed enough for most astronomers.

Paradoxically it might seem, it was the opponents of Gamow in the argument over the origin of the elements who next began to worry over helium. It was especially necessary for supporters of the steady-state theory to worry about helium. Because in that theory it was essential for the helium to be produced as an on-going process. It had become widely accepted, following a determination by Oort, that the average universal density of galactic material was about 3×10^{-31} g cm^{-3}. In the steady-state theory there should be no difference between galactic material considered locally, for which $Y \simeq 0.25$, and galactic material considered cosmologically. Thus $Y \simeq 0.25$ seemed a cosmologically significant number, implying a universal helium mass density equal to one-quarter of the baryonic density of visible matter, namely $7.5 \times 10^{-32}\text{g cm}^{-3}$. Since the energy yield from the transformation of one gram of hydrogen to helium is a little above $6 \times 10^{18}\text{erg}$, this meant that there should be a radiation background in the universe of 7.5×10^{-32} times 6×10^{18}, equal to $4.5 \times 10^{-13}\text{erg cm}^{-3}$. The embarrassment was that the energy density of starlight was observed to be not greater than $\sim 10^{-14}\text{erg cm}^{-3}$.

This embarrassment was not faced until 1955, when Bondi, Gold and Hoyle admitted the difficulty[13], pointing out that an escape from it required a universal radiation background of energy density $4.5 \times 10^{-13}\text{erg cm}^{-3}$ to be present in the far infrared. There was some difference between the three authors as to how the argument should be presented. Gold argued in favor of a thermalized background, because nature always turns out to be more efficient at degrading the quality of energy than we expect it to be. Whereas Bondi and Hoyle argued that because no adequate thermalizing agent seemed available, the more cautious position should be taken of simply placing the background generally in the far infrared. And on a vote of two-to-one, this was how the matter was published.

In subsequent years a great deal turned on this decision, which it should not have done. But when it comes to how the world decides broad issues in science, far too much depends on chance. Had Gold's picture of nature being an arch-degrader of energy been followed, a calculation of the resulting thermodynamic temperatures would have followed immediately.

Since
$$aT^4 = 4.5 \times 10^{-13}\text{erg cm}^{-3}, \tag{8.3}$$
and

$$a = 7.564 \times 10^{-15}\text{erg cm}^{-3}\ \text{K}^{-4}$$

$$T = 2.78\ \text{K}. \tag{8.4}$$

Had this elementary step been taken in 1955, it is easy to contemplate that when the cosmic microwave background was detected in 1965 to have a temperature somewhat below $3\,\text{K}$, this estimate of T made in 1955 would have seemed stronger than the estimates 'above $5\,\text{K}$' of Gamow, Alpher and Herman made in 1948. And as subsequent determinations of T lowered its observed values bit-by-bit towards the modern value of $2.73\,\text{K}$, big-bang supporters would not have been able to hold center-stage to the extent they actually have. And had Bondi, Gold and Hoyle had the wit to recall McKellar's determination of $2.3\,\text{K}$ for the thermal excitation of CN, it is likely that the big-bang theory would not have been on-stage at all. It is regrettable that the way things fall out should depend on such trivia, but experience shows that on a medium time scale of half a century they do.

Using a different approach in the same period, another of us (GB)[14] showed in 1958 that the luminosity of our galaxy over 10^{10} years could not have synthesized the observed He from H, and argued that either galaxies had been more luminous in the past, or that He was produced elsewhere, or that the radiation energy must be much greater than that in starlight generally, having the same value as that obtained by Bondi *et al.*[13].

History from 1965

In the course of determining the noise factor of a radiating horn at a wavelength of 7.35 cm, Penzias and Wilson in 1965 found they could not account for the whole noise factor in terms of instrumental effects. An unexplained small residue remained. This turned out to be the cosmic microwave background[15]. In the respect that a fundamental discovery came from the determination to pursue and explore, rather than to neglect, a small experimental discrepancy, the situation was similar to Kepler's discovery of the elliptic orbit of Mars, required to explain small discrepancies between Kepler's own improvements on Copernican theory and Tycho Brahe's observations of Mars.

This serendipitous discovery came just at a time when Robert Dicke and others at Princeton had developed a renewed interest in the physics of an early universe and had reworked much of the ground already covered by Gamow and his associates. They had further set out to detect the radiation which was predicted to arise in the early universe. Thus the Penzias–Wilson discovery was heralded by them as proving that the big-bang idea was correct.

With this discovery the study of cosmology was changed. It became headline news, with decisions on its future course made by a new generation of scientists at a technical level, and at a general level of belief by the writers of popular science. The time, 1965, coincided with the beginning of a great expansion in the number working in astronomy, an expansion begun by the intense interest at the time in space, which led to NASA being funded in billions of US dollars, rather than in the few tens of millions that had been previously available to astronomy, even in the most favorable circumstances. The number of attendees at meetings of the International Astronomical Union (IAU) jumped in only a few years from some 300 or 400 to more than 2000, an immense expansion which made it possible to claim that cosmology as a branch of physical science only began in earnest in 1965. All that was needed from the past was to know that Gamow had predicted a micro-wave background and had obtained a temperature of 5 K for it. Whence it was argued that the hot big bang, as it became called, was correct, and any other models were just as obviously wrong. To point to a prediction of $4.5 \times 10^{-13} \mathrm{erg} \, \mathrm{cm}^{-3}$ for some form of an infrared background, as discussed above, carried no weight in itself. To have carried weight the simple addenda contained in equations (8.3) and (8.4) were needed.

When issues come to turn on such inconsequential minutiae, one can feel that a degenerative moment has been reached. And so it proved. After a few years in which some progress was made, notably in the quality of the calcu-lations leading to results for the synthesis of the light elements in big-bang models, a quarter century of near stagnation set in. Of course, to those who live in real time, the impression is not one of stagnation. The impression is one of progress, frantically so at times of stagnation, according to historic examples. Yet we think that in future decades, this is what the period 1970–95 will be seen to have been.

The actual discovery of the microwave background prevented equations (8.3) and (8.4) being added in the casual way they might have been in 1955. If they were to be taken seriously post-1965, a thoroughly satisfactory therma-lizing agent had to be found. For many years this did not prove possible, although ironically the requisite metallurgical knowledge had already become available by 1960. The relevant facts simply remained hidden from those few astronomers who were interested in the problem. Only in the late 1980s, did what Bondi, Gold and Hoyle had been seeking in 1955 become at last avail-able. One can feel it should have been seen much sooner, and if the situation post-1965 in cosmology had been as measured as it was before 1960, perhaps it would have been.

The need was for a form of particle with a high absorption coefficient at micron wavelengths, significantly higher than at optical wavelengths, to pro-duce the absorptions and re-emissions required for thermalization in the far infrared, but without nearby intergalactic space becoming opaque to star-light, which would, of course, contradict observation. The dielectric particles responsible for the extinction of starlight in our own galaxy have properties, however, which are opposite from this. They contain little in the way of oscillators at microwavelengths.

By the early 1970s it was known that condensing carbon vapor forms long thin thread-like particles when cooled relatively slowly instead of being quenched (when quenched, it forms platelike particles of soot). In 1975, Narlikar, Edmunds and Wickramasinghe[16] realized that such whiskers had absorption properties going a step in the direction of converting optical radiation into the infrared, to the extent that carbon whiskers have an absorption coefficient in the far infrared that was not much less than at optical wavelengths. So one can contemplate that carbon whiskers would be capable, if present in the gas clouds within galaxies, of degrading starlight into the infrared generally; but not, it then seemed, of producing a very close approximation to a thermodynamic radiation field throughout the universe. The problem will be considered in detail in Chapter 15.

Metallurgical data available even as early as 1960[17] already showed that whiskers also form when metallic vapors are cooled slowly rather than quenched rapidly. These data were summarized by Hoyle and Wickramasinghe[18] as follows:

> The growth of a condensate begins as a more or less spherical cluster of atoms with a radius that increases to about 0.01μm. At this stage a dramatic change occurs, with the condensate growing linearly with great rapidity up to a length of ~ 1 mm or more, thus giving a ratio of length to diameter of the resulting whisker that is of the order of 10^5.

This was a result that one would not have guessed at without the evidence of experiment. It was an astonishing property which seems to arise for the following physical reasons. The initial condensate up to radius 0.01 μm is one in which the metallic atoms are fluid. The transition at this radius is to a solid crystalline state with a considerable energy output occurring. In the dissipation of this energy output, crystal dislocations arise, and a particular form of dislocation promotes whisker formation, with growth occurring linearly rather than more or less isotropically. Only a minority of dislocations would be expected to be of this special kind, however, but those that arise take the lion's share of the condensing vapor. A spherically growing particle has a mass m proportional to the cube of its radius and a growth rate proportional to its surface area, i.e. proportional to $m^{2/3}$, giving

$$\frac{dm}{dt} = (\text{constant}) \times m^{2/3}. \tag{8.5}$$

But a linearly growing particle has a mass m proportional to the square of the diameter, d^2 say, and is also proportional to the length l, $m \propto d^2 l$, while its growth rate – again proportional to the surface area – goes as ld. Hence if d is held constant during growth,

$$\frac{dm}{dt} = (\text{constant}) \times m. \tag{8.6}$$

If the constants in (8.5) and (8.6) are the same, implying equal attachment probability in the two cases, it is evident that the exponential growth in the linear case outstrips that in the spherical case. This is the second surprise in the process.

The physical reason for linear growth is that only at the ends of the particle are there sites with a sufficient energy of attachment to provide permanent binding in the slowly cooling vapor. Atoms are otherwise absorbed temporarily on to the long cylindrical portion of the surface, where they random walk until either they reach one of the ends and become permanently attached or they evaporate back into the surrounding vapor. The length of ~ 1 mm to which whiskers grow seems to be very reproducible, and seems to be determined by the random walk time along the particle. This random walk time increases with the length of the particle, until it becomes comparable with the residence time of absorption on the curved cylindrical surface. Growth then ceases.

In 1982, Wright[19] suggested that metallic whiskers might be the cause of large emissions of far infrared radiation from interstellar gas clouds. But he used the room-temperature conductivity of iron for the whiskers, $\sim 10^{17} s^{-1}$, at which metallic iron whiskers are not a great deal different in their properties from the carbon whiskers discussed by Narlikar *et al.*[16] in 1975. The resistivity of a metal is commonly regarded as the sum of two parts, $r_0 + r_i$ say, where r_0 is independent of temperature and r_i depends on electron scattering by lattice vibrations and is proportional to the temperature T. At temperatures below ~ 50 K for iron, r_0 dominates. Thus at room temperature the conductivity comes almost wholly from r_i, whereas mostly in interstellar space and in extragalactic space the conductivity comes from r_0, and is $6.9 \times 10^{18} s^{-1}$ for ordinary iron. It was first realized by one of us (FH)[20] in 1987 that at this much higher conductivity iron whiskers would have extraordinarily high absorption at microwaves. Table 8.1 shows the astonishing extent to which this is so.

These values imply a high cosmological opacity, say for an intergalactic iron whisker density of $10^{-34} g\ cm^{-3}$, at the important microwave center close to 1 mm, but little absorption at either optical wavelengths or at radio frequencies ~ 1 GHz. Thus over a cosmological distance of 10^{28} cm we would have a column density of $10^{-6} g\ cm^{-2}$ of iron whiskers, which if they are of length 500 μm, would have an optical depth of 57.3 at a wavelength of 1 mm. Yet at optical wavelengths the optical depth would be $\sim 10^{-3}$ and at 3 GHz it would be only 0.0232.

The values in Table 8.1 are not sacrosanct. The resistivity $r_0 = 0.13$ microhm-cm on which they are based is dependent on the impurity content of the iron used in the laboratory measurements leading to this value[21]. Iron whiskers produced in cooling vapor from supernovae could have variable

Table 8.1. *Mass absorption values for iron whiskers*

WL(MU)	Mass absorption coefficient (cm^2g^{-1})					
	Length					
	1000	500	300	200	100	50 μm
1.00E0	1.86E3	1.86E3	1.86E3	1.86E3	1.86E3	1.86E3
2.00E0	5.00E3	5.00E3	5.00E3	5.00E3	5.00E3	5.00E3
3.00E0	9.24E3	9.24E3	9.24E3	9.24E3	9.24E3	9.34E3
5.00E0	2.04E4	2.04E4	2.04E4	2.04E4	2.04E4	2.04E4
1.00E1	6.00E4	6.00E4	6.00E4	6.00E4	6.00E4	6.00E4
2.00E1	1.72E5	1.72E5	1.72E5	1.72E5	1.72E5	1.72E5
3.00E1	3.19E5	3.19E5	3.19E5	3.19E5	3.19E5	3.19E5
5.00E1	7.12E5	7.12E5	7.12E5	7.12E5	7.12E5	7.12E5
1.00E2	2.21E6	2.21E6	2.21E6	2.21E6	2.21E6	2.21E6
2.00E2	6.89E6	6.89E6	6.89E6	6.89E6	6.89E6	1.01E6
3.00E2	1.30E7	1.30E7	1.30E7	1.30E7	5.81E6	4.49E5
5.00E2	2.68E7	2.68E7	2.68E7	2.42E7	2.14E6	1.62E5
1.00E3	5.73E7	5.73E7	2.73E7	6.99E6	5.39E5	4.04E4
2.00E3	8.88E7	3.96E7	8.11E6	1.82E6	1.35E5	1.01E4
3.00E3	8.54E7	2.13E7	3.73E6	8.14E5	6.00E4	4.49E3
5.00E3	5.50E7	8.63E6	1.37E6	2.94E5	2.16E4	1.62E3
1.00E4	2.06E7	2.27E6	3.45E6	7.37E4	5.40E3	4.04E2
2.00E4	5.89E6	5.76E6	8.65E4	1.84E4	1.35E3	1.01E2
3.00E4	2.69E6	2.57E5	3.84E4	8.19E3	6.00E2	4.49E1
5.00E4	9.81E5	9.26E4	1.38E4	2.95E3	2.16E2	1.62E1
1.00E5	2.47E5	2.32E4	3.46E3	7.37E2	5.40E1	4.04E0

From Hoyle and Wickramasinghe (ref. 18)
d.c. conductivity = 6.90E18

resistivities and so could cover a wider range of values than those given in the table. For radiation in the far infrared, whiskers would be exposed to strong radiation pressure, being blown from regions of high radiation energy density to low, i.e. out of the parent galaxies into intergalactic space.

Thus by 1995, some 30 years after the discovery of Penzias and Wilson, it could be seen that the means to thermalize the background inferred by Bondi, Gold and Hoyle in 1955 was at last available. But, of course, it was 30 years too late to influence the big-bang advocates who remain unimpressed, either by the data shown in Table 8.1 or by the result $T = 2.78$ K to which the argument of 1955 leads. However, in our view such properties as iron whiskers are now seen to possess would, if they were of no relevance to the cosmic microwave background, be freakishly accidental.

If the iron whisker production is taken as 0.1 $M_\odot$ per supernova and if the supernova production rate is taken as 1 per 30 years per galaxy, the total production per galaxy in 10^{10} years is $\sim 2/3 \times 10^{41}$ g. By the term *galaxy* here is implied a major galaxy, of which the spatial density is about 1 per 10^{75} cm^3. Hence the expected whisker density, taking the whiskers to be expelled from their parent galaxies by radiation pressure, becomes

$$2/3 \times 10^{41} \times 10^{-75} \simeq 10^{-34} \text{g cm}^{-3}. \tag{8.7}$$

The calculated density is therefore seen to be $\sim 10^{-34}$ g cm^{-3}, as assumed above. If this had no relevance to the microwave problem, it would be yet another accident.

The isotropy of the cosmic microwave background

Figures 8.1, 8.2, 8.3 and 8.4 show the current COBE observations[22]. Since the temperature range is only $\pm 10^{-4}$ K, about three parts in 10^5 of the background temperature itself, the background is seen to approximate closely to an isotropic state. The well-known dipole component arising from the earth's motion relative to the Hubble flow of the galaxies has been removed from some of the maps. The strongest remaining deviations from uniformity are the dark regions along the Galactic plane, which come from the contamination of the map by our own Galaxy. From these data one can say that the background departs from an isotropic state by no more than about one part in 10^5 when measured with respect to the effective temperature, or a few parts in 10^5 with respect to energy density.

As we also show in Fig. 8.4, relative intensities of the background radiation measured for a wide range of frequencies, are very closely in accordance with a black-body distribution for a temperature of 2.73 K.

Since a black-body distribution is void of information concerning the manner of its origin, it might be surprising to an uncommitted physicist (if such could be found) that the COBE team claims so much from these results. But this was inevitable because the planning of the observations and the funding of COBE were carried out by individuals who were unaware of the history that we have just described, and who had been told that the microwave background was proof that a hot big bang occurred. The extravagant statements quoted in the preface to this book date from the announcement of these COBE results[23,24]. No-one who has spoken on this has been uncommitted. Here, for example, is part of the citation written by Margaret Geller for the award of the Rumford Prize of the American Academy of Arts and Sciences for 1996. John Mather is the leader of the COBE mission.

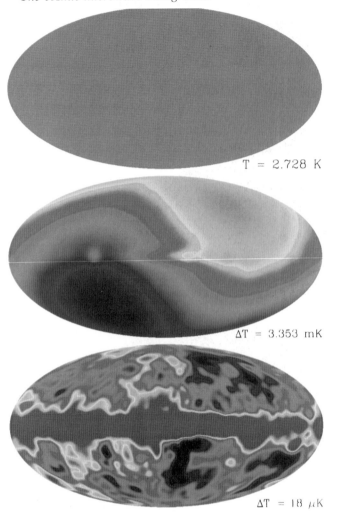

T = 2.728 K

ΔT = 3.353 mK

ΔT = 18 μK

Fig. 8.1. Maps based on 53 GHz (5.7 mm wavelength) observations made with the DMR over the entire four-year mission (top) on a scale from 0–4 K, showing the near-uniformity of the CMB brightness, (middle) on a scale intended to enhance the contrast due to the dipole and (bottom) following subtraction of the dipole component. Emission from the Milky Way Galaxy is evident in the bottom image. We are indebted to the NASA Goddard Space Flight Center and the COBE Science Working Group for this and Figs. 8.2, 8.3 and 8.4.

In 1950 Fred Hoyle christened the 'big-bang' model – to ridicule it. In 1965 the discovery of the cosmic microwave background – the whisper of the big bang – sounded the death knell for Hoyle's rival steady-state theory. Despite all efforts to the contrary, the name 'big bang' has stuck; in fact, the only revision is the addition of the word hot. It's the hot big-bang cosmology

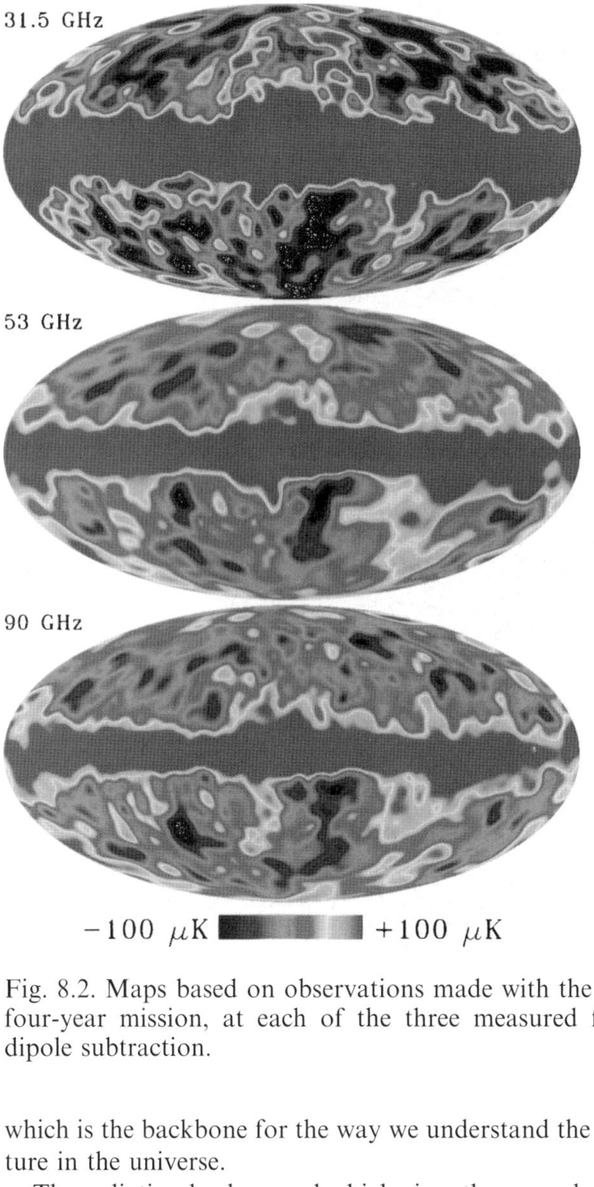
31.5 GHz
53 GHz
90 GHz
−100 μK
+100 μK

Fig. 8.2. Maps based on observations made with the DMR over the entire four-year mission, at each of the three measured frequencies, following dipole subtraction.

which is the backbone for the way we understand the development of structure in the universe.

The radiation background which gives the cosmological model its name – hot big bang – has a very nearly black body spectrum with a temperature of 3° Kelvin, and the temperature is the same in nearly every direction we look; we say that the background is highly isotropic. The best measurements by far of the spectrum of this background and the first measurements of tiny irregularities in this background – irregularities at the level of one part in 10^5 – are both fundamental results of the COBE (Cosmic Background explorer) mission. These measurements tell us about the thermal history of the universe, and they give us a picture of the lumps of matter which were the seeds of the structure we observe in the universe today. These tiny lumps and

Fig. 8.3. The 53 GHz DMR sky map (top) prior to dipole subtraction, (middle) after dipole subtraction, and (bottom) after subtraction of a model of the Galactic emission, based on data gathered over the entire four-year mission. The Galactic emission model is based on DIRBE far-infrared and 408 MHz radio continuum observations.

bumps, which were in the universe when it had an age of 300 000 years, are the origin of galaxies, and in some sense they are connected to our history because, of course, we live in a galaxy – the Milky Way.

With the microwave background coming from the thermalization of radiation from galaxies, there is no great significance to be attached to Fig. 8.1. The effects of the distributions of galaxies and of the thermalizing agents fade completely away as the thermodynamic state is reached; and with little of relevance remaining when the thermodynamic state is approximated as closely as it is shown to do in Fig. 8.4. We shall discuss this again in greater detail in

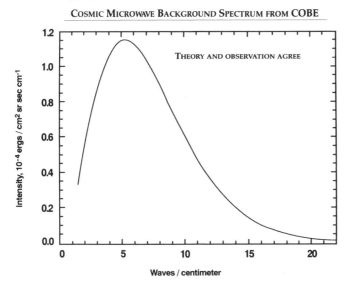

COSMIC MICROWAVE BACKGROUND SPECTRUM FROM **COBE**

Fig. 8.4. FIRAS measure cosmic microwave background radiation residual spectrum from Mather *et al.* 1994, *Ap. J.*, **420**, 439 (also see Fixsen *et al.* 1996, *Ap. J.*, **473**, 576). A Planck black-body spectrum and a small Galactic emission component have been subtracted from the measured spectrum in order to make the residuals visible. To a very good approximation, the cosmic microwave background spectrum is the same as that of a 2.738($\pm$ 0.004) K black-body.

Chapter 16. Why is so much made of these results by supporters of the big-bang models? In such models, there is an inversion of the usual time-order of thermodynamics. For in those models the initial situation is one in which the thermodynamic state is established already, the problem being to invent a sequence of events that led to small deviations from it. Quantum fluctuations are currently favored as the source of the deviations, the issue being to decide what form of fluctuations would lead to Fig. 8.1 at the present day. This indeed is a complicated affair, as it would be if we were to attempt to explain minute fluctuations of the fire heating a steam boiler in terms of the shapes of the coals which the stoker fed into it. Doubtless in the old days of steam someone could have tackled such a problem, but we doubt that anybody ever did. Because it would have been seen as irrelevant to the point of absurdity, as it is in our view in this way of looking at the microwave background. It is only if one starts by thinking one knows how the black-body radiation originated, that its very small real information content assumes apparent significance. In other words, the fluctuations in the figure could have been produced in very many ways, and it is only if one insists on one particular interpretation that anything of significance emerges. But the nature of this cosmological model is that there is no way at all of testing any particular assumption. We are in the realm of untestable theory which many would argue is not physics but metaphysics.

We shall return to this whole question, particularly to the technical details of the thermalization process in our theory, in a later chapter. Also we shall discuss the tremendous effort that is now going on to derive information from the fluctuations on the assumption that the conventional view is correct (Chapter 21).

References

Chapter 8

 1. Swings, P. & Rosenfeld, L. 1937, *Ap. J.*, **86**, 483.
 2. McKellar, A. 1940, *Pub. Astr. Soc. Pacific*, **52**, 307.
 3. Douglas, A.E. & Herzberg, G. 1941, *Ap. J.*, **94**, 381.
 4. McKellar, A. 1941, *Pub. Dom. Astrophys. Observatory, Victoria, B.C.* 7, 251.
 5. Boesgaard, A. M. & Steigman, G. 1985, *Ann. Rev. A & A*, **23**, 319.
 6. Gamow, G. 1946, *Phys. Rev.*, **70**, 572.
 7. Hoyle, F. 1946, *M.N.R.A.S.*, **106**, 343.
 8. Wagoner, R.V., Fowler, W.A. & Hoyle, F. 1967, *Ap. J.*, **148**, 3.
 9. Gamow, G. 1948, *Nature*, **162**, 680.
10. Alpher, R.A. & Herman, R. 1948, *Nature*, **162**, 774.
11. Alpher, R.A. & Herman, R. 1950, *Revs. Mod. Phys.*, **22**, 153.
12. Alpher, R.A., Fellin, J.W., Jr., & Herman, R. 1953, *Phys. Rev.*, **92**, 1347.
13. Bondi, H., Gold, T. & Hoyle, F. 1955, *Observatory*, **75**, 80.
14. Burbidge, G. 1958, *Publ. Astr. Soc. Pacific*, **70**, 83.
15. Penzias, A.A. & Wilson, R.W. 1965, *Ap. J.*, **142**, 419.
16. Narlikar, J.V., Edmunds, M. & Wickramasinghe, C. 1976, in *Far Infrared Astronomy*, Proc. 1975 Conf. of Roy. Astron. Soc. (ed. M. Rowan Robinson, Pergamon), p. 131.
17. Nabarrow, F.R.N. & Jackson, P.J. 1958, in *Growth and Perfection in Crystals* (eds. R.H. Doremus, P.W. Roberts & D. Turnbull, New York, J. Wiley, 65, 388. Gomez, R.J. 1957, *Chem. Phys.* 26, 1333; 1958, **28**, 457. Dittmar, W. & Neumann, K. 1958, in *Growth and Perfection in Crystals* (eds. R.H. Doremus, P. W. Roberts & D. Turnbull, New York, J. Wiley).
18. Hoyle, F. & Wickramasinghe, C. 1988, *Astrophys. & Space Sci.*, **147**, 248.
19. Wright, E.L. 1982, *Ap. J.*, **255**, 401.
20. Hoyle, F. 1987, in *Highlights in Gravitation and Cosmology*, Proc. International Conference on Gravitation and Cosmology at Goa, India (eds. B. R. Iver, A. Kembavi, J.V. Narlikar, Cambridge), p. 236.
21. White, G.E. & Woods, S.B. 1959, *Phil. Trans. R. Soc. Lond.* A**251**, 273.
22. Smoot, G. *et al.* 1992, *Ap. J.*, **396**, L1.
23. Mather, J.C. *et al.* 1990, *Ap. J.*, **354**, L37.
24. Mather, J.C. *et al.* 1994, *Ap. J.*, **420**, 439.

9

The origin of the light elements

Introduction

Historical considerations of the origin of the light elements overlap those of the microwave background, but with some shifts in emphasis. Thus there will be some overlap between the last chapter and this chapter. Where this is so, we have tended to proceed *de novo* rather than engage in a good deal of back referencing.

The question of the origin of the chemical elements had begun to be discussed already by the later months of 1945. One of the present authors (FH) had noticed that the 'natural abundances' of the elements exhibited a peak at mass numbers centred at $A = 56$. Since this peak coincides well with the range of the most strongly bound nuclei, the implication seemed to be that nuclear reactions had taken place somewhere with sufficient complexity to produce an approach to thermodynamic equilibrium among the basic constituents of nuclei, protons and neutrons. The temperature and density and time scale needed for such a situation could be estimated to be $T \simeq 5.10^9$ K, $\rho \simeq 10^9 \mathrm{g\,cm^{-3}}$, $t \simeq 1$–1000 s. These values suggested an astrophysical source, and it was proposed that supernovae were a likely possibility[1].

A very different view was preferred by many physicists, notably by R. Peierls in the UK, and by Fermi, Teller and Maria Mayer in the US. This was that the elements must have been formed by neutron addition, because otherwise Coulomb barrier problems would cause insurmountable difficulties. Peierls conceived of the process as taking place within 'neutron lumps', aggregations of upwards of a thousand neutrons, but without any definite site for the existence of such lumps being specified. It was Gamow who first suggested a site.

In his letter published in 1946, Gamow[2] pointed out that electron degeneracy in the early universe at high density would more than compensate for the mass difference between the neutron on the one hand, and a proton plus electron on the other. Whence Gamow argued that matter in the early moments of a Friedmann cosmology would be a single large neutron lump. Thus the existence of the chemical elements in general could be seen, according to Gamow's letter, as a verification of the Friedmann model.

It was quickly pointed out that there is a problem with this because there are gaps at atomic number $A = 5$ and $A = 8$ where there are no 'stable nuclei'. It was also soon realized that the presence of a radiation field in the early universe would lead to a more complicated situation with not all protons forced by degeneracy considerations into neutrons. These and other complications were taken into account in a more refined theory developed by Gamow and his collaborators[3,4,5].

The key point in these developments was the requirement for the process to synthesize an amount of helium in agreement with the observed value $Y \simeq 0.25$–0.30 by mass compared to hydrogen. Although helium was known to be produced from hydrogen inside stars, it was argued that stellar synthesis could only make a negligible contribution to Y. The Sun has an average energy production of about $2 \text{ erg g}^{-1} \text{ s}^{-1}$, while spiral galaxies such as our own are generally recognized to be less efficient in energy production per gram than the Sun by factors ranging from about 3 to 10; they may produce, say an average of $\sim 0.5 \text{ erg g}^{-1} \text{ s}^{-1}$. Since the complete transformation of hydrogen to helium yields somewhat more than $6 \times 10^{18} \text{ erg g}^{-1}$, the time needed to convert Y from zero to 0.25 typically in spiral galaxies would be $\sim 3 \times 10^{18}$ s, or about 10^{11} years. In 1950 the Hubble constant was still believed to be $\sim 550 \text{ km s}^{-1} \text{ Mpc}^{-1}$ or $H_{\text{o}}^{-1} \approx 2 \times 10^{9}$ years. Thus there seemed to be available much less time than was needed to explain $Y \simeq 0.25$ from astrophysical processes. Whence the decision was taken to explain Y from primordial synthesis in the early universe. For the moment we will pass this historic point by, but will return to it later in this chapter.

Setting aside the details of the calculation of Y, which took some years to clear up, it quickly became established that only if the energy density of radiation in the early universe was very large compared with the rest mass energy of matter can $Y \simeq 0.25$ be obtained primordially. Until then the opposite, that rest mass energy is much greater than radiation, had always been assumed in the Friedmann models. The immediate effect was to require the scale factor $S(t)$ of the universe to be proportional to $t^{1/2}$. Omitting electron–positron pairs, the radiation temperature T is inversely proportional to S, so that T in such a situation would be proportional to $t^{-1/2}$. With radiation alone, i.e. no neutrinos,

$$T_9 = \frac{15.2}{t^{1/2}}, \tag{9.1}$$

where T_9 is T measured in units of 10^9 K, and t is in seconds. However, the numerical coefficient in (9.1) is modified by the presence of electron–positron pairs and by neutrinos. For temperatures high enough for the electrons and positrons to be relativistic and for two mass-less neutrino types the modification is to

$$T_9 = \frac{10.4}{t^{1/2}}. \tag{9.2}$$

So long as the energy of the early universe is dominated by radiation, (9.2) can be considered to hold, or (9.1) in the more primitive theory of 1950. But the next step is completely *ad hoc*. And it remains just as *ad hoc* today as it was in 1950. The mass density of stable non-relativistic particles, explicitly neutrons and protons in the theory of 1950, decreases with the expansion of the universe proportionately to S^{-3}, i.e. as $t^{-3/2}$. So far so good. Alpher and Herman denoted this mass density by ρ_M. We will denote it by ρ_b (baryon). They took

$$\rho_b = 1.70 \times 10^{-2} t^{-3/2} \text{g cm}^{-3}, \tag{9.3}$$

the coefficient 1.70×10^{-2} being the *ad hoc* step. There is nothing in the big-bang theory, even in its most developed modern form, which fixes the value of this coefficient. It is a free choice that is hopefully adopted to make things come out right. In particular, it has been used to make the calculated value of Y agree with the observational scenario. This has meant that its value has changed over the years.

It is common to find that students emerge from a cosmology course in modern times believing that the big-bang theory *explains* the observed microwave background and that it also *explains* a cosmic helium value with Y close to 0.25. This is to distort the meaning of words. Explanations in science are normally considered to be like theorems in mathematics, to flow deductively from axioms and not to be mere restatements of the axioms themselves. As, for instance, the Dirac equation turned out to explain the fine structure of the hydrogen atom. Thus the radiation-dominated early universe is an axiom of modern big-bang cosmology, and the supposed explanation of the microwave background is a restatement of that axiom. When we return to the origin of the microwave background in Chapter 16, we shall offer what we hope is a true explanation. We shall also attempt to do this for Y.

Eliminating the time t between equations (9.1) and (9.3) gives

$$\rho_b = 4.84 \times 10^{-6} T_9^3 = 4.84 \times 10^{-33} T^3 \text{g cm}^{-3}, \tag{9.4}$$

which can be expressed as a baryon to photon ratio

$$\eta = 1.4 \times 10^{-10}. \tag{9.5}$$

A similar calculation for (9.2) and (9.3) gives

$$\eta = 4.5 \times 10^{-10}, \tag{9.6}$$

close to what is taken for η today. It gives $Y = 0.24$, close to the value preferred today. Thus already in 1950, Alpher and Herman had generated what is currently thought to be the best coefficient of proportionality to be inserted in (9.3).

The modern calculation of Y begins at $T_9 \gtrsim 20$ with equilibrium in the weak interactions connecting the protons and neutrons. Then follows an awkward integration down to $T_9 \sim 5$, in which both the weak interactions and the recombination of electron–positron pairs have to be followed in detail. Then down to $T_9 \simeq 1$ with the main effect to be followed being the decay of the free neutron. So far the coefficient of proportionality between ρ_b and T_9^3 in (9.4) has no effect, although the number of neutrino types – the difference between (9.1) and (9.2) – has to be correctly chosen. But then as T_9 falls below 1 the choice of the coefficient in (9.4), or equivalently the parameter η, becomes crucial. The first step in primordial helium synthesis is through the production of deuterium by the reversible reaction

$$p + n \rightleftarrows D + \gamma.$$

Until the temperature falls to $T_9 \simeq 1$, the reaction is strongly right-to-left because of the low binding energy of D (2.225 MeV). There is a short period as the temperature falls below $T_9 \simeq 1$ during which D tends to accumulate. Thereafter helium synthesis occurs through the following chains

$$D(D, n)^3\text{He}, \quad {}^3\text{He}({}^3\text{He}, 2p)^4\text{He},$$

$$D(D, p)T, \quad T(T, 2n)^4\text{He},$$

$${}^3\text{He}(D, p)^4\text{He}, \quad T(D, n)^4\text{He},$$

$${}^3\text{He}(T, D)^4\text{He},$$

which produce small quantities of D, ^{3}He as well as ^{4}He. The eventual abundance of ^{4}He by mass is given essentially by

$$Y = \frac{2n}{n+p} \; ,$$

where n, p are the neutron and proton densities as the temperature falls to $T_9 \simeq 1$.

A matter of outlook

When a theory is specifically adjusted to have a certain property, it cannot be given over-much credit for having that property. Which is how it is with the production of helium in the hot big bang. Examination of the papers cited earlier shows that the theory was quite explicitly constructed to fit the helium requirement. Consequently its ability to give $Y \simeq 0.25$ is not in itself worth a great deal as an indication of its correctness or otherwise. For that it is necessary to turn to other properties.

We remarked earlier on the problem of crossing the gaps at $A = 5$ and $A = 8$, originally in the theory for neutrons alone. When protons were further included, leading to the production of helium, reactions involving protons, neutrons and alpha particles became possible. And these were in principle capable of crossing both gaps through reactions such as

$$\alpha(\alpha, p)^7\text{Li} \quad \text{and} \quad \alpha(2\alpha, p)^{11}\text{B},$$

or $2\alpha(p, \gamma)^9\text{Be}$. But the low-density universe implied by the choice of the numerical factor in (9.4) produced yields from such reactions as these that were negligible. Thus in effect the gaps at $A = 5$ and $A = 8$ remained.

By 1950 a quite different difficulty had also emerged, a difficulty connected with Gamow's neutron-building chain of 1946,

$$A + n \rightarrow (A + 1) + \gamma.$$

The conditions in the early universe as envisioned by Gamow were those of rapid neutron addition, referred to some years later when adapted to stellar nucleosynthesis as the r-process[6]. Successive additions increase the atomic number, initially A, successively to $A + 1$, $A + 2$, ..., until the next neutron becomes too weakly bound to avoid its being lost again in an inverse reaction. At this stage the process must wait for a β-decay to occur, when further additions with increased binding energies become possible.

When one is dealing with large values of A, say $A > 70$, the details of the process show that the ultimately stable nuclei emerging from it correspond to the neutron-rich isotopes of the elements – the so-called r-process nuclei. But in nature, there are also comparatively neutron-deficient isotopes – the so-called s-process nuclides. Many of the latter cannot be reached through β-decays from the former. In 1950 the s-process nuclides were referred to as the shielded isotopes. They could not be produced in Gamow's model.

Despite these difficulties most physicists remained sympathetic to the cosmological explanation of the origin of the elements, demonstrating the extent to which the scientific community is willing to overlook difficulties, even severe difficulties, when it is dealing with a theory for which it has an emotional preference. The case of J. Robert Oppenheimer that we mentioned before could be said to typify this attitude.

It is generally agreed nowadays that most of the elements were not produced in the early universe. The modern claim is that while nuclides with $A \geq 12$ were produced inside stars, D, ^{3}He, ^{4}He and ^{7}Li were produced in the early universe, while ^{6}Li, ^{9}Be, ^{10}B and ^{11}B have been astrophysically generated. It is usually argued that these latter isotopes were made in spallation reactions of cosmic-ray protons on ^{12}C and ^{16}O, or through the inverse reactions – accelerated ^{12}C and ^{16}O on ambient hydrogen.

This complex argument does not appeal to a gambling man. A professional does not make odds based on emotions and impressions but by using past form. Thus a horse given odds of 100/1 may possibly win. But it rarely does. And successful professional gamblers generally live more prosperously than cosmologists. So if we put our trust in form, we would say that, since nuclides with $A \geq 12$ are produced in stars, all of the isotopes are produced astrophysically, and not cosmologically.

If this view is correct D, ^{3}He, ^{4}He and ^{7}Li must have been produced astrophysically. In a previous chapter we already had a startling indication that this might be so. We saw that production in stars, sufficient to give $Y \simeq 0.25$, will yield the observed microwave radiation background. The calculation was made using a mean universal density of galactic material of $\sim 3 \times 10^{-31} \mathrm{g\ cm}^{-3}$, a value very close to what is observed from the density of visible galaxies and the measured mass-to-light ratios. The resulting radiation when thermalized gives a temperature of ~ 2.78 K. If ^{4}He is made in a big bang, the fact that this temperature is so close to the observed temperature is a pure coincidence.

It was noted earlier in this chapter that research in this area was deflected away from this seemingly correct path in the early 1950s by what seemed to be a serious time scale problem. This was at a time when the Hubble constant was thought to be $\sim 550\ \mathrm{km\ s}^{-1}\ \mathrm{Mpc}^{-1}$. Now that the value is smaller by a factor of ten and the time scale ten times longer, the problem is less formidable. Moreover, in the quasi-steady-state model (QSSC) to be discussed in Chapters 15 and 16 (with oscillatory features set out explicitly in Table 16.1) the time scale is of the required order of magnitude. We also discuss in Chapter 16 the details of the thermalization of starlight to microwaves.

Thus as will be shown the QSSC is a sharper theory than big-bang cosmology, even in terms of its explanation of what are two of the latter's strongest points, namely helium production and the microwave background. In big-bang cosmology the radiation background was put into the Friedmann models by hypothesis, as we showed earlier. Whereas in the QSSC some form of background cannot be avoided, because we actually observe radiation being produced by stars. Thus even without refinements, we can say that the production of radiation by stars must inevitably produce a background with an energy density that is within an order of magnitude of what is observed. Whereas in big-bang cosmology the energy density of the background (in effect the parameter η) is a matter wholly of assumption.

It is frequently said that big-bang theory predicted the microwave background. This is only true if putting the factor 1.7×10^{-2} arbitrarily in equation (9.3) is called a prediction. If instead of eliminating the time t between (9.3) and the relation (9.2), we eliminate t between (9.3) and the relation (9.1) we get

$$\rho_b = 1.51 \times 10^{-32} T^3 \text{g cm}^{-3}, \tag{9.7}$$

which can be rounded off to $\rho_b \simeq 10^{-32} T^3 \text{g cm}^{-2}$. Alpher and Herman then put $\rho_b \simeq 10^{-30} \text{g cm}^{-2}$ for the present-day density of galactic material, and so obtained a value of about 5 K for T. This differs by a factor of 20 in energy density from the observed background, which is actually worse than the value obtained from helium production in stars without any refinements in the cosmological time scale from 10^9 to 10^{10} years.

Ten years after the 1949 paper of Alpher and Herman, the predicted temperature had been increased to about 15 K. This was the value suggested by both Gamow and Dicke separately in private conversations with one of us (FH). The latter value was obtained by putting ρ_b equal to the Friedmann closure value, $\sim 10^{-29} \text{g cm}^{-3}$ instead of $10^{-30} \text{g cm}^{-3}$.

Rather than use ρ_b to obtain the background temperature the opposite is done nowadays. The observed temperature 2.73 K requires ρ_b to be $\sim 10^{-31} \text{g cm}^{-3}$, which is only one-third of the estimate for the average density of visible matter. It leaves no room for the hot baryonic gas which is known to be present in large quantities within clusters of galaxies. Nor does it leave room for extending the stellar luminosity function to magnitudes fainter than $\sim +17$, which recent observations of gravitational microlensing by faint stars seem to require, stars that are believed to have masses $\gtrsim 0.5 M_\odot$ (cf. Chapter 20).

This estimate of $\rho_b \simeq 10^{-31}$g cm^{-3} is only about one percent of the closure density, $3H_\circ^2/8\pi G$, in the standard big-bang model. For $H_\circ = 60$ km s^{-1} Mpc^{-1} this is 6.8×10^{-30}g cm^{-3}, meaning that overwhelmingly most of the material in the universe in the standard model cannot be baryonic. This deduction was arrived at already in the 1960s, and so has been known for about 30 years.

Efforts have now been made over these decades to identify this 'dark matter'. The first idea was that neutrinos, which according to the hot big-bang model must be present universally with a density comparable to that of photons, have a small rest mass. More recently, dark matter has been identified by most supporters of the theory with a hypothetical particle produced in the very early universe at times t much less than the $t \simeq 10^2$–10^3 s at which helium production is supposed to have occurred. The hypothetical particle is often taken to possess a mass much larger than those of baryons. But like the neutrino it is uncharged and its interactions with baryonic matter are, again like neutrinos, supposed to be weak. Thus it hides its existence from normal astronomical observations, except that it acts normally with respect to gravitation. An elaborate theory of the formation of galaxies rests on the gravitational properties of these hypothetical particles.

None of these arguments are required if we take helium production to occur in stars, and not in the early universe. Then the whole of the mass can be baryonic, with ample provision for the number of dwarf stars of small mass to be very large, and with a time-scale extension in the QSSC from 10^{10} years to 10^{11} years and beyond, as will be discussed in Chapter 16. Also, two of us (GB & FH) have discussed this model in a recent paper[18].

Other light nuclei

In 1966, shortly after the discovery of the microwave background, Peebles[7] suggested that as well as ^{4}He, the two light nuclides D and ^{3}He might have been produced in the big bang. This possibility had been considered 15 years earlier by Gamow, Alpher and Herman in the papers already noted. Also in 1964, Hoyle and Tayler[8] had revived the question of ^{4}He in the big bang. In 1967, Wagoner, Fowler and Hoyle[9] repeated a network calculation for the entire range of light nuclei as had been done in 1950 by Fermi and Turkevitch (see ref. 5). The point of these repetitions was to update the parameters required in the nuclear reactions in order to get better estimates of the amounts of the various light nuclei to be expected in a range of big-bang models besides the so-called standard-closure model – and also with respect to variations of the numerical coefficient in (9.3), i.e. with respect to variations of the parameter η determining the ratio of baryons-to-photons emerging from the various models. It was shown in these later calculations that the light nuclei to be considered beside ^{4}He were D, ^{3}He and ^{7}Li. But no ^{6}Li or ^{9}Be, ^{10}B and ^{11}B, or any heavier nucleus, was produced in significant quantity in the big bang.

Thus the range of further possibilities that might be attributed to big-bang synthesis is small, three other nuclides only. And with ^{3}He there is the problem that it is produced in large quantities in dwarf stars, where in those of small enough masses it is not destroyed by ^{3}He(^{3}He, 2p)Be^8. Considering the low abundance values found in the interstellar medium, only a little ^{3}He needs to be ejected by such stars in order to give ^{3}He/H $\simeq 10^{-5}$, which is the abundance ratio observed in interstellar clouds. Big-bang supporters claim there has been insufficient time for the evolution of dwarf stars to produce this kind of contamination. But they do not *know* this to be true. It is only a self-promoting assumption, as it was in the case of ^{4}He. The stuff is there, and in considerable quantity. In our opinion it would need an improbably rigorous prohibition to prevent a little of ^{3}He becoming broadcast.

The story of ^{7}Li is somewhat similar, although more complicated. The interest in the cosmological significance of ^{7}Li comes from a sample of about 50 Type II stars with temperatures above about 5500 K and with metallicities less than [Fe/H] $= -1.5$ (i.e. less iron relative to hydrogen than in the sun by antilog -1.5, or a factor of about 30). The need for the temperature restriction is to prevent subsurface convection from carrying lithium down to depths where nuclear reactions destroy ^{7}Li by ^{7}Li(p, γ)^{8}Be. In the absence of such a destructive process, a star, even though it may be more than 10^{10} years old, will still have preserved the value of the ratio ^{7}Li/H possessed by its material at birth. An iron abundance [Fe/H] < -1.5 was taken to imply primordial ^{7}Li, which is found to go down in abundance to a certain 'plateau' at ^{7}Li/H $\simeq 1.7 \times 10^{-10}$, with only a few exceptions[10,11,12]. Out of ~50 stars, Bonifacio and Molaro[12] find seven that breach the 'plateau', with the ^{7}Li/H ratio going down to values significantly less than 1.7×10^{-10}.

Population I stars, on the other hand, have values of ^{7}Li/H that are often $\sim 10^{-9}$ or higher, an order of magnitude greater than the supposed primordial value of 1.7×10^{-10}. Meaning that, as with both ^{3}He and ^{4}He, some astrophysical means of producing ^{7}Li exists, a means that is *certain*, not a matter of hypothesis.

Two very different suggestions for producing ^{7}Li have been made. In 1970, Reeves, Fowler and Hoyle[13] found that collisions by cosmic-ray protons on ^{12}C and ^{16}O maintained at their present intensity, and integrated over an age of 10^{10} years for the galaxy, would produce $^7Li/H > 10^{-10}$ by spallation. But there was the difficulty with this proposal that it gave an isotope ratio $^6Li/^7Li$ of order unity, whereas ^7Li in the Solar System is more abundant by an order of magnitude in the Solar System than ^6Li. This suggested that while ^6Li might well have been produced by cosmic rays, most of the Solar System ^7Li was produced otherwise.

It might seem that susceptibility to destruction would prevent ^7Li from being synthesized inside stars. And this might have been accepted had not McKellar[14] discovered long ago an unexpected new class of lithium-rich supergiants, with the prototype WZ Cas.

An interesting idea for overcoming what seemed like a paradox was suggested by Cameron and Fowler[15]. Giant stars form after hydrogen is exhausted in an initially convective central core, with hydrogen burning continuing in a thin hot shell. Surrounding this shell is another consisting of convectively stable gas, which gives place to an extensive convective envelope going all the way out to the atmosphere of the star. As evolution proceeds the surface temperature falls, the convective envelope deepens, and the shell of material between the base of the envelope and the hydrogen burning regions becomes progressively thinner. The suggestion of Cameron and Fowler depends on the products of hydrogen burning, in particular ^7Be which forms from $^4He(^3He, \gamma)^7Be$ being able to penetrate this thinning shell of formally convectively stable material. If this can happen in some organized dynamical way so that ^7Be can reach the convective envelope, the lifetime of ^7Be against beta-decay of 53 days could be sufficient for a fraction of it to reach the outer atmospheric regions which are cool enough so that the reaction $^7Li(p, \gamma)^8Be$ will not take place when the ^7Be eventually decays to ^7Li. Because a class of lithium-rich giants exists one can feel some confidence that this multiply inter-linked process actually happens. In which case the ^7Li found in Type II sub-dwarfs could have been synthesized in previously existing supergiants with short lifetimes, only $\sim 10^6$ years. And then the relation of the ^7Li plateau in the Type II stars to primordial nucleosynthesis is gone, because yet another nuclide goes into the major group of nuclides formed inside stars.

Since ^9Be cannot be synthesized in the big bang, the small abundance of ^9Be also found in Type II sub-dwarfs is a further example of the subtlety of astrophysical processes. It has recently been shown by Boesgaard that the abundance ratio $^9Be/H$ is closely proportional to the metallicity[16]. Thus at $[Fe/H] = -1$ the logarithm of $^9Be/H$ is about -11. For $[Fe/H] = -2$, $\log {}^9Be/H = -12$, and for $[Fe/H] = -3$, $\log {}^9Be/H = -13$. The clear implication is that Fe and Be are synthesized together, in supernovae where $^8Be(p, \gamma)^9Be$ takes place at a rate proportional to the production of Fe. This three-particle reaction would need to arise from dynamic instability during the supernova explosion, with mixing currents stirring a small quantity of hydrogen into a helium-burning zone. The very low values of ^{9}Be/H found by observation require only a trace production of beryllium, with only a marginal event being required. The surprise is that such a minor process appears to be highly reproducible in even such a dynamic star as a supernova, well exemplifying the immense range of astrophysical processes in general. Boron is not an element that is kind to the spectroscopist. So that as far as we are aware there are no explicit problems that a process for the origin of boron is called-on to solve, the only issue being that we require the existence of some process, a requirement that is satisfied by cosmic-ray spallation, particularly on ^{12}C, from which only one or two nucleons need to be detached in order to produce either ^{10}B or ^{11}B.

As we have already noted there are many stable nuclides in the cosmos. Most of them are produced inside stars. Of the small group of only four – D, ^{3}He, ^{4}He, ^{7}Li – which are supposed by big-bang theorists to provide evidence of the correctness of their views, three now threaten to disappear from this small group. The lighter isotope of helium is produced so widely in dwarf stars and it remains largely unconsumed in those of small enough masses that there is a great deal of it around. And the extended time scale discussed in Chapter 16 within the framework of the QSSC gives a helium production with $Y \approx 0.25$, thereby generating a radiation background that in its intensity is almost precisely equal to that of the microwave background. Then with ^{7}Li synthesized in evolved supergiants, only D remains as the one nuclide for the big-bang enthusiasts.

The case of deuterium

The observed value of D/H in the gas clouds of our galaxy is about 10^{-5}, while a value two to three times larger than this has recently been measured by Tytler, Fan and Burles[17] from the absorption spectra of gas clouds in a high redshift QSO. Many clouds at such high redshifts only give null results however.

If the light isotopes are made in primordial nucleosynthesis a value of 2×10^{-5} for D/H is found for $\eta = 7 \times 10^{-10}$. With this value of η, $Y = {}^4$He/H $= 0.248$, when the calculation of Y is made for three neutrino types and for a neutron half-life of 10.3 min. When Y is determined observationally in giant HII regions, which are believed to have been least contaminated by helium produced in stars, some exceptional cases ranging down to $Y = 0.23$ are found. Margins are so fine that extrapolations down to zero metallicity are usually made; these extrapolations are taken to allow for contamination of primordial ^{4}He by helium production in stars. The extrapolation leads to what is judged to be a primordial value of about 0.235 for Y. Whether or not the difference between this observed value of 0.235 and the value of 0.248 deduced from D/H $= 2 \times 10^{-5}$ is a significant difference is a subject of debate at the time of writing.

That the numbers come out as close as this, thus linking the production of D and ^{4}He together, is a fair argument in favor of big-bang cosmology. In our opinion it is the *best* point to be made in favor of the big-bang theory. Already in 1950 it was realized by Fermi and Turkevich that such a connection could be made (see ref. 5). The circumstance that over the past 50 years the connection has improved, mostly through changes in the value given for the neutron half-life, can also be seen as an encouragement to the theory. The fact that nothing else as good for the theory has come along in half a century is not so encouraging.

However, it is an odd circumstance that the very objects which are taken by observers to have the least contamination from helium production in stars, namely giant HII regions, should be just the objects where helium is being synthesized at the fastest known rate. It is also curious that the massive stars which power the HII regions when they are at maximum activity as they begin to evolve have an internal helium mass fraction Y that is close to 0.25.

References

Chapter 9

 1. Hoyle, F. 1946, *M.N.R.A.S.*, **106**, 343.
 2. Gamow, G. 1946, *Phys. Rev.*, **70**, 572.
 3. Gamow, G. 1948, *Phys. Rev.*, **74**, 505.
 4. Alpher, R.A. & Herman, R. 1949, *Phys. Rev.*, **75**, 1084.
 5. Alpher, R.A. & Herman, R. 1950, *Rev. Mod. Phys.*, **22**, 153.
 6. Burbidge, E.M., Burbidge, G.R., Fowler, W.A. & Hoyle, F. 1957, *Rev. Mod. Phys.*, **29**, 624.
 7. Peebles, P.J.E. 1966, *Ap. J.*, **146**, 542; *Phys. Rev. Lett.*, **10**, 410.
 8. Hoyle, F. & Tayler, R. 1964, *Nature*, **203**, 1108.
 9. Wagoner, R.V., Fowler, W.A. & Hoyle, F. 1967, *Ap. J.*, **148**, 3.
10. Spite, M. & Spite, F. 1985, *Ann. Rev. A & A*, **23**, 225.
11. Thorburn, J.A. 1994, *A. J.*, **421**, 318.
12. Bonifacio, P. & Molaro, P. 1997, *M.N.R.A.S.*, **285**, 847.
13. Reeves, H., Fowler, W.A. & Hoyle, F. 1970, *Nature*, **226**, 727.
14. McKellar, A. 1940, *Pub. Astr. Soc. Pacific*, **52**, 407.
15. Cameron, A.G.W. & Fowler, W.A. 1971, *Ap. J.*, **104**, 111.
16. Vangioni-Flam, E., Cassé, M., Fields, B. & Olive, K. 1996. *Ap. J.*, **468**,199.
17. Tytler, D., Fan, X.-M. & Burles, S. 1996, *Nature*, **381**, 207.
18. Burbidge, G. & Hoyle, F.L. *Ap. J.*, **509**, 1998, 1.

10

A new primordial calculation of Y and of D/H

Introduction

In the previous chapter we discussed the origin of the light elements from two different points of view.

First we gave an historical account of the developments which led to the popular view that they were made in a very early phase of the universe. Secondly we discussed the view that we have developed and discussed elsewhere[1] that they, like all the other isotopes in the periodic table, have an astrophysical origin.

In the course of investigating all of the possibilities, we realized that within the framework of the quasi-steady-state cosmology there is a third possibility. This is that the lightest isotopes were generated in the creation process involving primordial fireballs which result from the creation and ejection of matter from galactic centers in the quasi-steady-state cosmology. The cosmogonical background to these ideas is given in Chapters 15–18.

An esoteric possibility for light-element synthesis

Big-bang cosmology requires all the matter in the observable universe to be created in one single moment occurring $\sim 10^{-36}$ s after the big bang itself. Our view relates matter creation to conditions within near-black holes. It is not a single process correlated throughout the whole visible universe, but a process that takes place in many locations. Otherwise there are points of similarity in the physics. We shall attribute the whole class of violent events to this process.

In Chapter 17 the gravitational theory is formulated in a scale-invariant way, with the usual Einstein form shown to be derivable from a more general equation, when a particular way of defining the scale is used. A feature of the more general theory is that it leads to a unique particle mass to be expected in a purely gravitational theory, namely the so-called Planck mass. When units are chosen so that $c = \hbar = 1$, the gravitational constant has the dimensionality of $(\text{mass})^{-2}$, and this mass is the Planck mass. In ordinary units it is about 10^{-5} g. A particle with this mass has the property that its Compton wavelength is equal to its gravitational radius. It is our view that at a larger mass, a particle has the ability to 'tear open', as it were, the structure of spacetime, and then it is from such a 'tearing open' that creation events emerge, leading to the generation of showers of particles with masses of the order of the Planck mass.

When interactions other than gravitation are included such particles decay. As they do so progressively the time scale of the decays increase until they come within the range studied by laboratory physics, with the particle identities ultimately evolving into quarks, as in the development of the post-big-bang physics. But now the physical ensemble is in the form of a rapidly expanding local cloud of quarks, instead of a sea of quarks making up the whole universe. We try in this chapter to determine the primordial nucleosynthesis that will eventually take place as the ensemble continues to expand apart.

The quarks eventually combine three at a time into baryons, which react among themselves provided there is a source of heat in the otherwise adiabatically expanding cloud. But for a finite number of particles there is a stage at which the baryons become too diffuse for any further reactions between them to be effective in producing D or ^{4}He. Thus for the $\sim 5 \times 10^{18}$ baryons produced by the decay of a single Planck particle an effectively expanded stage is reached before charmed quark decay occurs in $\sim 10^{-13}$ s. But a large cluster of particles, say $\sim 10^{24}$, maintains effective reactions for longer than $\sim 10^{-13}$ s, thus permitting charmed quark decay to occur, but not for as long as $\sim 10^{-10}$ s at which strange quark decay would occur.

It is this case that is investigated in some detail here. With the heat responsible for promoting the nucleosynthesis coming from the last of the charmed quark decays, and with the need for electrical neutrality paramount, giving a baryon content in which the up, down and strange quarks are equally represented, the particle content of the expanding cloud simplifies to P, N, 2Λ, Σ^{+}, Σ^{-}, Ξ^{0}, Ξ^{-}. The reason for the double representation of Λ is that Σ^{0} decays to Λ in a time scale of $\sim 10^{-20}$ s. Otherwise all eight members of the baryon octet would be equally represented.

Only the P and N contribute to the production of stable nuclides, and they do so by just the same reactions as in the big-bang case. Almost the whole of the P and N goes to ^{4}He, while the 2Λ, $\Sigma^{\pm}$, Ξ^{0}, Ξ^{-} all go in $\sim 10^{-10}$ s, after the cloud is effectively separated, to give H. If the P and N went completely to ^{4}He, the emerging value of Y would be 0.25. To the extent that P and N remain free, the value of Y falls below 0.25. We calculate all of this in detail below.

The calculation

In both the big-bang theory and the quasi-steady-state theory, creation of matter must occur. The difference in point of view is whether the products of creation expand as a universal sea (in the big bang) or in separated fireballs.

As well as the distinction between a universal sea of particles and a set of separated fireballs there is the difference that in the big-bang case the sea is a balanced particle–antiparticle system, whereas the fireballs are of particles only, a difference that in our view lies in their respective energies of formation. In our view, matter creation occurs when the Compton wavelength of the created particles is equal to, or comparable with, their gravitational radius, that is to say, at an energy of $\sim 5 \times 10^{18}$ GeV. Whereas for inflationary models the relevant energy is $\sim 2 \times 10^{15}$ GeV, when only particle–antiparticle creation is possible. This leads to the dilemma that, when applied strictly, only radiation remains, i.e. the baryon-to-photon parameter η is zero.

Our model contains a parameter, namely the number of baryons that emerge eventually from an individual fireball. Write this number as N. Energy from particle decay goes at first into kinetic motions and radiation, at some temperature T. But as expansion proceeds the temperature declines adiabatically after the decays have ceased, with the energy going into the expansion of the fireball at some appreciable fraction of the velocity of light, say $\frac{1}{3}c$. Thus the radius r of the fireball is related to the time t since its creation by

$$r \simeq \frac{1}{3}ct. \tag{10.1}$$

And the internal particle density n is given by

$$n \simeq 3N/4\pi r^3. \tag{10.2}$$

Nuclear reactions effectively cease when $nr\sigma$ becomes less than unity, where σ is a typical nuclear cross-section, say 10^{-26}cm^2. Thereafter, the fireball can be considered from a nuclear point of view to have completed its expansion.

Thus for nuclear reactions to persist for a time scale t it is necessary that

$$N \gtrsim \frac{4\pi}{3}r^2\sigma^{-1} = \frac{4\pi}{3} \times \frac{c^2 t^2}{9} \times \sigma^{-1}. \tag{10.3}$$

Putting $t = 10^{-13}$ s, the time scale for particles containing the charmed quark to decay, gives $N > 4 \times 10^{20}$, while for $t = 10^{-10}$ s, the time scale for particles containing the strange quark to decay, gives $N > 4 \times 10^{26}$. In what follows we shall assume that N lies between these values, meaning that all baryons containing the charmed quark and the top and bottom quarks will have time to decay before effective nuclear reactions cease. But there is not sufficient time from a nuclear point of view for baryons containing the strange quark to decay before the fireball becomes effectively expanded apart.

This condition on N is very coarse compared with the almost exact condition on the parameter η that is required in big-bang cosmology. Thus N can vary over several orders of magnitude while still maintaining the required condition. There is no requirement for fine tuning.

We expect the three heavier quarks – top, bottom and charmed – to decay in the order of their masses, with each decay serving to drive the thermal motions of the particles to energies of the order of 10 MeV or more. At the densities in question the radiation field carries only negligible energy compared to the particles. No nuclear association survives at such temperatures. Thus at each quark decay any previous nuclear association is destroyed and nuclear reactions start again *ab initio* as adiabatic expansion again lowers the temperature, with the effect that it is the last of the decays, that of the charmed quark in a time scale of $\sim 10^{-13}$ s, that decides the eventual nuclear content of the fireball. The earlier decays do nothing ultimately but endow the fireball with an increased expansion speed.

After the charmed quark has gone, we are left only with the up, down and strange quarks. Despite all the complexities experienced by the fireball, because of the near equality of the up and down quarks their numbers should be essentially the same. The necessity for electrical neutrality then forces the details of the decays to be such that the number of strange quarks is also the same as that of the up and down quarks. And from equal numbers of the three lighter quarks we get, on baryon formation, equal numbers of the baryon octet, with the exception, however, that Σ^0 decays to Λ, thus making for equal numbers of N, P, $\Sigma^\pm$, Ξ^0, Ξ^-, but with twice that number for Λ.

As was explained in Chapter 9, the complete association of N and P to form ^{4}He, with 2Λ, $\Sigma^\pm$, Ξ^0, Ξ^- all going eventually to H after the fireball has expanded, leads to a number ratio of helium to hydrogen of 1 to 12, or a mass fraction $Y = 0.25$ for the helium. Only to the extent that there is not complete association of N and P does the value of Y fall below 0.25. This is the ratio we have now to calculate, together with the small fraction of other nuclides formed in the process, in particular the ratio D/H.

Write y for the fraction of the N and P that remains free, reducing the value of Y to $0.25(1 - y)$. Denoting the number densities of free protons and neutrons by $n(P)$ and $n(N)$, a parameter ζ can be defined by

$$n(P) = n(N) = \frac{2(2\pi mkT)^{3/2}}{h^3} y\zeta, \tag{10.4}$$

where the temperature is T, h is Planck's constant and m is the proton, or neutron mass, considered here to be the same. Then the number density of alpha particles, $n(A)$ say, is given by

$$n(A) = \frac{(2\pi mkT)^{3/2}}{h^3}(1 - y)\zeta. \tag{10.5}$$

It was shown by one of us[2] that if free protons and neutrons are in statistical equilibrium with the alpha particles then by the application of the Saha equation,

$$\log\frac{1 - y}{y^4} = 0.90 + 3\log\zeta + \frac{142.6}{T_9}. \tag{10.6}$$

As T_9 falls below ~ 10 in the adiabatic expansion of a fireball the value of y given by (10.6) becomes very small, implying that in statistical equilibrium essentially the whole of the initial N and P would be combined into ^{4}He, i.e. $y \simeq 0$, $Y \simeq 0.25$. But nuclear reactions leading to ^{4}He become slowed as y falls towards zero and in the available time scale of $\sim 10^{-13}$ s there is only a limited period in which helium formation can take place. It is this limited period that determines the required value of Y, which must therefore be calculated with respect to the explicit nuclear reactions by which helium is formed.

The lifetime of a proton against conversion to a deuteron by $P(N, \gamma)$D is the inverse of

$$2.5 \times 10^4 n(N)/6.022 \times 10^{23}, \tag{10.7}$$

and similarly for the lifetime of a free neutron. As the temperature falls the resulting deuterons largely combine to ^{3}He and T, and then through further reactions involving ^{3}He and T, ^{4}He is formed. Half of the original protons and neutrons are returned in a free form, but the other half winds up as ^{4}He, and so goes to reducing y.

The smallest value to which $n(N)$, $n(P)$ can thus be reduced is given by requiring the inverse lifetime (10.7) to be 10^{13} s^{-1}, and is $n(P) \simeq n(N) \simeq 2.4 \times 10^{32} \text{cm}^{-3}$. Equating to (10.4) gives the required value of y when T, ζ are specified

$$\frac{2(2\pi mkT)^{3/2}}{h^3} y\zeta \simeq 2.4 \times 10^{32} \text{cm}^{-3}. \tag{10.8}$$

Taking logarithms

$$34.07 + \frac{3}{2}\log T_9 + \log y + \log \zeta = 32.38. \tag{10.9}$$

With ζ specified, (10.9) gives y as soon as T_9 is also specified, the sense of (10.9) being such that the higher T_9 the lower y, i.e. the more the N and P is combined into ^{4}He. But for specified y and ζ we cannot choose T_9 higher than the temperature satisfying (10.6). Otherwise the direction of the statistical balance towards ^{4}He is lost. This means that if (10.6) and (10.9) are solved simultaneously for y and T_9 we obtain the extent of the combination into ^{4}He and also the temperature at which the combination is completed. Further nuclear reactions with continuing falling temperatures are not able to lower y further. Critical values for $\log \zeta$ equal to -2 and to -2.5 are given in Table 10.1.

The temperatures in this table are critical in the sense that if they were higher the values of y could not be maintained, because y would increase owing to break-up of a fraction of the ^{4}He. At lower temperatures, smaller y cannot be reached because the nuclear reactions required for production of ^{4}He are not fast enough. The general order of y for this range of ζ is from 0.05 to 0.1, with resulting values of the helium mass fraction Y between 0.22 and 0.24.

It seems unlikely that all such expanding fireballs arising in creation events will be exactly similar to each other. Some degree of averaging with respect to ζ is to be expected, with a good agreement with observations occurring when this average lies in the range of Table 10.1.

The ratios D/H and ^{3}He/H

The parameter y remains at the value just calculated as the density in the expanding cloud continues to expand. Not because D production has stopped, or with helium production through such reactions as D(D, P)T and T(T, 2N)^{4}He having stopped. But such reactions are no longer able to affect y significantly, i.e. they are no longer happening fast enough to increase

Table 10.1. *Critical values with respect to ζ, which is determined by equation (10.4) when the number density of particles in the fireball is specified*

log ζ	T_9	log y	Y
−2	13.4	−1.38	0.24
−2.5	13.9	−0.91	0.22

significantly the fraction $(1 - y)$ of the original N and P that is converted to ^{4}He. Such reactions, following on $P(N, \gamma)$D will continue for a while, but as the density falls they too will slow down until in the end the flow of P and N, first into D and then into T and ^{3}He, effectively ceases. It is this cessation which has to be calculated in order to obtain the surviving abundances of D and ^{3}He.

The rate for D(D, P)T is: $4.17 \times 10^8 T_9^{-2/3} \exp(-4.258 T_9^{-1/3}) \times$

$$(1 + 0.098 T_9^{1/3} + 0.518 T_9^{2/3} + 0.355 T_9 - 0.010 T_9^{4/3} - 0.018 T_9^{5/3}),$$

$$(10.10)$$

with a similar rate for D(D, N)^{3}He. Writing n(D) for the number density of deuterons, the density declines until in a time scale of $\sim 10^{-13}$ s the reaction rate (10.10) plus its augmentation by D(D, N)^{3}He is able to convert no more than n(D) to T and ^{3}He. Thereafter, as the density is reduced, the deuterons remain largely undisturbed. We next calculate the critical value of n(D) at which this situation occurs.

The above reaction rate is in the form given by Fowler, Caughlan and Zimmerman[3] (FCZ). It is such that to obtain the number of reactions taking place per cubic centimeter per second the rate for a two-body reaction such as (10.10) has to be multiplied by the number density of each reactant and then divided by Avogadro's constant, 6.022×10^{23}, i.e. by n^2(D)/6.022×10^{23} for (10.10). The resulting expression is very insensitive to the choice of T_9 at the high values ~ 20 under consideration here, which is to say for an overall energy release into the expanding cloud of a few million volts per particle. For example, at $T_9 = 15$ (10.10) gives

$$1.76 \times 10^8 n^2(\text{D})/6.022 \times 10^{23} \text{cm}^{-3} \text{ s}^{-1}.$$

$$(10.11)$$

Doubling this rate to allow for D(D, N)^{3}He, the number of deuterons destroyed in 10^{-13}s is $1.18 \times 10^{-28} n^2$(D) cm^{-3}. When the reduction of density is to a value of n(D) such that

$$1.18 \times 10^{-28} n^2(D) \simeq n(D), \qquad (10.12)$$

the deuterons are reduced to about a half of the value of $n(D)$ given by (10.12), i.e. to about $4.25 \times 10^{27} \text{cm}^{-3}$, and from then onwards in the expansion the total number of deuterons is not much changed.

What is the value of $n(P) = n(N)$ at which this happens? The answer is obtained from the value of $n(D)$ just calculated, in the following way. The number of D nuclei formed in 10^{-13} s is from (10.7)

$$2.5 \times 10^4 n^2(P) \times \frac{10^{-13}}{6.022 \times 10^{23}}. \qquad (10.13)$$

Setting this equal to the value of $n(D)$ just calculated gives

$$n(P) = 1.4 \times 10^{30} \text{cm}^{-3}. \qquad (10.14)$$

Now the decline from $n(P) = 2.4 \times 10^{32} \text{cm}^{-3}$ to $n(P) = 1.4 \times 10^{30} \text{cm}^{-3}$ has not come from erosion due to nuclear reactions but from expansion. Moreover $n(P) = 2.4 \times 10^{32} \text{cm}^{-3}$ was the fraction y of the density of the original baryons (see equation (10.3)) six of which go to yield hydrogen. Thus $n(P) = 2.4 \times 10^{32} \text{cm}^{-3}$ corresponded to a hydrogen number density $H = 6 \times 2.4 \times 10^{32}/y$. Then after a density reduction by ~ 170 due to expansion we have

$$H = \frac{6 \times 2.4 \times 10^{32}}{170 y} = 1.7 \times 10^{32} \text{cm}^{-3} \qquad (10.15)$$

for $y = 0.05$. Thus using the value $D \simeq 4.3 \times 10^{27} \text{cm}^{-3}$ obtained above from (10.12) we get the ratio

$$\frac{D}{H} \simeq 2.5 \times 10^{-5}, \qquad (10.16)$$

in good agreement with the so-called low deuterium value obtained for the observed deuterium abundances[3,4].

The rate given by FCZ^2 for $T(T, 2N)He$ is

$$1.67 \times 10^9 T_9^{-2/3} \exp(-4.872 T_9^{-1/3}) \times$$

$$(1 + 0.086 T_9^{1/3} - 0.455 T_9^{2/3} - 0.272 T_9 + 0.148 T_9^{4/3} + 0.255 T_9^{5/3})$$
$$(10.17)$$

with a similar rate for $^3He(^3He, 2P)^4He$. At $T_9 = 15$ as before, this is $8.8 \times 10^8 \text{cm}^{-3}$, giving

$$8.8 \times 10^8 n(T)^2 / 6.022 \times 10^{23} \qquad (10.18)$$

for the number of reactions per cubic centimeter per second. Since the coefficient 8.8×10^8 in (10.18) is higher by a factor ~ 5 than that for D in (10.11) a similar line of reasoning to that given for D/H gives a value of ^{3}He/H lower than before simply by the factor $1/\sqrt{5}$, i.e.

$$^3\text{He/H} \simeq 10^{-5}, \tag{10.19}$$

again close to what is believed to be the primordial value of the ^{3}He abundance. This includes the eventual decay of T from D(D, P)T as well as ^{3}He from D(D, N)^{3}He.

Summary

In summary, if we assume that matter originates under the physical conditions within fireballs generating $\sim 10^{24}$ baryons leading to the production of D, ^{3}He and ^{4}He, we find that:

1. the helium fraction by mass Y is between 0.22 and 0.24 (Table 10.1);
2. D/H $\simeq 2.5 \times 10^{-5}$ (equation (10.16));
3. ^{3}He/H $\simeq 10^{-5}$ (equation (10.19)).

In Chapter 9 we considered the theory of the synthesis of the light elements according to big-bang cosmology. In a space defined by the density ρ, the temperature T_9 and the time scale t, synthesis according to the big bang is supposed to take place in a subregion close to the point

$$\rho = 10^{-5}\text{g cm}^{-3}, \ T_9 = 1, t = 10^2\text{s}. \tag{10.20}$$

The case for synthesis according to the big bang depends in a large measure on the *unproved* assumption that nowhere else in ρ, T_9, t space is there a subregion giving results that are as good or better than those given by (10.20). In this chapter we have shown that this assumption is incorrect. The results (1) to (3) above are better than those obtained in Chapter 9. They apply for a subregion close to the point

$$\rho = 10^9\text{g cm}^{-2}, \ T_9 = 10, \ t = 10^{-13}\text{s}. \tag{10.21}$$

This is a point in ρ, T_9, t space greatly different from (10.20).

The baryon-to-photon ratio at (10.20) is of order 10^{-10}. That ratio at (10.21) is of order unity. It has never been made clear why a ratio as small as 10^{-10} should be assumed, and this value of order unity seems to give a better result. This might suggest that if the lightest elements arise directly from a primordial process, an origin in fireballs is more likely.

References

Chapter 10

1. Burbidge, G. & Hoyle, F. 1998, *Ap. J.*, **509**, L1.
2. Hoyle, F. 1992, *Astroph. & Space Sci.*, **198**, 177.
3. Fowler, W.A., Caughlan, G. & Zimmerman, B. 1975, *ARA&A*, **13**, 69.
4. Tytler, D., Fan, X. & Burles, S. 1996, *Nature*, **381**, 207.

11

The new observational evidence and its interpretation: (a) quasi-stellar objects and redshifts

Introduction

By the early 1970s there had developed a general consensus that we live in a universe which began in a hot big bang. The evidence of Hubble, the attractiveness of the Friedmann–Lemaitre models, the discovery of the microwave background radiation and the strong belief that the light isotopes just discussed must have been made in a hot big bang, all contributed to the feeling that this model must in a general way be correct. In 1980, Guth introduced the idea of inflation which was also very warmly received. In addition to this, the general view had developed that the original steady-state model was ruled out by the observations.

This is why most modern books on cosmology have concentrated almost exclusively on the hot big-bang model, the general approach being that we must try to explain all of the new observational discoveries in terms of this model. In particular, the considerable amount of new evidence on the distribution of galaxies and QSOs in space have led to much detailed analysis of large-scale structure in the universe. All of this is based on the assumption that density fluctuations were present very early in the history of the universe and, as it expanded, individual galaxies and groups and clusters formed, while the microwave background, which was originally tightly coupled to the matter, cooled as the expansion continued.

While this observational evidence may be interpreted in terms of a hot big-bang model, we believe that other observational evidence that has accumulated since the 1960s provides the strongest argument showing that this approach is not correct. In this chapter and the next we outline that evidence. There are three major themes.

1. That not all redshifts arise wholly from the changing scale $S(t)$ of the universe.

2. That there is much direct evidence that coherent objects – galaxies, quasi-stellar objects, etc., do not originate from initial density fluctuations in the universe, but are generated and ejected from galactic nuclei as was originally proposed by Ambartsumian.

3. That the rapid release of large amounts of energy from galactic nuclei arise from creation processes. These are processes with some similarity to what is supposed to happen in the big bang, but they are taking place here and now in the modern universe.

The observational evidence concerning (1) and part of (2) will be discussed in this chapter, and (3) will be described in Chapter 12.

Redshifts

As was shown in Chapter 5 the identification of galaxies and quasi-stellar objects (QSOs) with radio sources and the measurement of their redshifts gave values of z much larger than anything that had been measured before. While these redshifts were, following Hubble, immediately interpreted as measures of distance, it was clear very early on in the investigations that this interpretation was not necessarily correct. This was obvious from the Hubble diagram. For normal galaxies there is a good Hubble relation, while for the QSOs the plot of redshift against apparent brightness does not show the $(m_v–5\log z)$ correlation discovered by Hubble.

Historically, following the identification of the redshift in the spectrum of 3C 273 in 1963 (cf. Chapter 5), QSOs identified first from radio positions and then from surveys of the colors of stellar objects, were found in increasing numbers and, by 1967, more than 100 redshifts were known[1]. However, a plot of the Hubble diagram for these objects already showed no correlation of apparent magnitude with Hubble's $5\log z$ dependence on redshift, but more accurately a correlation with $5\log(1 + z)$, which is simply the loss of energy due to the redshift z itself[2]. As the numbers increased there was still no sign of a $5\log z$ correlation[3,4]. Of course, as most people argued, this did not *prove* that the observed redshifts were not measures of distance. The diagram could be interpreted in the normal way provided there is a very large dispersion in absolute brightness of the QSOs.

We show the redshift–apparent magnitude diagram for about 7000 QSOs in Fig. 11.1. A careful examination of this figure shows that the apparent magnitudes of the QSOs are correlated with $5\log(1 + z)$, and not with $5\log z$. As we have said above this is caused by loss of energy due to the redshift term and not a distance effect.

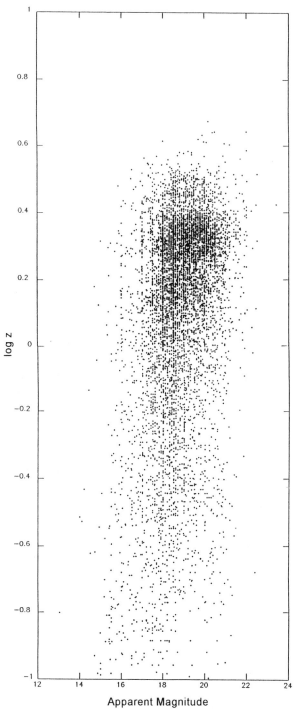

Fig. 11.1. Redshift–apparent magnitude plot for 7315 QSOs taken from the catalog of Hewitt and Burbidge[4].

Alternatively it could be argued that the QSOs have a component of intrinsic redshift z_i as well as a cosmological component z_c, so that the observed redshift z_0 is given by

$$(1 + z_0) = (1 + z_i)(1 + z_c),$$

with the first factor on the right-hand side of this equation producing a large scatter in the apparent magnitudes of QSOs.

Another remarkable property of these objects that was detected very early on in a subset of the QSOs, was that their optical and radio emission varies quite rapidly, on time scales of years or less. This was something that was unheard of in extragalactic astronomy. From causality arguments as usually applied it would mean that the size of the emitting region cannot be greater than the light travel time across the object.

It is interesting to point out that when the first rapid radio variations were found in a QSO, CTA 102, by Sholomitsky in the USSR in 1965, they were not believed by many in the US because it was already known that the QSO had a redshift $z_0 \simeq 1$, though that result had not yet been published. Only when observations of variability in 3C 273 were detected in the US by Dent[5] was the radio variability taken seriously. It turned out, of course, that Sholomitsky's observations were correct.

QSOs with large redshifts are not particularly faint. If the redshifts are measures of distance, the QSOs have bolometric luminosities ~ 100 times those of normal galaxies. This leads to luminosities $\sim 10^{46}$–10^{47}erg s^{-1} which in some cases would need to be generated in a region less than a light year in size. Moreover, it soon turned out that large variations in flux could take place on time scales much shorter than a year, and the current view, even in conservative circles (cf. Rees[6]) is that the sources are no greater than the size of the solar system (100 AU $\sim 10^{15}$cm).

It has been generally agreed for 40 years that the primary radiation is incoherent synchrotron radiation, and given the very high apparent luminosities and very small sizes, it was easy to show that the radiation density in these sources would be so high that the radiation would react on the radiating electrons, and we would reach what was called the Compton paradox[7], i.e. a simple physical model would not work. There were two ways out of the dilemma. The first was to assume that the variable QSOs lie much closer to us than the observed redshifts suggested. In this case the intrinsic luminosities are lowered, and thus the paradox can be at least partially relieved and the model may be viable. This requires that z_c is very much less than z_0 for a typical QSO.

The alternative way to go was suggested by Woltjer[8] and by Rees[9] who argued that the radiating surfaces must be moving relativistically, so that instead of the limit on the size R being that $R \leq c\tau$ where τ is the time scale for variation, $R < \gamma c\tau$ where

$$\gamma = \left[1 - \left(\frac{v}{c} \right)^2 \right]^{-1/2}.$$

This was the only way that could be thought of to make viable models of the rapidly variable optical sources *and* maintain the belief that their observed redshifts are wholly of cosmological origin.

A few years after the initial flurry of excitement and controversy over these results, radio astronomers began to measure structure in radio sources in QSOs on scales of milliarc seconds, and they showed that on time scales of years the apparent outward motion of radiating components from a center could be detected. If the distances of the sources were obtained from their redshifts this suggested that components were moving with apparent velocities from a few, to ~ 10 times the speed of light. Once again a paradoxical result had been obtained, and the alternatives were either to assume those QSOs are much closer than their redshifts indicate or argue that the radiating blobs are plasmoids moving at highly relativistic speeds with velocity vectors directed towards us. Values of γ in the range 3–10 are generally required and they correspond to values of $\beta = v/c$ often greater than 0.99.

To preserve the conventional redshift hypothesis the radio astronomical community claimed that they were (are) detecting *superluminal* expansion often requiring the sources to point closely toward us, and studies of such motions have turned into a very active field of endeavor. It is notable that in the cases where these motions can be detected in galaxies of stars whose distances are not in question (e.g. NGC 1275) the motion does not require a superluminal explanation though it does require fairly high speeds ($\sim 0.3c$).

These then are two examples of phenomena in QSOs which could have been understood if it were assumed that large intrinsic redshift components are present rather than highly relativistic motion pointing toward us, but always the high-velocity explanations have been possible, and have been preferred.

Direct evidence of the distances of the QSOs

In 1967 and 1968 Arp[10,11] began to identify radio sources in the vicinity of galaxies in the remarkable catalog of interacting galaxies which he had compiled[12]. He noticed that there were a number of radio sources which appeared to lie close to, and symmetrically about, some of those galaxies, suggesting physical association. However, the redshifts of the peculiar central galaxies are small, whereas the redshifts of the radio sources, mostly identified with QSOs, are very large. He argued that from both the geometry and statistics this showed that the systems are physically associated, so that the QSOs lie at the distances of the galaxies. Consequently most of the observed redshifts of the QSOs could not be of cosmological origin. His work was greeted with astonishment and disbelief, and his statistical arguments were heavily criticized, often very unfairly. In response he began an extensive observational program to detect radio-quiet QSO candidates from their colors in the vicinity of bright spiral galaxies with small redshifts, and many candidates were found. Almost without exception they were found to be genuine high-redshift QSOs.

One of us (GB) with others carried out the first really clean-cut statistical test, looking for correlations in position between a complete sample of QSOs taken from the 3C catalog of radio sources (50 objects), and the bright galaxies in the Shapley Ames catalog (about 1200 objects)[13]. A very strong effect involving four of the radio QSOs was found, and this was confirmed by a further extensive analysis by Kippenhahn and de Vries[14]. Actually it turned out that there were *five* QSOs lying very close to bright galaxies in this sample. The fifth case of 3C 455 had originally been identified with the galaxy NGC 7413, but it was shown in 1972[15] that the radio source was actually a QSO only 23 arcsec from that galaxy. So it had been demonstrated by statistical arguments using proper samples at a high level of significance (6σ) that there was a clear association between radio QSOs with large redshifts and galaxies with very small redshifts[a].

While these five QSOs had been picked out by statistical methods, it was noticed that when the angular separation θ between QSO and the nearby galaxy was plotted against the redshift of the galaxy, which is a measure of its distance d, there was a proportionality,

$$\theta \propto d^{-1}$$

or $\theta d = \text{constant},$ (11.1)

---

[a]There were statistical investigations using other samples[16] where this result could not be confirmed, but it was shown that there were selection effects that were responsible for this failure[17]. Later Webster[19] attacked Arp's statistics of association (see also Browne[20] and Arp[21]). The work by Webster was used to cast doubt on Arp's association for many years.

i.e. the projected linear separation between galaxy and QSO was approximately the same in each case[16,17,18]. This then added to the statistical evidence that the QSOs and galaxies are physically related. This result is shown in Fig. 11.2.

As time went on and more QSOs were identified, larger numbers of QSOs and apparently associated galaxies were found. Thus it eventually became possible to look at the $\theta - d$ relation for a much larger number of pairs.

In 1989, a study was begun to look for all pairs of QSOs and galaxies known at the time separated by 10 minutes of arc or less. To do this a computer search was made using the catalog of all QSOs known in 1987[22], about 3000 QSOs, and *The Revised New Catalog of Nonstellar Astronomical Objects*[23]. The computer gave about 400 pairs with $\theta \leq 600''$ and of these about 315 involved galaxies with known redshifts z_G. The distribution of the separations of the pairs is shown in Fig. 11.3, together with a number of pairs with larger θ which had been reported in the literature. This represents the

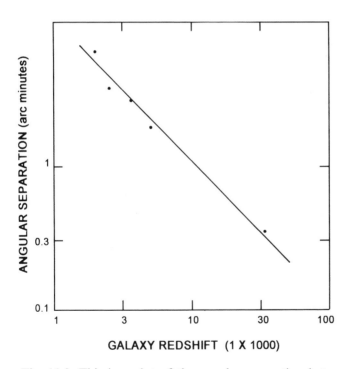

Fig. 11.2. This is a plot of the angular separation between the QSOs and galaxies against the galaxy redshifts (proportional to the galaxy distances) for the first five QSO galaxy pairs which were singled out as being likely to be physically associated by statistical analysis[13,15,16]. The line that is drawn has a slope of 1 corresponding to the relation $\theta z_G = $ constant.

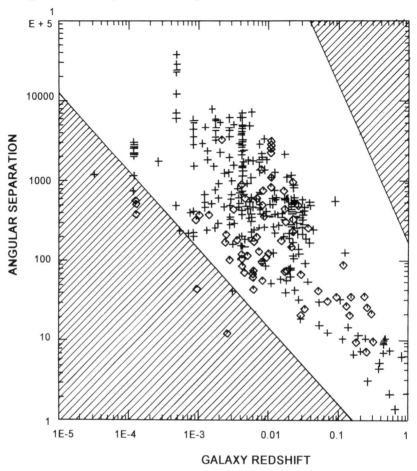

Fig. 11.3. This is a plot similar to that in Fig. 11.2. This is taken from the paper by Burbidge *et al.*[24] and shows θ against z_G for 392 pairs of galaxies and QSOs. Although there is a much larger scatter it is clear on average $\theta z_G \simeq$ constant. The hatched areas are regions which are excluded by selection effects.

complete set of pairs with available z_G, as they were known in 1990[24]. Thus there is no question of these data being deliberately biased in any way through human intervention. It shows without any possibility of doubt that the overwhelming majority of the pairs in question are physically associated. And hence that the QSO in each case is at the distance of the associated galaxy, requiring the large observed redshift of the QSO to have a dominant intrinsic property.

In Fig. 11.4 we show the plot of numbers of cases against z_G.

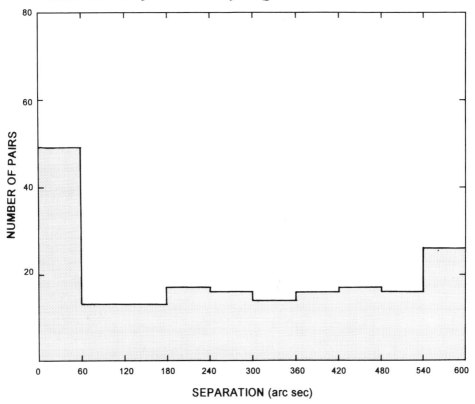

Fig. 11.4. A histogram of the number of QSO–galaxy pairs as a function of their separations also taken from Burbidge *et al.*[24]. Here 197 pairs within $600''$ are included. In all cases $z_{QSO} \gg z_G$. It is clear that there is a large excess of pairs with galaxy/QSO separations $\leq 60''$. This is completely contrary to what would be expected if the pairings were accidental.

The pairs in Fig. 11.3 with $\theta > 600''$ were obtained in a different way[24]. The QSOs in these cases were discovered from an examination of the fields surrounding nearby bright galaxies, and the figure includes all QSOs which had been discovered in this way at the time it was constructed. The fact that this different sample fits on to the computer sample, again indicating an average galaxy to QSO separation of about 100 kpc, re-enforces what was already an overwhelming result.

The correlation seen in Fig. 11.3 (taken from ref. 24) is very strong evidence that much of the redshifts of QSOs are not of cosmological origin. In addition to these studies, a number of other statistical investigations showing strong correlations between samples of QSOs and galaxies with much lower redshifts have been published in recent years[25–28].

In 1970, Weedman[29] noted that the bright QSO or AGN (Active Galactic Nucleus), object Markarian 205, which has an observed redshift $z_0 = 0.07$, lies only 43″ from the nucleus of the spiral galaxy NGC 4319. He pointed out that the chance that this was an accidental configuration was very small. Deep optical imaging of the pair was then carried out by Sulentic and Arp[30], and they showed that there is a luminous bridge connecting these objects, which have very different redshifts. Criticisms were made of this work on the ground that the connection was due to a photographic effect[31], but that argument was not sustained, and the data are now generally accepted (Fig. 11.5). In a number of other situations where QSOs with high redshifts are found to lie very close to galaxies with much lower redshifts, connections of various kinds are seen. We show some of these in Figs. 11.6, 11.7, 11.8 and 11.9.

These examples are nearly all associated with very bright, comparatively closeby galaxies with distances $\gtrsim 300$ Mpc. In the 1970s and 1980s Arp investigated the fields around many bright comparatively nearby galaxies

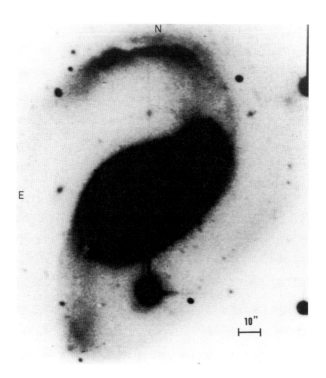

Fig. 11.5. This image of NGC 4319 ($z_G = 0.0057$) and Markarian 205 ($z_Q = 0.07$) was obtained by Sulentic and Arp[30]. It clearly shows the 'bridge' joining the galaxy and Mk 205, nearly parallel to and 3 arcsec E of the diffraction spike.

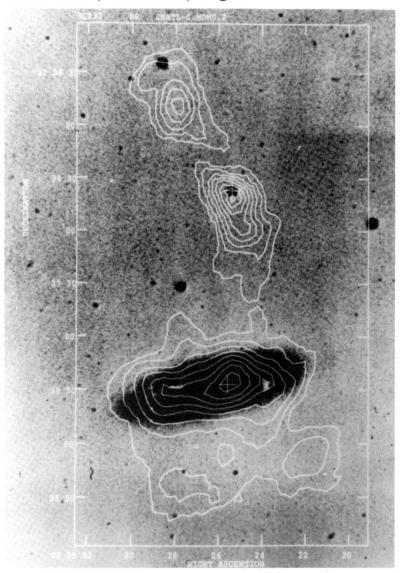

Fig. 11.6. This shows the 21 cm radio contours connecting the bright nearby galaxy NGC 3067 ($z_G = 0.0047$) with the QSO 3C 232 ($z_Q = 0.533$). 3C 232 is the stellar object at RA 09^h 55^m 25.4^s Dec $32°38'23''$ at the center of the 21 cm contour map immediately north (above) the galaxy. In addition absorption is seen in the spectrum of 3C 232 at the redshift of NGC 3067. This shows that the absorbing gas envelops the QSO. It should not be forgotten that this pair is one of the five pairs singled out as being associated in the original statistical study of Burbidge *et al.*[13]. The observers who looked for and found absorption have never chosen to point this out. They would have the reader believe that this is an accidental pairing and that the absorption shows that the QSO lies behind the galaxy. Of course the association based on statistical evidence shows that this is not true.

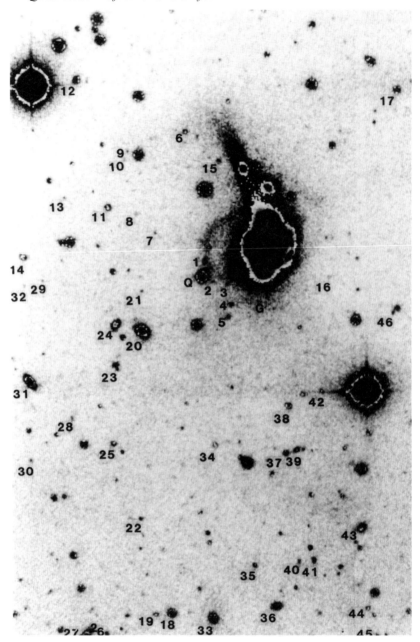

Fig. 11.7. This QSO galaxy pair, ESO 1327-2041 ($z_G = 0.018$) and Q 1327-206 ($z_Q = 1.170$) is reproduced from a figure published by Bergeron *et al.*[81]. The QSO is marked **Q**. *The bridge joining it to the galaxy is very clearly present.* In the text of this paper Bergeron *et al.* described it as an arm which passes roughly on the sight line to the QSO and bends back toward the galaxy in an EW direction at about the QSO declination. Absorption at the redshift of the galaxy is seen in the spectrum of the QSO.

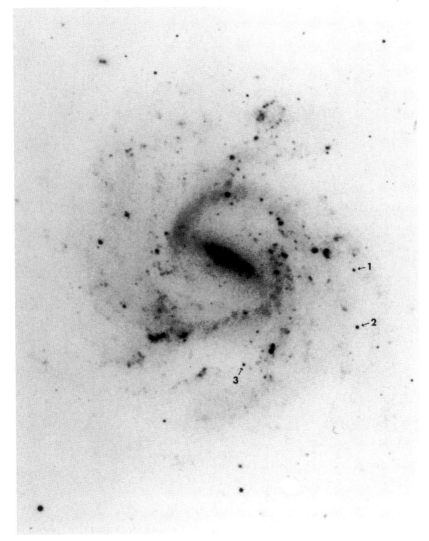

Fig. 11.8. This shows the nearby spiral galaxy NGC 1073 which has three QSOs lying within 2 arc min of its nucleus. They were first identified by H.C. Arp, redshifts were measured by M. Burbidge. The redshifts of these QSOs which are marked 1, 2, and 3, are $z_Q = 1.945, 0.599$ and 1.411 respectively. Since the density of QSOs of this magnitude on the sky is less than 20 per square degree, the chance that this is a random occurrence is less than one part in 50 000.

and he found that QSOs were much more frequently found near to the galaxies than in the general field. Also, several cases were found where *two* or *three* QSOs with very different redshifts lie very close (within 2′) to the center of the same bright galaxy. These are NGC 1073 (Fig 11.8) and

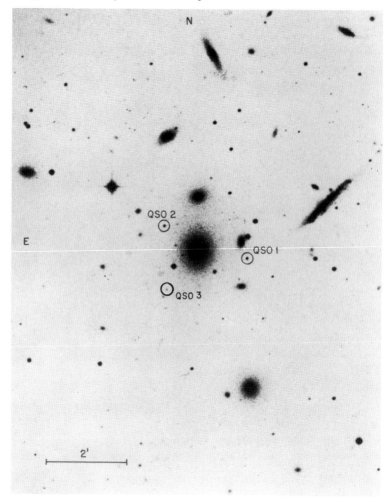

Fig. 11.9. Here are three more QSOs lying within 2 arc min of the center of NGC 3842. Two of them were also first identified by H.C. Arp based on X-ray data and the third by Arp and Gavazzi. The redshifts of the QSOs 1, 2 and 3 are 0.335, 0.946 and 2.205 respectively. NGC 3842 has a very small redshift and lies in a nearby cluster of galaxies. Once again the chance that this is an accidental configuration is incredibly small.

NGC 3842 (Fig 11.9). In all of these cases the statistical argument clearly shows that the QSOs and galaxies lie at the same distances.

A list of more than 40 QSOs lying within $3'$ or less of bright galaxies with $m_v < 15.5$ is given in Table 11.1. This is taken from the paper by Burbidge[32]. If we use the measured surface densities of QSOs to different magnitude limits, it is easy to calculate the number of pairs we would expect to find due to accident. This number is given by

Table 11.1. *QSOs close to bright galaxies* ($m_v \leq 15.5$)

Galaxy	m_v	QSO	m_v	z_Q	sep$^{(")}$	Remarks
UGC 0439	14.4	PKS 0038-019	16.86	1.674	72	
NGC 470	12.5	(0117 + 0317g)	19.9	1.875	93	
NGC 470	12.5	(0117 + 0317g) 68D	18.2	1.533	95	
NGC 622	14.0	0133 + 004 (UB 1)	18.5	0.91	71	
NGC 622	14.0	0133 + 004 (UB 1)	20.2	1.46	73	
IC 1746	14.5	0151 + 048 (PHL 1226)	17.5	0.404	6.4	
NGC 1073	11.3	BS0 1	19.8	1.945	104	
NGC 1073	11.3	BS0 2	18.9	0.599	117	
NGC 1073	11.3	RSO	20.0	1.411	84	
NGC 1087	11.5	0243 − 007 (UB 1)	19.1	2.147	170	
ZW 0745.1 + 5543	15.3	0745 + 557	17.84	0.174	100	
IC 2402	13.5	844 + 319 (4C 31.32)	18.87	1.834	30	QSO in direction of radio jet
NGC 2534	14.0	0809 + 358 (UB 1)	18.7	2.40	121	
NGC 2693	13.1	0853 + 515 (UB 1)	19.5	2.31	188	
UGC 05340	14.8	0950 + 080	17.69	1.45	103	

Table 11.1. (*contd*)

Galaxy	m_v	QSO	m_v	z_Q	sep$^{('')}$	Remarks
NGC 3067	12.8	0955 + 326 (3C 232)	15.8	0.533	114	21 cm contours connect galaxy to QSO Absorpt. in QSO at z of galaxy; active galaxy
NGC 3073	14.1	0958 + 558 (UB 1)	18.8	1.53	144	
NGC 3079	11.5	0958 + 559	18.4	1.154	114	Extremely active galaxy; absorption in QSO at z of galaxy
ZW 1022.0 − 0036	15.5	PKS 1021-006	18.2	2.547	122	
NGC 3384	10.8	1046 + 129	20.6	0.497	149	
NGC 3407	15.0	1049 + 616 (4C 61.20)	16.3	0.422	173	
NGC 3561	14.7	1108 + 289	20.0	2.192	66	Extremely disturbed galaxy
NGC 3569	14.5	1109 + 357	18.1	0.91	31	
NGC 3842	13.3	QSO 1	18.5	0.335	73	
NGC 3842	13.3	QSO 2	18.5	0.946	59	
NGC 3842	13.3	QSO 3	21.0	2.205	73	
NGC 4138	12.1	3CR 268.4	18.1	1.400	174	
NGC 4319	13.0	Mk 205	14.5	0.070	43	Luminous bridge joining QSO to galaxy; absorption in QSO at z of galaxy
ZW 1210.9 + 7520	15.4	1219 + 753	18.16	0.645	94	
NGC 4380	12.8	1222 + 102 (Wdm 6)	17.6	cont.	88	
NGC 4550	12.6	1233 + 125 (Wdm 8)	17.2	0.728	44	Galaxy in the Virgo cluster

NGC 4651	11.8	3CR 275.1	19.0	0.557	210	Active galaxy with jet and counterjet
NGC 5107	13.8	1319 + 38	19.5	0.949	40	Jet or bridge pointing to QSO; absorption at z of galaxy
ESO 1327 − 2041	13.2	1327−206	17.0	1.169	38	
ZW 1338 + 0350	14.9	1333 + 0.35	17.98	0.85	41	
NGC 5296	15.0	1342 + 440 (BSO 1)	19.3	0.963	55	
NGC 5406	13.1	1358 + 932	17	3.30	95	
NGC 5682	15.1	1432 + 489	19.2	1.940	95	
ZW 1640.1 + 3940	15.2	1640 + 396	18.16	0.54	180	
NGC 5832	13.3	3CR 309.1	16.8	0.905	372	
NGC 5981	13.9	1537 + 595	19.0	2.132	10.7	
IC 1417	13.6	2158-134	17.8	0.73	76	
Anon	15	2237 + 0305	17.3	1.41	≤ 0.3	
NGC 7465	13.3	2259 + 157	19.2	1.66	128	
NGC 7413	15.2	3CR 455	19	0.543	24	
NGC 7714 − 15	13.1	2333 + 019 (UB 1)	18.0	2.193	120	Pair of interacting galaxies

$$n = 8.73 \times 10^{-4} \Gamma \theta^2 N \qquad (11.2)$$

where Γ is the surface density of QSOs brighter than a prespecified magnitude per square degree, and N is the number of cases that have been investigated. θ is the separation in arc minutes. The numbers of pairs for different values of the brightness of QSOs using this system are given in Table 11.2. While this is a very simple calculation, it could not be seriously in error unless the general surface density of QSOs has been grossly underestimated. The fact is that the rarity of QSOs on the sky has all along suggested that there should be far fewer apparent associations than we see, if they are all accidental. Arp and a few others have paid a very heavy price in their professional careers for discovering this fact.

The community has remained sceptical of these results apparently because the implications are so great. Apart from claims that the statistical arguments of Arp were never made correctly, one argument made against the reality of the associations by a leading observer was that if these results were correct, we had no explanation of the nature of the redshift! In other words, if no known theory is able to explain the observations, it is the observations that must be in error!

Arp's own colleagues at the Mount Wilson and Palomar Observatories became so disturbed and so disbelieving of the results he was getting that, in the early 1980s, they recommended to the directors of the two observatories that his observational program should be stopped, i.e. that he should not be given observing time on the Palomar or Carnegie telescopes to carry on with this program. Despite his protests, this recommendation was implemented, and after his appeals to the trustees of the Carnegie Institution were turned down, he took early retirement and moved to Germany where he now resides, working at the Max-Planck-Institut für Physik und Astrophysik in Munich. Arp's account of this whole episode is described in his book *Quasars, Redshifts and Controversies*[33]. Thus, Arp was the subject of one of the most clear cut and successful attempts in modern times to block research which it was felt, correctly, would be revolutionary in its impact if it were to be accepted.

Given that far more QSOs with large redshifts are found to lie close to galaxies with much smaller redshifts than are expected by chance, is there *any* way that the statistical evidence could be correct, and yet the cosmological redshift hypothesis could be preserved? In 1980, Canizares[34] came up with a very ingenious idea to explain the statistical effect as follows. We shall discuss later the possibility that spiral galaxies may have large halos of dark matter. This dark matter may be made up of very large numbers of dark stars or

Table 11.2. *Comparison between numbers of QSOs found within 180″ of bright galaxies (N_0) and the numbers expected by chance $\langle n \rangle$*

Apparent magnitude	N_0	$\langle n \rangle$
≤ 17	6	0.69
≤ 18	14	2.38
≤ 19	31	6.93

stellar remnants. The theory of gravitational lensing tells us that there is a small probability that the image of a background QSO at its redshift distance will be amplified by gravitational lensing due to a halo object so that it will appear much brighter than it actually is. If this mechanism is to explain the apparent associations of QSOs and galaxies, all of the spiral galaxies and also galaxies of other types must have these extended dark-matter halos; but more important, a very large density of QSOs fainter than 20^m must be present so that there are always enough QSOs in the field to explain the number of close associations. Now we have very good data on the density of QSOs on the sky. We show in Fig. 11.10 a plot of the surface density of QSOs as a function of apparent brightness. Using this, it has been shown by several authors[35,36] that

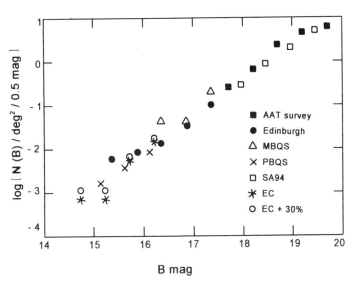

Fig. 11.10. This figure shows the differential surface density of QSOs over the sky plotted against *B* magnitude. It is taken from a recent paper by Kilkenny *et al.*[82].

this mechanism will not explain the observed effect essentially because the density on the sky of QSOs begins to flatten off as the QSOs get fainter ($> 21^m$).

There is yet another type of argument that indicates that QSOs are physically related to galaxies with much lower redshifts. It comes from studies of the geometry of QSOs and galaxies.

In their studies, Arp and others found many examples of alignments of QSOs about galaxies, and we show cases in Figs. 11.11, 11.12, 11.13 and 11.14. Several of them are discussed in Arp's book[33]. In addition to this, two of the brightest QSOs are 3C 273 and 3C 279, and these lie in the direction of the nearest cluster to us, the Virgo cluster, which is only $\sim$ 20 Mpc away. If it were not for the very large redshifts of these QSOs compared with that of the Virgo cluster, the association would have been realized long ago, as it was by

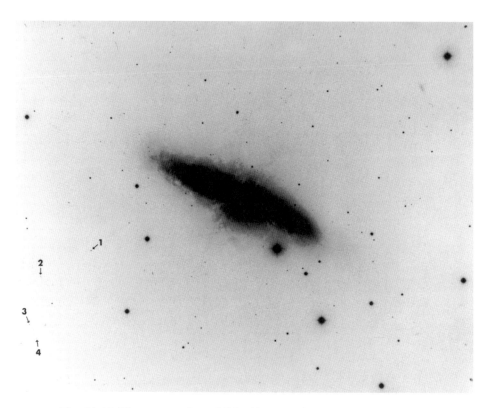

Fig. 11.11. Here we see four QSOs discovered serendipitously near the active galaxy M82 (NGC 3034) which is one of the closest galaxies to the Milky Way. The redshifts of the QSOs are z_Q = 2.053, 2.058, 2.033 and 0.85. The QSOs lie in the quadrant in which most of the ejection of gas from M82 is directed.

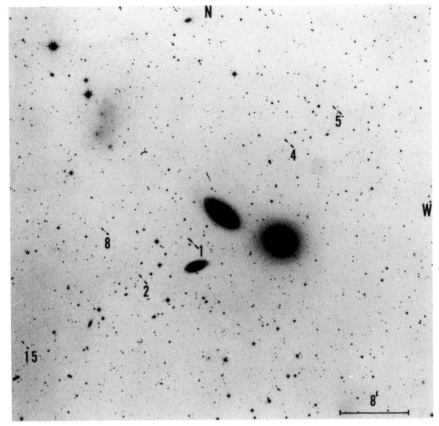

Fig. 11.12. An elongated group of high-redshift QSOs marked with numbers 15, 8, 2, 1, 4, 5 discovered near the triple system of bright comparatively nearby galaxies by H.C. Arp.

Arp. One of the brightest active galaxies in the Virgo cluster is M87, which has a well-known optical and radio synchrotron jet. In the position angle of the jet there are a number of QSOs and further away exactly in this alignment (the PA of the jet and the line joining M87 to M84 are the same to $1°$) is a radio galaxy M84. This was pointed out as long ago as 1960 by Wade[37]. We show this in Fig. 11.15.

In the past two or three years surveys of X-ray sources have led to the identification of X-ray-emitting QSOs and to further evidence of the same kind. In the ROSAT survey many active galaxies have been studied, and also many compact X-ray sources have been identified. In particular Arp[38] and Radecke[39] have made a number of identifications of compact X-ray sources which from their geometry appear to have been ejected from comparatively

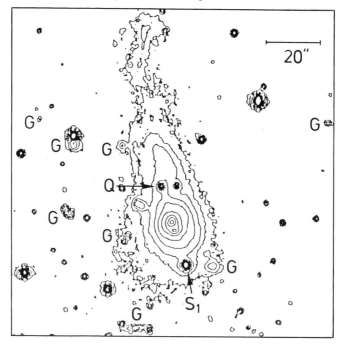

Fig. 11.13. This shows the galaxy–QSO pair 0248 + 430. The galaxy which has a double nucleus has $z_G = 0.051$ and the QSO marked Q has $z_Q = 1.311$. This contour map is taken from a paper by Borgeest *et al.*[83]. S_1 is a foreground star. The objects G are other galaxies in the field. The separation between QSO and central galaxy is only $14''.5$.

nearby active galaxies. In each case what is seen are two or more X-ray sources lying along an axis which passes through the central regions of these galaxies.

For example, the active galaxy NGC 4258, which is only about 7 Mpc away from the Milky Way, was shown many years ago to be ejecting hot gas and relativistic particles from its nucleus in cones of ejection roughly perpendicular to the plane of rotation[40]. Studies of its X-ray emission showed that in addition to the X-rays from the main body of the galaxy there are a number of compact X-ray sources outside the main body, and in particular two sources roughly equidistant and paired across the nucleus of the galaxy, each of which has a faint blue stellar object at its center[41]. The X-ray astronomers suggested from the geometry that these sources had been ejected from the galaxy[41]. Optical investigation has shown that they are both QSOs with redshifts $z = 0.39$ and 0.65 respectively[42]. This system is shown in Fig. 11.16.

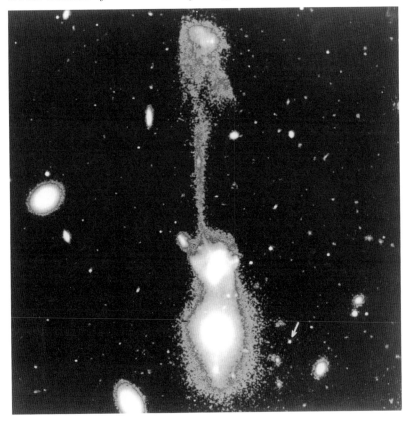

Fig. 11.14. This is a famous interacting galaxy pair NGC 3561 A & B. The plate is reproduced with the kind permission of Dr L. Duc and Dr F. Mirabel. Many years ago V.A. Ambartsumian gave this as an example of a galaxy in the process of ejecting a new galaxy. A high-redshift QSO very close to this pair is marked with an arrow. While we and Ambartsumian believe that this is an excellent example of one galaxy ejecting another, Duc and Mirabel believe that it is one galaxy disrupting another. This to them is an example of a violent merger. We believe that the existence of much star formation and the high-redshift QSO shows that the creation process is at work.

Several other active galaxies have been found to exhibit the same effect. They include NGC 1068 (one of the classical Seyfert galaxies) in which three ejected QSOs appear to have been identified, NGC 2639 where two X-ray emitting QSOs have been identified[43] and NGC 3516 (another classical Seyfert galaxy) where five X-ray emitting QSOs have been identified[44]. In Table 11.3 we give a list of all of the X-ray emitting QSOs which have now been discovered close to their 'parent' galaxies.

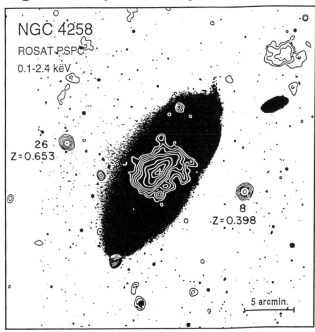

Fig. 11.16. This shows the galaxy NGC 4258 which is only about 7 Mpc away. X-ray studies made by ROSAT led the X-ray astronomers to conclude from the geometry that the two compact X-ray sources had ejected from the galaxy[41]. Optical studies show that they are both QSOs with large redshifts[42].

These conclusions about QSOs are not generally accepted by the astronomical community. Therefore before turning to other evidence concerning the ejection of compact objects from galaxies, we must discuss the arguments against the hypothesis.

Arguments against the non-cosmological redshift hypothesis

Three objections based directly on observations have been raised against the non-cosmological redshift hypothesis. They are as follows.

1. That all QSOs lie in the centers of host galaxies and the QSO components and the galaxy components have the same values of the redshift.
2. That many QSOs lie very close to galaxies with the same redshifts as would be expected if their redshifts are wholly of cosmological origin.
3. That the discovery of gravitational lensing involving variable QSOs allows us to determine the difference of the light travel paths of the different images

and hence a distance scale. In the case of $0957 + 561A\&B$ this leads to a distance scale of the order of gigaparsecs.

In addition to these objections there is the question of the nature of the absorption which is found in the spectra of all high-redshift QSOs. Apart from the fact that there is usually an absorption redshift system $z_{abs} \approx z_{em}$, meaning that absorption and emission is taking place in or near the same object, the bulk of the absorbing systems have $z_{abs} < z_{em}$. And the community has chosen to interpret these systems' z_{abs} as being due to matter lying between us and the QSO. So there is an apparent conflict with the view that much of z_{em} is intrinsic. We shall discuss this question after covering objections (1), (2) and (3).

Before dealing with the critical points (1), (2) and (3), it is necessary to mention several questionable techniques which have been employed to enforce the standard ideas. They include casting doubt on statistical methods used to reach the conclusion that non-cosmological redshifts are present, by generally criticizing data, but never being specific (cf. Peebles *et al.*[45]) or by calling all QSOs active galactic nuclei, thus planting in the mind of the reader that a QSO is *known* to be part of a galaxy, or by writing monographs in which none of the evidence in favor of non-cosmological redshifts is ever even mentioned (e.g. Peterson[46]). Here we concentrate on dealing with *evidence* concerning the arguments (1) to (3).

Much of the scepticism over the existence of intrinsic redshifts is due to the fact that while the evidence began to appear already in the 1960s it was not very clear cut. At the same time, the initial discovery that QSOs in general have redshifts very large compared with galaxies led to the very popular belief that they must be much more luminous than galaxies, and that they could be used to carry out cosmological investigations. This was a situation in which, in our view, the wish was father to the thought.

Host galaxies

That all QSOs lie in the centers of host galaxies is consistent with the continuity argument in which it is supposed that in galaxies with very small z we see normal spiral structure surrounding Seyfert nuclei, which spectroscopically are very similar to QSOs. These are the original galaxies identified by Seyfert in 1943[47]. It is then clear that as we go to larger redshifts and intrinsically brighter QSOs, it will be harder and harder to detect the surrounding galactic structure, or what is called, without proof, the host galaxy. When Kristian[48] originally made a plot of angular size of the image against z for the few objects identified as QSOs in the early days, it did appear that the apparent size declined as z^{-1}, but Kristian had already left out of his plot the brightest QSO with the smallest redshift, 3C 273.

There are two ways to establish the nature of the fuzz around QSOs. The first is by direct imaging. Unfortunately nearly all of the imaging work until recently has been made on QSOs with very small redshifts. This is because of the belief of the observers that the continuity argument is correct, in which case the best chance of detecting host galaxies would be to look at low-redshift QSOs. The proper way to test the hypothesis would have been, from the start, to examine a sample of QSOs with a large range of redshifts.

The best imaging work[49] has been carried out by using the Hubble Space Telescope on QSOs with $z \lesssim 0.3$. This work has shown that some QSOs do lie in galaxies (as established from their morphology), some appear to be interacting with galaxies, but do not appear to lie in host galaxies, and some appear to be bare. We show a selection of the images in Figs. 11.17–11.20. It is also clear that the faint QSOs apparently ejected from NGC 4258[42] (distant only 7 Mpc) are bare.

Some imaging has recently been done of high-redshift radio-emitting QSOs. For these it is stated that both UV continuum and Lyα emission are resolved around radio-loud QSOs, but only in a minority of radio-quiet QSOs which, of course, make up the majority of the QSO population[50]. Those who have reached that conclusion state that their results are consistent with the idea that radio-quiet QSOs lie in less luminous systems, but of course they do not prove it. There is still no evidence at all that the luminous fuzz detected around these QSOs is caused by stars. It is as likely that it is from hot gas.

The way to establish that QSOs lie in normal host galaxies is to obtain the spectra of the QSO and the fuzz separately, and show that the redshifts from the QSO emission lines and the stellar absorption spectrum agree. The first spectra obtained of the fuzz surrounding several well known QSOs[51,52,53] showed narrow emission lines at the same redshift as the QSOs. Thus there was no evidence for starlight. In only one case, 3C 48, were absorption features attributable to stars found at the emission redshift[54]. In this case the absorption was not symmetrically distributed about the QSO. It only appears on one side of the QSO and indicates the presence of a young A-type stellar population, which is not typical of disk or elliptical galaxies. More recently Miller[55], using the Keck telescope, has shown that at least one host galaxy in the group imaged by the Hubble Space Telescope has a genuine stellar spectrum with the redshift the same as the QSO ($z < 0.2$).

Thus there *is* evidence in a few cases that normal host galaxies surround QSOs. But this is certainly not true in all cases, and is much less than over-

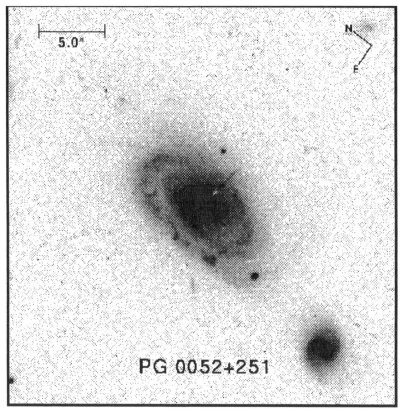

Fig. 11.17. This figure and the succeeding three figures are taken from the work of Bahcall *et al.*[49] who have taken direct images of a number of low redshift ($z < 0.2$) QSOs to look for direct evidence for host galaxies with the Hubble Space Telescope. We are indebted to John Bahcall for making these images available. This figure clearly shows that PG $0052 + 251$ lies in a host spiral galaxy.

whelming. Thus if we rely on evidence rather than belief, the continuity argument cannot be accepted as a general rule. Nor can it be used as strong evidence against non-cosmological redshifts, especially since it is confined to low-redshift QSOs.

There is one very rare class of Active Galactic Nuclei, called BL Lac objects after the prototype BL Lac (originally thought to be a variable star), which undoubtedly do lie in galaxies. These are rapidly variable AGN at radio, optical and sometimes X-ray energies, with optical spectra showing no traces of emission lines, or only very weak emission lines, and absorption features such as the H and K lines of Ca II and the G band, which are due to starlight. A plot of the redshifts of these objects against their apparent magnitudes at minimum light give a good Hubble relation, *demon-*

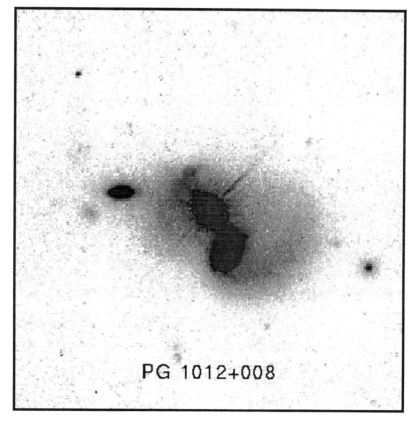

PG 1012+008

Fig. 11.18. This figure shows that PG 1012 + 008 may have a host galaxy and is clearly interacting with another nearby galaxy.

strating that they are giant elliptical galaxies with absolute magnitudes ~ -23, and with a component of variable non-thermal radiation coming from their nuclei (Fig. 11.21). Thus there is strong observational evidence that these BL Lac objects *do* lie in the nuclei of elliptical galaxies.

Galaxies with adjacent QSOs at the same redshift

Soon after Arp began to show that QSOs with very large redshifts could be found very close to bright nearby galaxies, strenuous efforts were made to find QSOs associated with galaxies having the same redshifts. The first sample found was due to Stockton[56]. Later, other samples were found, and in the paper by Burbidge *et al.*[24] a considerable number of such results are listed. For some, these results have apparently cast doubt on the reality of other pairings where the redshifts are quite different, although the statistical argu-

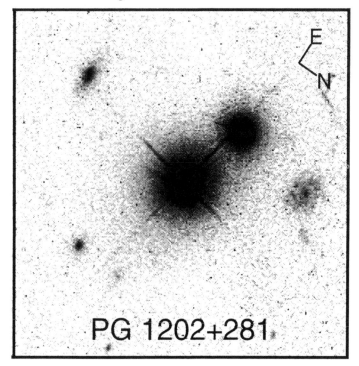

Fig. 11.19. This figure shows that PG 1202+281 has symmetrical fuzz around it which may be an elliptical galaxy, and a second galaxy nearby.

ments are equally good in both cases. Even in Stockton's sample there are 12 discrepant pairs, but these are ignored for no stated reason. The conclusion we draw from this is rather obvious. We suppose that for all pairs,

$$(1 + z_Q) = (1 + z_G)(1 + z_i)$$

where z_Q is the redshift of the QSO and z_G is the redshift of the adjacent galaxy. If the intrinsic redshift component $z_i = 0$, $z_Q \approx z_G$. If not, $z_Q = z_G + z_i + z_i z_G$, and for many of the pairings involving nearby galaxies, $z_i \gg z_G$, so that $z_Q \simeq z_i$. Consequently z_i can range from 0 to z_Q. Thus these observations do not give a disproof of the existence of the intrinsic redshift hypothesis, which is evidently required in many of the cases that are quite clearcut.

Gravitational lensing

The principles underlying the gravitational lens effects are well known. Such an effect will be present in any situation in which two masses lie at very

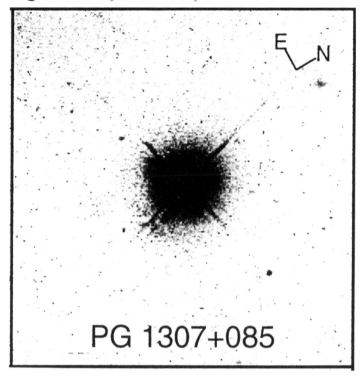

Fig. 11.20. This figure shows PG 1307 + 085, which does not appear to have a host galaxy.

different distances from us but are close enough together on the sky that light from the background source is gravitationally deflected by the mass closer to us. Cosmologically distant QSOs are good candidates to show lensing effects, though their rarity suggests that the number of cases will remain small. Following the serendipitous discovery of what most people believe is a gravitational lens, 0957 + 561A&B by Walsh, Carswell and Weymann[57], systematic searches have been undertaken involving large numbers of sources. However, only a comparatively small number of QSOs (~ 40) with multiple images with the same redshift have so far emerged as strong candidates. Those with nearly the same spectra, but not quite the same, are called physical pairs.

Since gravitational lensing is a phenomenon which is expected, because the theory on which it is based is acceptable and well understood, the criteria required for believing that the effect has been discovered are considerably less than those which are demanded of discoveries of what appear to be intrinsic redshift effects.

For gravitational lensing involving QSOs, the primary criterion should be to show that a lensing object at least of galactic mass lies very close to the

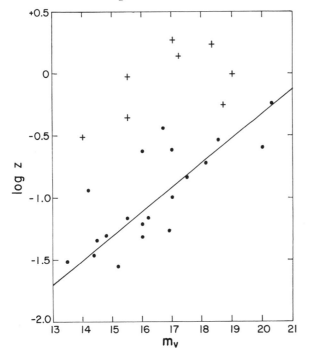

Fig. 11.21. Hubble plot for BL Lac objects with underlying elliptical galaxies (black dots) and objects called BL Lac which show QSO characteristics. It can be seen that the black dots follow a Hubble relation with a slope of 5. This is taken from a paper by Burbidge and Hewitt[84].

light path ($\leq 10''$) and is in front of the QSO. However, the discovery of $0957 + 561$A&B and the assertion that it is a gravitationally lensed QSO was based on the fact that the two images A and B are separated by only $6''.1$ and have identical spectra though they have different apparent magnitudes. *After* this discovery was made, the search began for a lensing galaxy. When a suitable candidate was found with a redshift $z = 0.36$ it was concluded that a satisfactory model of the deflecting potential could be made to explain both optical and radio images if the gravitational effect of a cluster of galaxies around the primary deflecting galaxy was also invoked. Such a cluster is present.

However, in general in only a fraction of the gravitational lens candidates, all based on the detection of multiple images with the same redshifts, have the lensing galaxies been detected.

On the other hand, if pairs of QSO images with identical spectra are found with separations $\geq 10''$ they are also called physical pairs *because* in these cases their separations are too large for them to be gravitational lens candidates, and they are not studied further. Also when very close pairs with very different redshifts are found they are usually assumed simply to be accidental configurations, and ignored by those interested in gravitational lenses. In fact there are four very close pairs of QSOs with separations of $\leq 5''$ and very different redshifts known[58]. The probability that these are all accidental is very small indeed, and it has not been possible so far to make a realistic lensing model to explain them[58]. Thus they may be added to evidence for non-cosmological redshifts.

The most remarkable instance of a multiply imaged system is 2237 + 031 A B C D where four QSO images with $z = 1.69$ have been found symmetrically placed within $0''.3$ of the center of a galaxy with $z = 0.039$. This system was found serendipitously[59] and it was announced to be a gravitationally lensed system *before* the multiple images were detected though it was immediately shown that the probability of this configuration being accidental was incredibly small[60]. After the four components had been detected Arp and Crane, using HST images[61], proposed that this was actually a galaxy ejecting four QSOs with non-cosmological redshifts, similar to many of the QSO-galaxy associations described earlier in this Chapter. However this paper was turned down by all of the leading astronomical journals on the ground that as more than one referee put it, 'It *must* be a gravitational lens' and it was only finally published in a physics journal.

Given the uncertainties associated with the gravitational lensing hypothesis, how seriously should we take the argument that lensing clearly proves that at least some QSOs lie at cosmological distances? We should mention that a given lensing model in which the lens galaxy has not been identified can be scaled down if the same QSO is closer than its measured redshift distance. The only way of then bringing an absolute measure of distance into the model is through the measurement of time delays.

The argument for this is straightforward. Once the gravitational field due to the lensing mass(es) is well determined it is possible to measure the distance of the system from the delay in the light travel time between the different images. In 0957 + 561A&B both components vary in time in both optical flux and in radio flux. Thus by measuring the delay in the light curves between components A and B a distance can be obtained. It is now believed that a good value for this delay is about 420 days[62]. The light curves matched for a 417 day delay by Kundic *et al.*[62] are shown in Fig. 11.22.

At first sight this matching of solid points and open circles looks good, and to the believer it probably looks very good indeed. But the sceptic will notice that the filled circles for A have been adjusted by 0.12 mag., in order to make the first drop of the two components essentially equal, meaning that both coordinates have been adjusted. Thus the first drop is an artifact which has no option but to look good. Then when one examines the subsequent oscillations the situation becomes scrappy, not better than one would expect for independent oscillations, meaning that both happened to fluctuate up and down in a period of a little more than two months. In particular, the situation at approximately JD 1170 is out of phase for the two components.

The time delay between A and B can be expressed as a product of three factors, provided the redshifts of 0957 + 561 and of the lensing object are interpreted as cosmological with $z = 1.413$ and 0.36 respectively. One factor depends on the detailed structure of the lens, the second one on the redshifts, and the third is simply proportional to the Hubble constant H_o, making H_o determinable from the time delay provided the other factors are considered known and provided 417 days for the time delay is correct. From this, the calculated value of H_o is 61 km s^{-1}Mpc^{-1}, close to the value estimated from

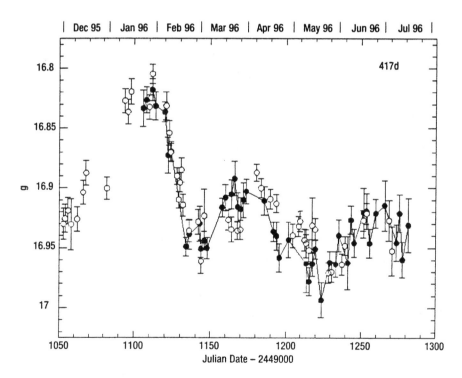

Fig. 11.22. This shows the light curves of the two images of the gravitationally lensed quasar, Q0957 + 561; filled and empty circles represent images A and B respectively. The light curve of image A has been advanced by the optimal value of the time delay, 417 days, and offset by −0.12 magnitudes, overlaid on the 1996 image B data. This figure is taken from work by Kundic *et al.*[62] reproduced in *Astronomy and Geophysics*, **38**, p. 13 (1997).

traditional procedures earlier in this book. It is this result, based on one example, that has led Kundic *et al.*[62] to claim with Dar[63] and Peebles *et al.*[45] that 'gravitational lensing events (note the use of the plural) confirm that high redshift quasars are at cosmological distances'.

Because this method of determining H_0 does not depend on the usual step-ladder distance indicators, e.g. the Cepheid variables, it is viewed by many as being likely in the not so distant future to provide the most reliable determination of H_0, provided that there is no intrinsic contribution to the redshift of the QSO. We are inclined to think that this might be so. However, it would be possible with $z_{obs} = 1.413$ for $0957 + 561$ to contain an intrinsic redshift component z_i, while still maintaining the gravitational lens interpretation, the effect being to bring the QSO closer to the lens, causing the calculated value of H_0 to be too low.

Thus we would not be surprised if the gravitational lensing arguments lead to the conclusion that some QSOs with small intrinsic redshift terms do lie at great distances. There is a selection effect that needs to be mentioned, however.

When looking for time delays, the observers may be subtly influenced by the theoretical model. Thus if the cosmological hypothesis predicts a time delay of N days and the intrinsic hypothesis predicts a delay of $N/100$ days, in either case there being no lensing object seen, will the observer seriously look into the second case?

To summarize our conclusions concerning objections (1), (2) and (3).

1. Some QSOs lie close to galaxies with the same redshift, so that for them $z_i \approx 0$. For many this is not true, however.
2. While there are many puzzles in the gravitational lensing models these may ultimately show that a minority of QSOs at least lie at cosmological distances.

None of these arguments removes or detracts from the evidence we have presented earlier in this chapter, which shows that many QSOs, particularly radio-emitting objects, have large intrinsic redshift components.

The problem of the absorption line spectra

In the past decade much of the research in the QSO area has centred around the absorption-line spectra. While the primary characteristic through which the initial discoveries were made was the emission-line spectrum, it soon became clear that absorption lines are present and are found in essentially all high-redshift QSOs. The absorption is of several types. In nearly all objects there is an absorption line system with $z_{abs} \approx z_{em}$ and multiple absorption line systems with $z_{abs} < z_{em}$. This immediately suggests three possible origins for the absorption.

(i) Gas with $z_{abs} \approx z_{em}$ must be physically associated with the QSO.
(ii) Systems with $z_{abs} < z_{em}$ could be due to gas ejected from the QSO with the difference between z_{abs} and z_{em} being a measure of the ejection speed.
(iii) The absorbing gas could be due to clouds or halos of galaxies which lie along the path between the QSO and the observer.

In addition to these and from the evidence described earlier in this chapter, we can also consider the possibility that some part of these redshifts, both from emitting and from absorbing gas, may have an intrinsic origin.

The opinion of most astronomers at present is that, apart from the 10–15% of QSOs showing very broad absorption features, which it is generally agreed are intrinsic and due to shells of gas being driven out of the QSO with speeds in the range $0.1–0.2c$, and also apart from the narrower 'associated absorption systems' recognized from their line profiles and time variability to be intrinsic, the absorption is of type (iii), i.e. it is due to intervening matter.

On this basis it is widely assumed that cosmologically distant QSOs can be used as probes of the intergalactic gas. The narrow absorption line spectra with $z_{abs} < z_{em}$ are of several types. Often the lines are very narrow and many different redshifts due both to hydrogen and the commoner heavier elements, C, N, O, Mg, Si, Fe and a few other elements are found. In a small fraction of the absorption line systems Lyα is extremely strong. These are known as damped Lyα systems with $N_H \gtrsim 10^{20} \mathrm{cm}^{-2}$. Wolfe *et al.*[64] have argued that such strong absorption indicates that the absorber must be an intervening halo or thick disk of a forming galaxy. For absorbing clouds with $N_H < 10^{20} \mathrm{cm}^{-2}$, smaller, less massive intervening clouds are indicated. The least absorption is found when only weak Lyα or Lyα, Lyβ pairs are identified. These lines often occur in profusion and were christened by R. Lynds the *Lyα forest*[65]. They are thought to arise in hundreds of small clouds, or parts of galactic halos, lying along the light path from the QSO.

Given this scenario a large research effort has now been mounted to use the absorption spectra as probes of the intergalactic medium. It is supposed that we are looking at intergalactic clouds or protogalaxies and the like, randomly distributed, in an era of galaxy formation in the cosmological redshift range $2 \leq z \leq 5$.

How strong is the case that all of the absorption is caused by intervening matter at cosmological distances, and that none of the systems have either redshift components involving ejection from the QSOs or intrinsic redshift components? In other words, how reasonable is it to ignore evidence for ejection of some absorption systems, as well as the evidence for intrinsic redshifts? What are the reasons for ignoring all the evidence for intrinsic redshifts described earlier in this chapter?

There are several possible answers to these questions.

1. To study the intergalactic medium at high redshifts using absorption from intervening clouds is such an attractive idea that evidence the absorption is not due to intervening clouds can be disregarded.
2. The evidence for intrinsic redshifts is not really known or accepted by those who are working on the absorption phenomenon.
3. There *is* strong evidence *for* the absorption being intervening, and none against.
4. Some QSOs with large emission-line redshifts and many absorption redshifts both lie at cosmological distances and *also* have intrinsic redshift components.

Arguments (1) and (2) are certainly operating in the real world, but they cannot be countered by scientific evidence, since they are manifestations of the band-wagon approach to science. Only arguments (3) and (4) can be discussed using normal scientific criteria.

As far as (3) is concerned, Sargent *et al.*[66] investigated the distribution of Lyα absorption redshifts in a small sample of QSOs and concluded that the distribution of z_{abs} was compatible with a Gaussian distribution of clouds or galaxy halos. This, together with later studies, originally led to the general view that the absorbers are individual clouds lying at smaller distances from us than the QSO. This conclusion was strengthened in the eyes of most astronomers by the results of Bergeron and her associates[67,68], who investigated a sample of high-redshift QSOs each of which had a much lower absorption redshift system identified in its spectrum, and were able to find a galaxy with this redshift close to the line of sight.

However, there is evidence which is in conflict with the intervening cloud hypothesis. For example, Borgeest and Mehlert[69] have shown that the number of absorption redshifts per unit interval of z is proportional to the emission redshifts of the QSOs, suggesting that the absorption redshifts are related to the emission redshifts, which would not occur if the absorption was due to intervening clouds. Secondly, there are a number of QSOs such as PKS 0237-23 with many absorption redshifts widely distributed in z which cannot be understood simply in terms of a random distribution of intervening clouds or halos[70]. In addition to these arguments, Richards *et al.*[71] have shown a large number of the absorption systems in many QSOs are intrinsic to the QSOs and suggest that gas is being ejected at speeds up to 75 000 km s^{-1}.

In a few of the bright galaxy–QSO pairs in Table 11.1 absorption is seen in the spectrum of the QSO at the redshift of the galaxy. There are also pairs of fainter galaxies and QSOs with very small separation. In Table 11.4 we show an incomplete list of QSOs with very close faint galaxies. Some of them were found by the method of Bergeron. Thus a far higher fraction of these show absorption at the redshift of the nearby galaxy than do the pairs involving the brighter galaxies in Table 11.1.

For the pairs in Table 11.1, there is often strong statistical or morphological evidence for physical association. For those involving fainter galaxies in Table 11.4, there is also sometimes morphological evidence for physical associations. There is therefore a conflict between this evidence and the evidence based on the fact that the absorption redshift agrees with the galaxy redshift and is much smaller than the emission redshift of the QSO. However, for all of the pairs, when absorption at the redshift of the galaxy has been reported, astronomers have ignored the statistical or morphological evidence for physical association, and have claimed that the presence of absorption means that the absorbing galaxy must lie far in front of the QSO. Ignoring the morphological evidence, they have tended to attribute such clustering to gravitational lensing, arguing that there must be much dark matter associated with the (foreground) galaxies. Of course, this will not work for the bright galaxies in Table 11.1, but only for the faint galaxies of Table 11.4.

If we accept the evidence for physical association, it must be supposed that the QSO, with an intrinsic redshift component, *lies in the halo of the galaxy.* In this case, if a fixed optical path length for absorption is chosen for a sample of pairs with similar galactic halos we would expect that on the average 50% would show absorption, and 50% would not. In the studies mentioned above the authors only chose QSOs in which *absorption was already detected.* If our hypothesis is correct, for a sample of QSOs without low-redshift absorption we would expect to see a similar distribution of galaxies as far as $\Delta\theta$ and apparent magnitude are concerned. This is apparently now being found[72,73].

Another important point is that in some of the best cases where interaction between ejected QSO and galaxy are implied, e.g. 3C 232 and NGC 3067 (Fig. 11.6), and the QSOs near NGC 3079, there is every indication that the gas which is doing the absorbing is not quiescent halo gas, but is gas that is being ejected from the galaxy.

Since we see evidence for activity in many of the galaxies listed in Table 11.4, it is possible that much or all of the absorption is associated with gas

Table 11.4. *Very close QSO – galaxy pairs (m > 15.5)*

Galaxy	z	m	QSO	m_v	z_Q	sep$^{(")}$	Remarks
Anon	—	19.7	0109 + 176	18.0	2.157	8	bridge joining galaxy to QSO
Anon	0.535	—	0109 + 200	17	0.746	7.1	abs. at galaxy redshift
Anon	0.133	20.0	PKS 0119–046	16.47	1.948	14	abs. at galaxy redshift
Anon	0.4176	—	PKS 0229 + 131	17.0	2.065	6.8	abs. at galaxy redshift
Anon	0.051	—	GC 0248 + 430	17.45	1.311	3.5	QSO aligned with double nucleus galaxy; abs. at galaxy redshift
Anon	0.0669	—	0446 – 208	17	1.896	13	abs. at galaxy redshift
Anon	0.726	—	0453 – 423	17.1	2.661	30.3	abs. at galaxy redshift
Anon	0.071	—	PKS 0454 + 036	16.53	1.345	4	abs. at 0.8596 and1.154 not at galaxy redshift
Anon	—	20.2	0809 + 483 (3C R196)	17.8	0.871	1.6	
Anon	0.153	—	PKS 0952 + 179	17.2	1.478	22.8	
Anon	0.441	21	1038 + 064	16.81	1.27	9.4	abs at galaxy redshirt
Anon	0.359	—	1101 – 264	11.0	2.148	12.2	abs. at galaxy redshift
Anon	0.030	16.3	1107 + 036	19	0.964	20	
Anon	0.313	—	1127 – 145	16.9	1.187	8	abs at galaxy redshift
Anon	0.392	21.9	1209 + 107	17.76	2.191	7.1	abs at galaxy redshift
Anon	0.450	21.7	1441 + 522 (3C 303C)	19.97	1.57	4	abs. at galaxy redshift
3C 303	0.414	17.3	1441 + 522 (3C 303C)	19.97	1.57	20	QSO in lobe of radio emission
Anon	0.199	—	PKS 1229–021	16.7	1.045	8.6	
Anon	0.4359	—	1511 + 103	17.73	1.546	6.9	abs. at galaxy redshift
Anon	0.121	18.5	PG 1522 + 101	15.7	1.321	5	

Name			Object	mag	z		Comment
Anon	0.076	—	1543 + 489	16.1	0.400		
Anon	0.434	—	1548 + 114B	19	1.901	10	second QSO 1548 + 114A 5″ from 1548 + 1148
Anon	0.091	—	1704 + 607 (3C 351)	16.1	0.371	34	abs. at galaxy redshift also other absorption
Anon (Several galaxies)		—	1622 + 238 (3C 336)		0.927	2–10	abs. at z of one galaxy Steidel & Dickinson (1992) (ref. 83)
Anon	0.366	—	1632 + 391 (4C 39.46)	18	1.082	4.2	abs. at galaxy redshift
Klemola 31 A&B	0.029	16	PKS 2020 − 370	17.5	1.048	20	abs. at galaxy redshift
Anon	0.430	21.5	PKS 2128 − 123	15.46	0.501	8.6	abs. at galaxy redshift; axis of galaxy aligned with QSO
Anon	0.075	—	PKS 2135 − 147	15.19	0.200	49	abs. at galaxy redshift also other absorption
Anon	0.790	—	PKS 2145 + 067	16.47	0.990	5.5	abs. at galaxy redshift
3C 771	0.707	—	2203 + 292B	22	4.399	51	QSO lies along optical radio axis of 3C 771
Anon	0.2	—	2203+292B	22	4.399	7	
Abell 2854 (2 galaxies) G_1 & G_2	0.12	—	2319 + 272 (4C 27.50)	18.6	1.253	17(G_1) 25	QSO on line joining centers of galaxies

that has been ejected from the galactic disks. After all, we know that even in our own Galaxy much of the neutral gas at high latitudes is due to ejection from the Galactic disk.

It is also possible that at least some of the absorbing material is *ejected* from the QSO and its speed is determined by radiation pressure from the QSOs. In this case, whatever the velocity turns out to be, it would prove that the gas is intrinsic to the QSO. Such outflowing gas absorbs both continuum and line radiation from the QSO. Ions in the outflow of any particular species will, if they have an appropriate velocity, 'see' redshifted radiation from the central source that may have a resonance line or ionization absorption edge falling at the frequency of their particular resonance lines. Such ions will experience less radiation pressure than ions at neighboring velocities. With gravity acting inward a balance can be set up since both scale as $1/r^2$ so that ions will accumulate in velocity space at appropriate velocities where they experience no net inward or outward acceleration. At the velocities where balance occurs, resonance lines will 'lock on' to each other, or on to any point of strong intensity–wavelength gradient in the radiation flux such as an ionization edge.

This is the phenomenon of line locking or edge locking. A typical ratio $(1 + z_{em})/(1 + z_{abs}) = 1.11 = \lambda 1025(\text{Ly}\beta)/\lambda 912(\text{Ly limit})$. Burbidge and Burbidge[74] looked for this and other ratios in an early sample of QSOs with absorption redshifts and reported positive results. Sargent and Boroson[75] claimed to have rebutted this result, blaming it on poor statistics. More recently, however, Wampler *et al.*[76] and Barlow[77] have also reported results which suggest that line locking is present in some objects.

An extensive new investigation of this phenomenon is required since we now have many thousands of absorption redshifts available. We suspect that the evidence is there, and that line-locking and edge-locking will demonstrate beyond doubt that many absorption systems are intrinsic to the QSOs.

However, as long as most astronomers in the field adhere to the belief that the absorption is due to intervening gas and analyse their spectra accordingly, progress on this topic will not be made.

We turn now to a way in which it is possible to show that some QSOs have redshift components due to intervening absorption clouds as well as intrinsic and cosmological components.

Intrinsic redshifts and intervening absorption redshifts

Evidence that galaxy halo and/or disks are responsible for some of the absorbing systems in the spectra of QSOs comes not only from the direct detections of galaxies very close to the QSOs[67,68] but also from the identification in QSOs having redshifts in the range $z \approx 1.8$–4 of the Lyα damped systems where Wolfe and others[78] have made a good case for supposing that these systems are due to gaseous disks of spiral galaxies.

Such observations are compatible with the view that the QSOs in question have both an intrinsic redshift component z_i and a cosmological component z_c. In a model in which the QSO and galaxy are physically associated, a damped Lyα absorption at a redshift z_{abs} must define the cosmological redshift of the parent galaxy. Thus the damped Lyα redshift will be the cosmological redshift z_c of the system, and the difference between it and the emission redshift of the QSO must be due to the intrinsic redshift term, i.e. $z_i = [(1 + z_{em})/(1 + z_{abs})] - 1$.

For the systems in which a galaxy with $z_g = z_{abs}$ has been detected, cf. Bergeron and Boissé[67], the values of z_i turn out to lie in the range 0.3–0.7. For the systems in which the galaxy is identified only through the identification of the Lyα damped systems $z_{abs} \approx z_c$, and the values of z_i lie in the range 0.05–0.3.

In all of those cases $z_c = z_{abs} > z_i$ and in the Lyα systems $z_c = z_{abs} \gg z_i$.

Thus in this interpretation, the investigations of galaxies at high redshifts being made by studying the absorption systems in high-redshift QSOs are still valid and can be used to set limits to cosmological models.

The damped Lyα systems are comparatively rare among all of the absorption systems. However, in the spectra of high-redshift QSOs the frequency of absorption is high. In this model the Lyα damped systems are associated with the parent galaxies of the QSOs and when they are not present we conclude that the QSOs have moved sufficiently far away from their parent galaxy so that they are outside the thick disk. What about the situation in which more than one absorption redshift, each of which has all of the spectral characteristics associated with either disk or halo absorption, is found? In this case we must relax the condition that it is the lowest value of z_{abs} that is due to the parent galaxy, and argue instead that it is the highest value of z_{abs} compatible with absorption in a galactic disk that defines the cosmological redshift.

How do we account for the other absorption systems, both those containing metals, and those which make up those so-called Lyα forest systems which are commonplace in the spectra of high-redshift QSOs? Since we have accepted that the damped Lyα systems are galaxies at high redshifts, these QSOs also must lie at large cosmological redshifts. Thus some of the Lyα systems are galaxies at high redshifts, and absorption involving metals could be due to intervening gas as is commonly assumed. However, the problems raised by Borgeest and Mehlert[69] previously referred to, and the difficulty of explaining the very wide range of z_{abs} seen in some QSOs, make us believe that at least some of the absorption is due to ejection of gas from the QSOs at the appropriate velocities, so that the redshift is made up of a cosmological component, an intrinsic component due to the QSO, and a Doppler component due to ejection towards us.

In cases that are often found, where $z_{abs} \approx z_{em}$, the absorption must be due to gas associated with the QSO with nearly the same intrinsic component as the QSO. It is curious that the ejection hypothesis has been neglected. It clearly needs further study. In some cases the velocities must be quite high, but they are never as high as those tacitly accepted by those who believe in the existence of superluminal motions in the compact radio sources. A good example of the manifestation of ejection may be associated with the Lyα forest system seen in the spectrum of 3C 273. Several Lyα forest systems ranging all the way from $z_{em} = 0.158$ to the redshift of the Virgo cluster have been detected[78,79,80] and the search for galaxies at those (cosmological) redshifts has given null results. With 3C 273 in the Virgo cluster as Arp has suggested, ejection of gas from it at quite modest velocities would explain these systems. Even if it turned out that 3C 273 were at a cosmological distance, it is still possible that the Lyα absorption could be due to ejection.

References

Chapter 11

1. Burbidge, G. & Burbidge, M. 1967, *Quasi-Stellar Objects* (San Francisco and London, W. H. Freeman and Co.).
2. Hoyle, F. & Burbidge, G.R. 1966, *Nature*, **210**, 1346.
3. Hewitt, A. & Burbidge, G. 1987, *Ap. J. S.*, **69**, 1.
4. Hewitt, A. & Burbidge, G. 1993, *Ap. J. S.*, **87**, 451.
5. Dent, W.A. 1965, *Nature*, **205**, 487.
6. Rees, M.J. 1984, *ARA&A.*, **22**, 471.
7. Hoyle, F., Burbidge, G. & Sargent, W.L.W. 1966, *Nature*, **209**, 751.
8. Woltjer, L. 1966, *Ap. J.*, **146**, 597.
9. Rees, M. 1967, *M.N.R.A.S.*, **135**, 345.
10. Arp, H. 1968, *PASP*, **80**, 129.
11. Arp, H. 1967, *Ap. J.*, **148**, 321.
12. Arp, H. 1966, *Atlas of Peculiar Galaxies* (California Institute of Technology).
13. Burbidge, E.M., Burbidge, G., Solomon, P.M. & Strittmatter, P.A. 1971, *Ap. J.*, **170**, 223.
14. Kippenhahn, R. & de Vries, H.L. 1974, *Astroph. Space Sci.*, **26**, 131.
15. Arp, H., Burbidge, E.M., Mackay, C. & Strittmatter, P. 1972, *Ap. J. L.*, **171**, L41.
16. Burbidge, G., O'Dell, S.L. & Strittmatter, P.A. 1972, *Ap. J.*, **175**, 601.
17. Bolton, J. & Wall, J. 1970, *Aust. J. Physics*, **23**, 789.
18. Bahcall, J.N., McKee, C.F. & Bahcall, N. 1971, *Astroph. Letters*, **10**, 147.

19. Webster, A. 1982, *M.N.R.A.S.*, **200**, 47.
20. Browne, I.A.W. 1982, *Ap. J.*, **263**, L7.
21. Arp, H.C. 1983, *Ap. J. L.*, **271**, L41.
22. Hewitt, A. & Burbidge, G. 1987, *Ap. J. S.*, **63**, 1.
23. Sulentic, J.W. & Tifft, W.G. 1973, *The Revised New General Catalog of Nonstellar Astronomical Objects* (Tucson, Arizona, The University of Arizona Press).
24. Burbidge, G., Hewitt, A., Narlikar, J.V. & Das Gupta, P. 1990, *Ap. J. S.*, **74**, 675.
25. Chu, Y., Zhu, X., Burbidge, G. & Hewitt, A. 1984, *A&A*, **138**, 408.
26. Bartelmann, M. & Schneider, P. 1993, *A&A*, **271**, 421.
27. Bartelmann, M. & Schneider, P. 1994, *A&A*, **284**, 1.
28. Zhu, X-F. & Chu, Y.-Q. 1995, *A&A*, **297**, 300.
29. Weedman, D. 1970, *Ap. J. L.*, **161**, L113.
30. Sulentic, J. & Arp, H.C. 1987, *Ap. J.*, **319**, 687.
31. Lynds, R. & Millikan, A.G. 1972, *Ap. J.*, **176**, L5.
32. Burbidge, G. 1996, *A&A*, **309**, 9.
33. Arp, H.C. 1987, *Quasars, Redshifts and Controversies* (Berkeley, Interstellar Media).
34. Canizares, C.L. 1981, *Nature*, **291**, 620.
35. Arp, H.C. 1990, *A&A*, **229**, 93.
36. Osriker, J. 1989, in *BL Lac Objects: Lecture Notes* (ed. L. Maraschi, Berlin, Springer Verlag), Vol. 334.
37. Wade, C.M. 1960, *Observatory*, **30**, 235.
38. Arp, H.C. 1997, *A&A*, **319**, 33.
39. Radecke, H.D. 1997, *A&A*, **319**, 18.
40. Burbidge, E.M. & Burbidge, G. 1997, *Ap. J. L.*, **477**, L13.
41. Pietsch, W., Vogler, A., Kahabka, P., Jain, A. & Klein, V. 1994, *A&A*, **284**, 386.
42. Burbidge, E.M. 1995, *A&A*, **298**, L1.
43. Burbidge, E.M. 1997, *Ap. J. L.*, **484**, L99.
44. Chu, Y., Wei, J., Hu, J., Zhu, X. & Arp, H. 1997, *Ap. J.*, **500**, 596.
45. Peebles, P.J.E., Schramm, D., Turner, E. & Kron, R. 1991, *Nature*, **352**, 769.
46. Peterson, B. 1997, *An Introduction to "Active Galactic Nuclei"* (Cambridge U. Press).
47. Seyfert, C.K. 1943, *Ap. J.*, **97**, 28.
48. Kristian, J. 1973, *Ap. J. L.*, **179**, L11.
49. Bahcall, J.N., Kirhakos, S. & Saxe, D.H. 1997, *Ap. J.*, **479**, 642.
50. Heckman, T. & Lehnert, M. 1997, *The Hubble Space Telescope and the High Redshift Universe* (eds. N.R. Tanvir, A. Aragon-Salamanca and J.V. Wall, World Scientific Pub. Singapore), p. 377.
51. Wampler, E.J., Robinson, L., Burbidge, E.M. & Baldwin, J. 1975, *Ap. J. L.*, **198**, 49.
52. Boroson, T. & Oke, J.B. 1982, *Nature*, **296**, 397.
53. Richstone, D. & Oke, J.B. 1977, *Ap. J.*, **213**, 8.
54. Boroson, T. & Oke, J.B. 1984, *Ap. J.*, **281**, 535.
55. Miller, J. 1998, private communication.
56. Stockton, A. 1978, *Ap. J.*, **223**, 747.
57. Walsh, D., Carswell, R. & Weymann, R. 1979, *Nature*, **279**, 381.
58. Burbidge, G., Hoyle, F. & Schneider, P. 1997, *A&A*, **320**, 8.

59. Huchra, J., Gorenstein, M., Kent, S., Shapiro, I., Smith, G., Horine, E. & Perley, R. 1985, *A. J.*, **90**, 691.
60. Burbidge, G. 1985, *A. J.*, **90**, 1399.
61. Arp, H.C. & Crane, P. 1992, *Phys. Lett. A.*, **168**, 6.
62. Kundic, T. *et al.* 1997, *Ap. J.*, **482**, 75.
63. Dar, A. 1991, *Ap. J. L.*, **382**, L1.
64. Wolfe, A.M., Turnshek, D., Smith, H.E. & Cohen, R.D. 1986, *Ap. J. S.*, **61**, 249.
65. Lynds, R. 1971, *Ap. J. L.*, **164**, L73.
66. Sargent, W.L.W., Young, P., Boksenberg, A. & Tytler, D. 1980, *Ap. J. S.*, **42**, 41.
67. Bergeron, J. & Boissé, J. 1991, *A&A*, **243**, 344.
68. Le Brun, V., Bergeron, J., Boissé, P. & Christian, C. 1993, *A&A*, **279**, 23.
69. Borgeest, V. & Mehlert, D. 1993, *A&A*, **275**, L21.
70. Duari, D. & Narlikar, J.V. 1995, *Int. Mod. J. Phys.*, **D4**, 367.
71. Richards, G., York, D., Yanny, B., Kollgaard, R., Laurent-Muehleisen, S. & Vanden Berk, D. 1999, *Ap. J.*, **513**, 576.
72. Drinkwater, M., Webster, R.L. & Thomas, P. 1993, *A. J.*, **106**, 848.
73. Lanzetta, K.M., Wolfe, A.M., Turnshek, D.A. 1995, *Ap. J.*, **440**, 435.
74. Burbidge, E.M. & Burbidge, G.R. 1975, *Ap. J.*, **202**, 287.
75. Sargent, W. & Boroson, T. 1977, *Ap. J.*, **212**, 382.
76. Wampler, E.J., Chugai, N.N. & Petitjean, P. 1995, *Ap. J.*, **443**, 586.
77. Barlow, T. 1997, private communication.
78. Wolfe, A.M., Turnshek, D., Lanzetta, K.M. & Lu, L. 1993, *Ap. J.*, **404**, 480.
79. Bahcall, J.N., Jannuzi, B.T., Schneider, D.P., Hartig, G.F., Bohlin, R. & Junkkarinen, V. 1991, *Ap. J.*, **377**, L5.
80. Morris, S.L., Weymann, R.J., Savage, B.D. & Gilliland, B.D. 1991, *Ap. J. L.*, **377**, L21.
81. Bergeron, J., D'Odorico, S. & Kunth, D. 1987, *A&A*, **180**, 1.
82. Kilkenny, D., O'Donoghue, D., Koen, C., Stobie, R.S. & Chen, A. 1997, *M.N.R.A.S.*, **287**, 867.
83. Borgeest, V., Dietrich, M., Hopp, V., Kollatschny, W. & Schramm, K.-J. 1991, *A&A*, **243**, 93.
84. Burbidge, G. & Hewitt, A. 1987, *A. J.*, **93**, 1.

12

The new observational evidence and its interpretation: (b) ejection phenomena and energetics

Ejection of coherent objects from galaxies

As we have already shown in the previous chapter, in the galaxy M87 relativistic particles and optical synchrotron radiation are ejected in a preferred direction (the jet), and QSOs are ejected in the same direction, and much further out is a radio galaxy M84, which also appears to have been ejected (Fig. 11.15).

In addition to the galaxy–QSO pairs discussed in the previous chapter and galaxies like M87, there are also many very peculiar galaxies which have complex morphology and lack of symmetry. Such galaxies were originally designated simply as irregular systems by Hubble, and were called highly peculiar or interacting systems in the catalogs of Vorontsov–Velyaminov and Arp. In some cases the velocity fields have been measured, and other studies have been made, and a few of these galaxies have been interpreted as two previously separate systems which have collided. Thus the convention has developed that in almost all cases we are seeing collisions or 'mergers', and much modeling has been made on the assumption that two galaxies are coming together and the only forces operating are gravitational.

The original argument along these lines comes from work by Toomre and Toomre[1] who showed that the bridge and tidal tails found in the well known galaxy pair NGC 4038-39 could be well modelled by supposing that this system is two previously separate galaxies interacting through their gravitational fields.

This result is completely valid. However, it has led to a situation in which almost all complex galaxies are *assumed to be merging* systems whether or not tidal tails can be discerned. For example, a small number of the luminous infrared galaxies, such as Arp 220 and NGC 6240 which contain more than one nucleus are called merging systems. Whole conferences are now devoted to studies and analysis of merger activity. But there is no *direct* evidence that *any* of these systems are previously separate systems coming together. There are obviously three possible interpretations – they are coming together, they evolve *in situ*, or they are coming apart. Presumably the reason why only the first interpretation has been accepted is that gravity is an attractive force, and it has been impossible without new concepts to consider any mechanism that could lead to large-scale ejection.

We believe that it is more likely that some of these are systems which are splitting apart – i.e. they may well be systems similar to the galaxies which eject QSOs, at an earlier stage in their evolution. We believe that it is the energy generated in the ejection process which gives rise to the high luminosity. One remarkable example is the configuration NGC 3561A & B[2]. This is shown in Fig. 11.14. It is sometimes described as a spiral and elliptical galaxy in collision with a tidally bound dwarf galaxy created in the collision[2]. The embedded nebulosity is attributed to tidal effects. But as Ambartsumian originally pointed out[3] this system may equally well be the result of the ejection of one galaxy from another, along the lines of M87 and M84 with this debris extending over ~ 100 kpc with blue dwarf galaxies and star-forming regions emerging out of the debris. It is notable that a QSO with $z \approx 2.2$ (marked in Fig. 11.14) has probably also been ejected from the brighter galaxy.

This evidence, and this point of view are directly related to the proposal made by Ambartsumian[4,5] that many galaxies are to be found in expanding associations with positive total energy, i.e. $(KE + PE) > 0$. We shall discuss this further when we survey the large-scale structure and dark matter. Ambartsumian's view was that the galaxies which are expanding away from a parent galaxy have been generated by processes in the nucleus of that galaxy. He gave as examples some of the small groups of galaxies, and highly irregular clusters like the Hercules cluster, in which the kinetic energy of the visible galaxies is much greater than the potential energy.

His conclusion was that the phenomena all pointed to ejection from what he speculated was a superdense state.

The rapid release of very large amounts of energy in galaxies

In the late 1950s, after the mechanism of generation of radio emission from galaxies had been identified as incoherent synchrotron radiation, it was a straightforward matter to calculate how much energy must be present in the form of relativistic electrons and magnetic flux. The problem was then, and still remains, that the total energy E_e (particle energy) + E_m (magnetic energy) could not be directly evaluated, because although the critical frequency ν_c at which the bulk of the radiation is emitted from an individual electron with energy E is proportional to $E^2 B_\perp$, where $B_\perp$ is the perpendicular magnetic field component, and the total energy radiated is proportional to $E^2 B_\perp^2$, no independent measurements of B could be made. What was known was the range of frequencies over which the source is radiating, and its angular size and, if it was optically identified, the redshift of the source. Since the first identifications were with galaxies, this meant that we could determine the distances from the redshifts and hence the sizes and volumes of the sources.

For a given source, the total energy $(E_e + E_m)$ was then calculated for a series of assumed values of B, and the most conservative assumption was made, that $(E_e + E_m)$ was a minimum. It was found for the simplest models that under these conditions $E_e \approx 4/3 E_m$; i.e. rough equipartition must be present. Typical extended radio sources consist of two large emitting lobes lying symmetrically about the central galaxy each with dimensions ~ 100 kpc. For the most powerful sources the total energy (minimum) amounted to about 10^{59}–10^{60} erg, with the values of $B_\perp$ lying between 10^{-5} and 10^{-6} G[6,7].

Apart from the fact that we do not have any direct measurement of the magnetic field strength in those regions far outside galaxies, there is the question of the proton flux. Everywhere in nature where fast particles are generated, most of the energy is carried by the protons. In the primary cosmic ray flux in our own Galaxy the ratio of energy in protons to energy in electrons $E_p/E_e \simeq 100$, in part because the electron component is more vulnerable to radiation and collisional losses, while the proton flux will only be removed by nuclear collisions or escape. Thus while the proton flux cannot be detected in radio sources, we have every reason to believe that with the electron flux there is a (more energetic) proton flux. Assuming a ratio of the proton to electron energy of $\sim 100{:}1$ (the value in the primary cosmic rays of our own Galaxy) the total energy is increased to 10^{61}–10^{62} erg and the magnetic fields are slightly increased. Thus total energies for powerful sources amount to $\sim 10^6 M_\odot c^2$, and such values are conservative.

At the Solvay conference in 1964[7], one of us (GB) gave the first presentation of this theory and the results to a number of the leading physicists with little astronomical background. Several of them, in particular E. Segre and Robert Oppenheimer, were obviously very concerned or even taken aback by the very large numbers.

A major issue that was raised in the discussion was how efficiently such large energies can be generated. The characteristic time scale for the large sources was estimated from $\tau \sim E_e/L_r$, where L_r is the radio luminosity; for $L_r \simeq 10^{44}$erg s^{-1} and $E_e = 10^{60}$erg, $\tau \simeq 3 \times 10^8$ years. But how much energy is required to generate such huge fluxes of relativistic electrons (and protons) and how can so much magnetic flux be arranged in regions where the density of the gas is very low compared with that in galaxies?

What is the efficiency with which this can be done on the earth? The obvious analogy is with man-made accelerators. Here the efficiency factor is given by the ratio of the power output (in the beam) to the total power input. It turns out that for a linear accelerator, the efficiency factor is $\sim 10^{-2}$, while for a synchrotron it is $\sim 10^{-3}$ or less. As thermodynamics requires, the bulk of the energy is degraded and lost as heat energy in the magnets.

Is there any reason why such low efficiency factors should not apply in nature? In solar flares, in discharges, and the like, apparently the efficiency factors can be much larger[a]. Thus if the energy is generated in discharges and in bursts perhaps this is true. But for the large radio sources, the currently popular models based on the accretion disk–black hole paradigm or the rotating black hole scheme require the following sequence of events.

Gravitational energy is released from collapsing mass at an efficiency of say 10% or less[8], and at high efficiency, it is transformed into relativistic particles. Let us suppose that an efficiency for transformation from the gravitational mode to relativistic particles is f_1. This energy is now transported, via a relativistic beam of particles assumed to be stable, through a distance ~ 100 kpc into a radio-emitting lobe. Let us suppose that the efficiency factor with which it is transmitted is f_2. This energy is then dumped into the low-density plasma in the lobes, and generates shock waves that are used to re-accelerate electrons which now emit the radio waves. Suppose that this final efficiency factor is f_3.

If rest mass is converted to gravitational energy with a 10% efficiency, then the total energy in the final radiating particles requires a rest mass of

$$M = \frac{10^{61}}{M_\odot c^2} \times \frac{1}{f_1 f_2 f_3} \quad \text{solar masses.}$$

Thus even if $f_1 = f_2 = f_3 = 0.1$

$$M = 10^{10} M_\odot$$

---

[a] But of course it should be remembered that while the efficiency of conversion of energy in a solar flare may be very high, the total energy released in all non-thermal processes in the Sun is a minute fraction of the energy released by thermonuclear processes ($\sim 10^{-6}$).

and somewhere, the bulk of this energy which has been released in other forms, i.e. $9999 \times 10^6 M_\odot c^2$ is lurking, undetected. Not only is this a requirement for such a model, but we also have to explain how such strong magnetic fields, as strong as the interstellar field in our own Galaxy, can have been generated outside the radio galaxies. Can the fields really be as strong as this? How has the flux been carried out without being reduced in strength? If the fields are weaker than the equipartition value, the value of E_p is rapidly increased, meaning that M must be much greater. But clearly there is an upper limit on M – the mass of the whole galaxy.

In Chapter 5, we pointed out that following the discovery of these phenomena, gravitational energy release was only one of *several* mechanisms proposed to generate the great energies. Unfortunately all of the theoretical efforts in modern times have been devoted to the black hole–accretion disk paradigm or the release of energy from a rotating black hole, but the problems associated with the efficiency, etc., which we have raised above, have been ignored for 30 years. We believe that the difficulties associated with the black hole–accretion disk paradigm, and in particular the efficiency problem and the inability of that mechanism to explain the ejection of coherent objects, make it likely that the observed phenomena are due to mass creation in nuclei of galaxies.

To summarize, the evidence described in this and the previous chapter suggests the following to us.

1. Violent events in galaxies are manifestations of mass creation.
2. From these events huge fluxes of relativistic particles and rapidly moving gas clouds are generated.
3. QSOs with intrinsic redshifts are also ejected, and they are probably related to the mechanisms through which new galaxies are formed.
4. Following Ambartsumian, it is reasonable to consider the possibility that *systems* of galaxies with positive total energy also originate in this way.

Thus the evidence strongly suggests that condensed objects are generated in the nuclei of galaxies and are ejected periodically along with hot gas and relativistic particles.

They are obviously not formed by density fluctuations in an early universe which then collapse into galaxies. It is ejection that we are seeing, and not accretion and collapse long ago that is the basis of the real cosmogony leading to the formation of galaxies and QSOs.

It is the evidence summarized in these two chapters, together with the cosmological evidence described earlier which has led us to the conclusion that a cosmological model which replaces the classical big-bang approach and can also point us in the direction of the correct cosmogony is desirable. The observational evidence tells us directly that the centers of galaxies are the places where new objects are generated and the fact that the activity can be found in galaxies with a wide range of distances strongly suggests that the activity goes on at all epochs.

This is the model which we describe in the following six chapters, beginning in Chapter 13 with a short restatement of the conventional position; and then continuing in Chapter 14 with a more refined discussion of the big-bang theory. After that in Chapter 15 we introduce the creation of matter from a mathematical point of view, showing how a wide range of new possibilities is then opened up, including the old steady-state theory discussed earlier in Chapter 7 as a special case. Then we go on in Chapter 16 and thereafter to relate the new theoretical possibilities to observation.

References

Chapter 12

1. Toomre, A. & Toomre, J. 1972, *Ap. J.*, **178**, 623.
2. Duc, P.-A. & Mirabel, I.F. 1991, *A&A*, **289**, 83.
3. Ambartsumian, V.A. 1961, *A. J.*, **66**, 536.
4. Ambartsumian, V.A. 1958, *Solvay Conference on Structure and Evolution of the Universe* (R. Stoops, Brussels), p. 241.
5. Ambartsumian, V.A. 1965, *Structure and Evolution of Galaxies, Proc. 13th Conf. on Physics, University of Brussels* (New York, Wiley Interscience).
6. Burbidge, G. 1959, *Ap. J.*, **129**, 849.
7. Burbidge, G. & Burbidge, E.M. 1964, *The Structure and Evolution of Galaxies* (ed. Professor Prigogine) *Proc. 13th Solvay Conf. on Physics*, Sept. 1964, p. 137.
8. Rees, M.J. 1984, *ARA&A*, **22**, 471; Blandford, R. & Rees, M.J. 1974, *M.N.R.A.S.*, **169**, 395.

13

Modern Friedmann cosmology

Preliminaries

There is no theoretically determinate way of choosing coordinates of cosmological relevance. In general relativity the choice has a large measure of
arbitrariness, and indeed it is possible to gain the impression from general
relativity that apparently interesting properties of spacetime are of physical
relevance when in fact they are nothing but the products of special choices of
the coordinate frame. This awkward aspect of the theory was a serious
problem for cosmology in the 1920s, and it continued to be so until the
early 1930s when H.P. Robertson and A.G. Walker independently intro-
duced a system of coordinates r, θ, φ, t, where t was proper time as measured by the observer, and r, θ, φ were spherical polar coordinates with
respect to the observer. Provided the universe was considered to be homogeneous and isotropic, it could be shown for observers attached to so-called
comoving particles (galaxies) that the values of r, θ, φ were constants, and
that t could also be taken synchronously for such observers. Then the metric
(line element) necessarily had the form

$$ds^2 = dt^2 - S^2(t)\left[\frac{dr^2}{1 - kr^2} + r^2(d\theta^2 + \sin^2\theta d\varphi^2)\right], \qquad (13.1)$$

where ds was proper distance between the neighboring spacetime points
$(r + dr, \theta + d\theta, \varphi + d\varphi, t + dt)$ and (r, θ, φ, t). We have taken the speed of
light $c = 1$, so that the time unit is the same as the space unit. Symmetry
considerations do not determine the value of k, which can take any real but
constant value. However, by suitably adjusting the unit of r, k can be made
equal to $+1$ if it is positive, -1 if negative, otherwise zero. The three cases
respectively indicate whether the curvature of hypersurfaces given by $t =$
constant are positive, negative or zero. This system of coordinates, in which

galaxies have constant r, θ, φ values, has been used widely and indeed almost exclusively in modern times. Any object with r, θ, φ constant is often said nowadays to *belong to the Hubble flow*. One also sometimes hears of the 'cosmological rest frame', which implies that the galaxies in this coordinate frame have constant spatial coordinates. The immense simplification from the full generality of Riemannian geometry to the simple form (13.1) has been brought about by the far-reaching assumption of a homogeneous and iso-tropic universe. It was the correct choice of $k(0, \pm 1)$ that Hubble attempted to decide through his program of galaxy counts discussed in Chapter 3.

The function $S(t)$ determines the scale at time t of the universe. The func-tion $S(t)$ also determines the cosmological redshift z of light emitted at time t_{em} by a Hubble-flow galaxy of coordinates r, θ, φ and received by a Hubble-flow observer at $r = 0$ at time t_0 according to

$$1 + z = \frac{S(t_0)}{S(t_{em})}, \tag{13.2}$$

with t_0 and t_{em} related by

$$\int_{t_{em}}^{t_0} \frac{dt}{S(t)} = \int_0^r \frac{dr}{\sqrt{1 - kr^2}}, \quad k = 0, \pm 1$$

$$\begin{aligned} &= \sin^{-1} r \quad \text{for } k = +1; \\ &= r \qquad\;\; \text{for } k = 0; \\ &= \sinh^{-1} r \;\, \text{for } k = -1. \end{aligned} \tag{13.3}$$

Determining $S(t)$ is, however, a matter of physics, not of geometry. In the standard theory, the equations of general relativity are used

$$R_{ik} - \frac{1}{2} g_{ik} R = -8\pi G \, T_{ik}, \tag{13.4}$$

where T_{ik} is the energy-momentum tensor of matter and radiation. The coordinates are denoted numerically by $x^1 = r$, $x^2 = \theta$, $x^3 = \varphi$, $x^4 = t$, and with respect to this numerical notation T_k^i takes the following forms in certain special cases.

(i) Isotropic fluid

$$T_1^1 = T_2^2 = T_3^3 = -p, \quad T_4^4 = \rho_0 + 3p, \tag{13.5}$$

where ρ_0 is the rest mass proper density of the fluid and p is its internal isotropic pressure, the off-diagonal elements of T_k^i being zero.

(ii) Black-body radiation

$$T^1_1 = T^2_2 = T^3_3 = -\frac{u}{3}, \quad T^4_4 = u, \tag{13.6}$$

where u is energy density of the radiation. The off-diagonal components are again zero.

Since the divergence of the left-hand side of (13.4) is identically zero, the need for T_{ik} to satisfy equation (13.4) requires energy and momentum to be conserved automatically. But the need to satisfy (13.4) does not take account in general of physical processes occurring between particles and fields. By reducing the metric (13.1) locally to flat spacetime, the processes to be inserted additionally to the equations of general relativity are those of normal physics. This follows from the strong principle of equivalence, assumed in general relativity. As for instances under adiabatic conditions, the proportionalities

$$\rho_0 \propto S^{-3}, \quad p \propto S^{-5}, \quad u \propto S^{-4}. \tag{13.7}$$

As we have stated earlier, all coordinates have the same unit, t as well as r. It will be convenient to think of the unit as being determined by the Compton wavelength of some chosen particle, say the electron. In which case a familiar unit such as the centimeter is given by

$$1 \text{ cm} = 2.589\ 605 \times 10^{10} \text{ Compton wavelengths.} \tag{13.8}$$

It would be possible to signal to an inhabitant on a distant planet what we mean by a centimeter by giving the information on the right-hand side of (13.8), and in this sense we can speak of the Compton wavelength of the electron as being a fundamental physical quantity. But imagine trying to signal how we determine 1 cm or 1 m in terms of an arbitrary lump of matter. Those who work oppositely, as do the authors of the so-called SI system of units, read (13.8) in the opposite sense, viz.

$$\text{Compton wavelength} = 3.861\ 592 \times 10^{-13} \text{m.} \tag{13.9}$$

For our purpose, the SI system is not a good one and we shall not use it.

When the metric has the form (13.1), the 16 components of the left-hand side of (13.4) have only two non-zero independent forms

$$R^1_1 - \frac{1}{2}R = R^2_2 - \frac{1}{2}R = R^3_3 - \frac{1}{2}R = -\frac{2\ddot{S}}{S} - \frac{\dot{S}^2 + k}{S^2}, \tag{13.10}$$

$$R^4_4 - \frac{1}{2}R = -3\frac{\dot{S}^2 + k}{S^2}. \tag{13.11}$$

The standard model

Instead of the 10 independent equations (13.4) in the most general case, there are only two non-zero equations for a homogeneous isotropic universe. Where there is a mixture of non-interacting relativistic particles (photons, neutrinos) and of non-relativistic particles with the internal pressure p neglected, the two equations are of the form

$$\frac{2\ddot{S}}{S} + \frac{\dot{S}^2 + k}{S^2} = \frac{-8\pi Gu}{3}, \tag{13.12}$$

$$3\frac{\dot{S}^2 + k}{S^2} = 8\pi G(\rho + u), \tag{13.13}$$

where ρ, u are the energy densities of the non-relativistic and relativistic particles respectively. With the two not considered to interact together, $\rho \propto S^{-3}$, $u \propto S^{-4}$ and the equations can be written in the form

$$2\frac{\ddot{S}}{S} + \frac{\dot{S}^2 + k}{S^2} = -\frac{1}{3}\frac{A}{S^4}, \tag{13.14}$$

$$3\frac{\dot{S}^2 + k}{S^2} = \frac{A}{S^4} + \frac{B}{S^3}, \tag{13.15}$$

where A, B are positive constants depending on what the densities of the particles happened to be at a particular moment of time.

Eliminating $\dot{S}^2 + k$ between (13.14) and (13.15) gives

$$\ddot{S} = -\frac{A}{3S^3} - \frac{B}{6S^2}, \tag{13.16}$$

whence for $A > 0$, $B > 0$, $\ddot{S}$ is negative, meaning that in this case the universe necessarily decelerates, $\dot{S}$ becomes less with increasing t. It is already interesting that general relativity requires both the gravitational effect of the galaxies contained in the non-relativistic B term in (13.16) *and* the effect of radiation contained in the A term to decelerate the universe, whereas intuitively one might have expected the positive pressure of the radiation to increase the expansion rate $\dot{S}$ of the universe. This means that the observed expansion of the universe cannot have been caused by the exercise of any positive form of pressure, as in the sense of a conventional explosion which is the outcome not of pressure, but of a pressure gradient.

In general relativity (as opposed to Newtonian gravity), the introduction of any positive pressure brings positive energy density, and the attractive effect of this causes deceleration. To explain the observed expansion of the universe in terms of the operation of physical fields, it is necessary that the dominating pressure contribution shall be *negative*, a circumstance first realized in 1948 by one of us (FH)[1] and subsequently much emphasized in inflationary cosmological models.

Multiplying (13.16) by $\dot{S}$ and integrating gives

$$\frac{3\dot{S}^2}{S^2} = \frac{A}{S^4} + \frac{B}{S^3} + \frac{\text{constant}}{S^2}, \tag{13.17}$$

simply recovering (13.15), so that really we have only one independent equation, the constant in (13.17) being simply $-3k$.

When the constant in (13.17) is < 0 $(k > 0)$ it is possible, whatever the values are of A and B, to choose a value $S = S_0$ such that $\dot{S} = 0$. However, this is not a stable resting position of the universe. Because (13.16) shows that $\ddot{S} < 0$ at any S, the universe implodes, with the terms on the right-hand side of (13.16) ultimately diverging as $S \to 0$. Indeed, whatever the values of A, B, so long as $A > 0$, for S small enough the A term dominates on the right-hand side of both (13.16) and (13.17), giving

$$\frac{3\dot{S}^2}{S^2} \simeq \frac{A}{S^4}, \tag{13.18}$$

which integrates to yield

$$S \simeq \left[-\sqrt{\frac{4A}{3}} \, t + \text{constant} \right]^{1/2}. \tag{13.19}$$

Writing $S = S_0$ at $t = t_0$, the constant of integration in (13.19) is $S_0^2 + t_0\sqrt{4A/3}$, and the collapse from $S = S_0$ into the singularity $S = 0$ occurs in a time

$$t - t_0 = \sqrt{\frac{3}{4A}} \, S_0^2. \tag{13.20}$$

The situation we have calculated here can be contemplated for the actual universe simply by inverting the time sense, i.e. by considering time to run backwards. Then as S decreases a stage is reached at which the relativistic contribution to the energy density eventually dominates, leading to (13.18), and with the negative square root of (13.19) applying. Whence from $S = S_0, t = t_0$ there is collapse into a singular, unknown state in the time interval (13.20).

If the time sense of the world was like this, if in cosmology the observed sign of $\dot{S}$ was negative, with the Hubble constant H_0 negative

($H_o = -60$ km s^{-1}Mpc^{-1}, say) our conclusion would thus be that starting from now ($S = S_0, t = t_0$) the standard model requires the universe to disappear in all aspects at present known to us. For large enough S, the contraction is dominated by the B term of (13.17), and the time involved could easily be shown with a little more attention to the equations to be $2/3H_o^{-1} = 10.9 \times 10^9$ years.

Reference

Chapter 13

1. Hoyle, F. 1948, *M.N.R.A.S.*, **108**, 372.

14

Standard cosmology

General relativity

The rules of general relativity in (13.4) can be modified in two ways. First, through a change from the usual forms of the energy-momentum tensor T_{ik}. This is sufficient to yield new cosmological models, including the steady-state theory and its developments. It is also the approach adopted in inflationary scenarios. But it is not sufficient to include the intrinsic redshifts shown unequivocally to exist in QSOs in Chapter 11. For this, it will be necessary to extend the theory in the direction indicated in Chapter 17. In Chapter 17, the action principle is that of the Machian theory of inertia and gravitation which was first proposed by Hoyle and Narlikar in 1964 (see Chapter 17). In this theory the inertia of a particle is supposed to arise from a large-range scalar interaction of that particle with all other particles in the universe. Explicit rate of matter creation (and annihilation) is taken through the possibility that the typical worldline can have a beginning. In this present chapter and the next we shall show what can be obtained simply from changing the usual T_{ik}, beginning in the standard way. We begin with a discussion of the principle of stationary action that plays a key role in any physical theories, including general relativity.

The action $\mathcal{A}$ is a dimensionless number constructed from the parameters that define a configuration of the universe, not necessarily the configuration of the universe as we conceive of it in cosmology. But a configuration as it might be, with the paths of the particles being allowed to be different from the way they are in the 'actual' universe and the spacetime geometry given by

$$ds^2 = g_{ik}\, dx^i dx^k \tag{14.1}$$

being different, but a configuration that would be kinematically possible. Then in general relativity

$$\mathcal{A} = \frac{1}{16\pi G} \int R\sqrt{-g}d^4x - \sum_a m_a \int da \,, \qquad (14.2)$$

where R is the fully contracted Riemann tensor constructed from (14.1), i.e. from the kinematically possible spacetime geometry, and $\sum_a$ is the sum over the kinematically possible paths for the particles $a, b, ...,$ each particle being considered to possess an inherent mass, with m_a for a, m_b for b, etc.

For each such configuration we then obtain $\mathcal{A}$, and hence $\exp i\mathcal{A}$, which can be represented as a point on the unit circle in the complex plane. Next we make the quantum statement that $\exp i\mathcal{A}$ is the probability amplitude possessed by the universe for the particular configuration with respect to which $\mathcal{A}$ has been calculated. Then the probability amplitude for specified small ranges of the parameters defining the configuration, small ranges for the functions $g_{ik}(x)$ in the metric and small variations of the particle paths, is given by the basic summation rule of quantum mechanics. Namely, that we add the amplitudes $\exp i\mathcal{A}$ over the distribution of parameter values in the specified small ranges. Should it be the case that $\mathcal{A}$ is a large number (as it is for the universe) that varies markedly even for small changes of the parameters, then the mathematical problem of adding the amplitudes $\exp i\mathcal{A}$ will be that of adding complex numbers distributed essentially at random around the whole unit circle, the result being effectively zero. But if $\mathcal{A}$ has almost the same value for all values of the parameters in the ranges in question, then the points $\exp i\mathcal{A}$ will fall together on the unit circle, and the values of $\exp i\mathcal{A}$ will add cumulatively. And if there is but one neighborhood in parameter space for which this second situation applies, this will be the only infinitesimal region that survives the addition of points on the unit circle. Consequently the parameter values in question will be the only ones with any appreciable probability of existing in the universe. In other words, this is the way the universe will be found to be, except with infinitesimal probability of its being otherwise.

This is the explanation given by Paul Dirac of the so-called principle of least action that was discovered in the nineteenth century, because it was found empirically to yield equations that were the same as those already known from Newtonian physics. But it was only really to be understood much later in terms of quantum physics.

It is usual when this explanation is presented to divide the classically computed action $\mathcal{A}$ by a constant $\hbar$ before the exponential is taken $\exp i\mathcal{A}/\hbar$, the constant being Planck's constant. Using appropriate units we can put $\hbar = 1$. We use units such that a mass unit has been specified, and the Compton wavelength associated with that mass is taken as the unit of length (and time). This is more convenient than taking the gram as the unit of mass and the centimeter as the unit of length. In that case, $\hbar$ has the value 3.515×10^{-38}g cm. This is because the gram and the centimeter have been arbitrarily chosen. Then the Compton wavelength associated with 1 g is small compared to 1 cm by the factor 3.5105×10^{-38}, and this is nothing more than a scaling constant between the arbitrarily chosen centimeter and what the length should properly be when the mass unit is taken as 1 g.

Varying the form of $\mathcal{A}$ in (14.2) with respect to the metric tensor g_{ik} leads to the equations of general relativity (see Chapter 13, equation (13.4)). But if we believe the equations of general relativity lead properly to the singularity $S(t) \to 0$ as $t \to 0$ in big-bang cosmology, should we not write

$$\mathcal{A} = \frac{1}{16\pi G} \int_{t>0} R\sqrt{-g}d^4x - \sum_a m_a \int_{t=0} da \qquad (14.3)$$

instead of (14.2)? If we could prove that $\mathcal{A}$ calculated from (14.3) is finite and stationary with respect to slight variations of the moment $t = 0$ there would be chances for such a step. But this is not the case and no such simple maneuver seems possible. Instead the best hope seems to be to argue that carrying general relativity back literally to $t = 0$ goes beyond the validity of the principle of least action which led to general relativity. And this is to lose definition of the problem. Particularly now we no longer can be assured that we know the parameters determining the possible kinematic configurations for the universe. Is it correct, for example, to go on using the spacetime coordinates? Is it correct to even go on using spacetime at all? What we might hope to do is consider evaluating $\Sigma \exp i\mathcal{A}$ (the wave-function) in terms of a generalized parametric description of the configuration of the universe. We then have to argue that in the region of the universe, which in a spacetime description we would characterize by $t < \epsilon$ with ϵ some small positive number, the summation of points on the unit circle leading to the evaluation of $\Sigma \exp i\mathcal{A}$ for some small specified ranges of the generalized parameters, does not condense around a particular point on the unit circle – we do not have the simplification occasioned by the principle of least action. We have a summation of points over the unit circle determining the wave-function for each small range of the parameters.

If we imagine that such a quantum state of the universe possesses a causal parameter in the description of its configurations, as the time is a causal parameter in the usual description, then at some value of the causal parameter we have to imagine a transition occurring, to a state with spacetime as its parameters and with the principle of least action applying to an action of the form

$$\mathcal{A} = \frac{1}{16\pi G} \int_{t>\epsilon} R\sqrt{-g}d^4x - \sum_a m_a \int_\epsilon da. \tag{14.4}$$

The universe thus emerges from an unknown state, to which no constraints can be applied because we have no observational evidence relating to it.

Inflationary scenarios

The concept of inflation was introduced into cosmology by Guth[1] in 1981, although Kazanas[2] and Sato[3] had independently discussed similar ideas slightly earlier. Today, after a decade and a half, the concept is still around with typical responses to it ranging between fanatical belief and cynical scepticism. Let us briefly examine it in the light of our discussion so far.

When proposing a new idea experience shows that it is a good maneuver to begin by identifying new weaknesses of the existing paradigm and then to show how the proposed new idea removes them. If you simply point out the weaknesses and ask for the revision of the standard paradigm nobody will take you seriously, because doing so would imply a threat to the existing beliefs. Thus we believe that the success enjoyed by the inflationary model rests not so much on its intrinsic merit but on how it was presented to the cosmological community.

In the years immediately prior to 1981 three 'defects' of the standard model were highlighted under the headings (i) the horizon problem, (ii) the flatness problem, and (iii) the entropy problem. All these problems were known at an earlier date but were largely ignored. Now they had been stressed (prior to 1981); it became clear that if one could show that by having inflation grafted onto the standard model these problems disappeared, there was every chance that the cosmological community would recognize the problems and the solution as one packet. And indeed this is what happened.

Briefly, the horizon problem tells us that given the particle horizon of a typical observer at epoch t in the early universe $R_H \sim ct$, one immediately runs into difficulty with the postulate of homogeneity. For, if one uses the epoch of grand unification when a major phase transition is believed to have occurred in the very early universe, then the time scale is 10^{-36} s and the distance over which homogeneity could be established was of the order of 3×10^{-26} cm. Taking the big-bang relationship of temperature decreasing inversely with the square root of the epoch, this distance would expand to ~ 30 cm at the present epoch. In other words, the expectation is that at the present epoch the universe should be homogeneous only on scales of no more than 30 cm! Clearly it was necessary to get round this severe restriction of the limited particle horizon, or to postulate *ad hoc* that the universe started with homogeneous conditions already in place.

The flatness problem relates to the curvature parameter k. The Friedmann models can either contract to a singularity or expand out to infinity in the two cases $k = 1$ or -1 respectively. The characteristic time scale for their so doing is linked with the initial conditions. If the time scale of the initial phase was determined by grand unified physics, then it follows that the characteristic time is of the order of 10^{-36} s. That is, within this time scale the universe would have contracted back to a singularity ($k = 1$) or diffused out to infinity ($k = -1$). Why then do we see the universe today in a state of modest density at all? Unless it were created almost flat, i.e. very close to the state $k = 0$, when its initial state would have to be extremely finely tuned, with its initial density lying within one part in 10^{50} of the closure density!

The entropy problem is essentially the statement of the large photon to baryon ratio of the order of 10^9. The entropy per unit proper volume is given by $(4\pi/45)T^3$ at temperature T, and as the proper volume increases as S^3 with $TS = $ constant, the total entropy in a proper volume in an expanding universe is conserved, its value in the present observable universe being as large as 10^{86}. It is hard to explain such a large dimensionless number except as an input of initial conditions.

Inflation was put forward as a cure for these 'ailments'. Its rationale was that at the time of the phase transition the nature of the vacuum changes if one defines it quantum mechanically as the state of lowest energy of a scalar field ϕ. Such scalar fields intervene in the process of phase transition wherein the grand unified interaction splits into the separate components of strong interaction described by the group $SU(3)$ and the electro-weak interaction described by the group $SU(2)_L \times U(1)$. Like super-cooling of steam to below the standard boiling point of water, the steam eventually condensing into droplets of water, the universal phase transition may have a transient false vacuum of the earlier state with higher energy than the true vacuum appropriate for the final state. Like the condensed steam the changeover of vacuum releases energy which is dumped into the system and causes the entropy to rise. During the changeover process, however, the energy-momentum of the vacuum reacts back on the spacetime geometry via the Einstein field equations. The process is as follows.

To proceed to details, the mass term in (14.4) is omitted and two terms involving a postulated field $\phi(x)$ are inserted in its place

$$A = \frac{1}{16\pi G} \int_{t>\epsilon} R\sqrt{-g}d^4x + \frac{1}{2}\int_{t>\epsilon} \phi_i\phi^i\sqrt{-g}d^4x - \int_{t>\epsilon} V(\phi)\sqrt{-g}d^4x,$$

(14.5)

and the principle of least action is considered as applying to variations of ϕ as well as of g_{ik}.

The latter gives

$$\frac{1}{8\pi G}\left(R_{ik} - \frac{1}{2}g_{ik}R\right) + \varphi_i\varphi_k - \frac{1}{2}g_{ik}\phi_l\phi^l + V(\phi)g_{ik} = 0,$$

(14.6)

while $\delta A = 0$ with respect to $\varphi \to \varphi + \delta\varphi$ gives

$$\Box\varphi + \frac{dV}{d\varphi} = 0.$$

(14.7)

Again using the Robertson–Walker metric we get from (14.6)

$$\frac{2\ddot{S}}{S} + \frac{\dot{S}^2 + k}{S^2} = -4\pi G\dot{\varphi}^2 + 8\pi GV,$$

(14.8)

$$3\frac{\dot{S}^2 + k}{S^2} = 4\pi G\dot{\varphi}^2 + 8\pi GV,$$

(14.9)

when in an isotropic–homogeneous situation the field φ is taken to be a function only of the time. And from (14.7),

$$\frac{1}{S^3}\frac{d}{dt}(S^3\dot{\varphi}) + \frac{dV}{d\varphi} = 0.$$

(14.10)

The concept of inflation now depends on choosing the initial values of $\phi, V(\varphi)$ in a suitable way. Writing the initial values of φ, $\dot{\varphi}$, $V(\varphi)$ as $\varphi_0, \dot{\varphi}_0, V_0$ the aim is to choose these quantities so that the $\dot{\varphi}^2$ terms on the right-hand sides of (14.8) and (14.9) are small compared with the V term. As S increases, the k/S^2 term in these equations also becomes negligible, thereby solving the so-called flatness problem, with the consequence that S increases with t approximately as $\exp[(8\pi \frac{G}{3}V_0)^{1/2}t]$. If V is chosen to make $dV/d\varphi$ also small and negative then $\ddot{\varphi}$ in (14.7) can be neglected and

$$\dot{\varphi} \simeq -\frac{1}{3H}\frac{dV}{d\varphi},$$

(14.11)

where $3H^2 = 8\pi GV_0$. Thus φ increases as V declines.

A popular form of V with the required properties is

$$V(\varphi) = \frac{1}{2} B \sigma^4 + B\phi^4 \left[\ln \frac{\varphi^2}{\sigma^2} - \frac{1}{2} \right], \tag{14.12}$$

the so-called Coleman–Weinberg potential[4] giving for ϕ_0 small compared with σ

$$V_0 \simeq \frac{1}{2} B\sigma^4, \ 3H^2 = 4\pi GB\sigma^4. \tag{14.13}$$

Here B and σ are initially specified constants. The constant σ, taken to be $\sim 2 \times 10^{15}$ GeV, is the maximum value to which φ can grow, with $V(\varphi)$ declining to zero at $\varphi = \sigma$. The time required for this to happen is obtained from integrating (14.11), and the number N of exponentiations of S occurring in this time is

$$N \simeq 3H^2 \int_{\varphi_0}^{\sigma} \frac{d\varphi}{|dV/d\varphi|}. \tag{14.14}$$

The number N given by (14.14) can be of the order of 100, raising S by an enormous factor $\exp(100)$. Thus the original difficulty of an unacceptably small particle horizon is resolved because the earlier size of 30 cm for a homogeneous region will be inflated by this factor. This of course makes the universe homogeneous on a scale far larger than the presently observable size. This is probably as unsatisfactory on aesthetic grounds as the non-inflated value of 30 cm. But it does not cause any conflict with observations. (An analogy may be relevant: a badly prepared tennis court may have lots of ridges and bumps. To improve it a groundsman prepared a court suitable for playing billiards.) Anyway, the horizon problem is solved this way.

The constant B is chosen to be $\sim 10^{-3}$, which makes H, given by (14.13) with $\sigma = 2 \times 10^{15}$ GeV, equal to $\sim 2 \times 10^{10}$ GeV. This is in the energy units used by those who believe in inflation. In conventional units, the resulting time scale H^{-1} is $\sim 3 \times 10^{-35}$ s, and the whole inflationary phase of $\sim 100H^{-1}$ lasts for $\sim 3 \times 10^{-33}$ s.

In order that the integral (14.14) shall not diverge at its lower limit the initial value ϕ_0 of the field cannot be zero, but ϕ_0 is taken sufficiently small that the ϕ-dependent term in $V(\varphi)$ is very small initially compared to the constant $\frac{1}{2}B\sigma^4$ term in V, $\phi_0 \simeq 2 \times 10^{11}$ GeV, $\sigma \simeq 2 \times 10^{15}$ GeV.

The end of inflation results in the dumping of the excess energy of the false vacuum into matter and radiation, and this reheating is what raises the entropy of the universe. The original model of Guth failed to achieve this

although the 'new inflationary model' using the Coleman–Weinberg potential
has a better success but at the cost of a fine tuning of its parameters.

All this is obtained without reference to the physical properties of the field
φ, which is considered to decay eventually into particles, ultimately yielding a
radiation field of very high temperature

$$T \simeq 10^{14} \text{ GeV}. \tag{14.15}$$

Since a temperature of 1 eV in energy units is about 10^4 K, (14.15) is a
temperature of $\sim 10^{27}$K in conventional units. Inflationists have apparently
become convinced that the decay of φ into other particles did not occur until
the N exponentiations had taken place, until ϕ had risen to σ and V had died
away to zero. There are two reasons for wanting the inflationary phase to
take place without being troubled by the decays of ϕ.

The large increase of the scale $S(t)$ of the universe by $\exp N$, with N given
by (14.14), ensures that the k term in the dynamical equations (14.8) and
(14.9) becomes completely negligible. This fixes the subsequent behavior of
the universe as being determined by the case $k = 0$, the so-called standard
model. The cases $k = \pm 1$ are thus not permitted in an inflationary model.

Because it is hardly credible that the initially unknown state of the universe
should have led to a state at time $t_0 = \epsilon$ that was homogeneous and isotropic,
inflation is called on to reduce any initial deviations from uniformity by a
factor $\sim \exp{-N}$. The initial neighborhood that becomes the present-day
observed universe was reduced by this factor. But with N finite this is not
a strict mathematical proof of homogeneity, since the initial non-uniformity
could be postulated to be $\exp N$ times larger than before. It is an argument
aimed at what, at best, might be termed our sense of the fitness of things.

The ultra-high temperature (14.15) leads to a sea of esoteric particles. But
as $S(t)$ increases and T declines, things eventually ease off. Indeed, after
an easing off in T by a factor $\sim 10^{-12}$, things at last come into the range
that high-energy physicists actually know about from experiments in the
laboratory.

We are not yet quite done with our explanation of what goes on. Baryon
pairs will be created, leading at last to the appearance of known forms of
matter. But the symmetry between baryons and anti-baryons, if perfect,
would lead, as the universe continued with its expansion, to the ultimate
annihilation of all baryonic matter. So it is an easy deduction of convenience
to argue that the matter–anti-matter symmetry cannot be quite perfect. A
small asymmetry must exist leading to one type in comparatively low concentration, after annihilation has wholly removed the other type. When this
stage is reached, the energy density of radiation dominates that of matter

with η, the baryon-to-photon ratio, a small number. We do not know what η should be, because the small asymmetry that led to it was an *ad hoc* assumption. But it is another easy deduction to arrange that η has just the value required for the synthesis of the light elements discussed in Chapter 9. A few years ago this led to the assumption that $\eta \simeq 2 \times 10^{-10}$, when this was thought to be the right value. At present it is supposed that $\eta \simeq 6 \times 10^{-10}$, now that determinations of D/H show this to be a better number.

The calculation of the abundances of the light isotopes in the classical big-bang cosmology is sometimes given the rather grand title Baryogenesis. However, it is not stressed enough that extrapolations beyond what is known in physics are being used and there are no independent tests of the correctness of these procedures. But what do you do when there are no independent tests at the energy level of 10^{14} GeV to decide how particle physics operates? Unfortunately, rather than admitting that we really do not know, we are often told by those who believe, simply, 'That this is the way it is'. We briefly repeat the canonical argument below.

First, the following three ingredients are needed to produce the baryon asymmetry found in the present universe:

(a) microscopic violation of baryon number,
(b) *CP*-violation,
(c) departure from thermal equilibrium.

Thus suppose that there are massive particles at high energy, which we designate as X particles with their antiparticles $\bar{X}$. The Xs decay into two channels with branching ratios $\mathcal{R}$ and $1-\mathcal{R}$ and baryon numbers B_1 and B_2, while the anti-X decays similarly with the respective rates $\mathcal{F}$ and $1-\mathcal{F}$, and decay products with baryon numbers the exact negative of the above. The average baryon number produced by decays of X and $\bar{X}$ is

$$\triangle B = 0.5(\mathcal{R} - \mathcal{F})(B_1 - B_2).$$

The average baryon asymmetry is then determined by the $\triangle B$ and the number of Xs at the time of decay:

$$\frac{n_B}{n_\gamma} = \left(\frac{N_X}{N}\right) \triangle B.$$

Here N is the total number of helicity states and N_X is the number of type X.

The point is that none of these quantities follow uniquely from a fundamental theoretical framework. As in the case of inflation itself, these quantities have to be put in by hand to justify the ultimate value of n_B/n_γ. So far as the *CP*-violation is concerned, it has not been possible to establish a connection between what is proposed here and what has been actually observed in the baryon decay. While not denying the possibility of everything assumed here turning out to be right, we would question the definitive authority claimed for it.

The upshot of the whole exercise is to arrive at a post-inflation universe that has the right relic density dominated by matter over anti-matter, and a relatively large entropy density.

Let us summarize what is now required in this currently popular cosmological scenario.

(i) We have to postulate a universal quantum state that undergoes transition to one dominated by the potential $V(\varphi)$.

(ii) The potential $V(\varphi)$ is chosen so that it decreases with φ increasing. It has to contain suitably chosen constants, whose properties are determined by the initially unknown state.

(iii) The decay of φ is held off until V falls to zero.

(iv) The usual baryon–anti-baryon symmetry is broken so that baryons emerge from the subsequent mélé just so as to give the value of η required by light element synthesis.

(v) Although the quantum transition thus generating the early universe produces an isotropic–homogeneous symmetry, the symmetry is broken by fluctuations which are chosen to permit galaxies to form once radiation and matter become decoupled in the subsequently expanding universe. The deduction $k = 0$ is important and indeed considered crucial to success in this endeavor. The hope is that by adjusting the initial fluctuations appropriately, the observed slight variations in the temperature of the microwave background can be explained.

(vi) This scenario with all its *ad hoc*ism and fine tuning, has sometimes been described as the most important scientific achievement of the century, more important than the Dirac equation.

(vii) With the best current determination of $60 \text{ km s}^{-1}\text{Mpc}^{-1}$ for the Hubble constant, the standard model with $k = 0$ (necessary for (v)) leads to an elapsed time since the quantum transition in (i) that is somewhat less than 11 billion years. Whereas, the ages of the oldest stars may be greater than 15 billion years. It is argued against this impasse that the observed value of the Hubble constant remains doubtful (but see Chapter 4). Either that, or those who have calculated the ages of the oldest stars have still not got accurate results.

(viii) The best determinations of the relative abundances of the light elements are not quite right. When η is chosen to correspond to what seem good determinations of D/H, the $^4\text{He}/(\text{H}+^4\text{He})$ mass ratio is calculated to be close to 0.25, whereas the best observational determinations give 0.235. Thus the position today in this regard is hardly better than it was in the days of George Gamow – after almost 50 years of substantial effort. But it is argued that the correspondence of the numbers is near enough. This despite the experience of all who have attempted to fit a wrong theory to a limited number of facts, namely that provided a sufficient number of parameters of choice are available, it is usually possible to make the numbers fit to what seems like a moderate measure of accuracy.

It has turned out that the original expectations of the inflationary scenario have not been matched by the performance of the model in its various incarnations. The following difficulties have appeared.

(i) The prediction that $\Omega = 1$ has led to a severe age problem, mentioned above, and which we have discussed elsewhere in this book. So now it is being relaxed by some to the more vague $\Omega + \Lambda = 1$.

(ii) But this is to jump in an important respect from the frying pan into the fire! For now it is necessary to explain how a lambda term came about from the very early physics. *Prima facie*, the Λ term comes from the potential $V(\phi)$ but its value from V is about 108 orders of magnitude greater than what is needed by the present-day cosmologist. So one has to introduce a fine tuning that saves about 10^{-108} of the primordial Λ for the present epoch.

(iii) One of the powerful dividends of inflation has been claimed to be that it naturally delivers a scale-invariant power spectrum for initial perturbations. However, keeping in view the primordial significance attached to the COBE-discovered inhomogeneities, all attempts to evolve a structure formation scenario within the big bang have proved disappointing, given the added latitude provided by the choices of dark matter (CDM, HDM or MDM), biasing/anti-biasing, cosmological constant, etc. The façade has been maintained that with a few iterations on the basic idea the problem is all but solved. This is another way of saying that the fine tuning necessary to fit all constraints has not been discovered as yet.

An earlier introduction of a scalar field

The theory developed over the years 1948–60 led to new cosmological models, of which the steady-state theory was one. It was based on a scalar field with some analogies to the scalar field of inflationary theory, except the field was written as $C(X)$ instead of $\phi(x)$.

The action formula of general relativity,

$$A = \frac{1}{16\pi G} \int R\sqrt{-g}d^4x - \sum_a m_a \int da, \qquad (14.16)$$

assumes that the line integrals for the particles extend backwards indefinitely in time. This leads to the contradiction that general relativity does not permit this supposition to be true, because general relativity requires the entire universe to plunge into the singularity of the big bang, as was explained earlier in this chapter. It can be argued therefore that it would be more logical to write the line integrals in (14.16) with explicit lower limits, as in

$$A = \frac{1}{16\pi G} \int R\sqrt{-g}d^4x - \sum_a m_a \int_{A_0} da, \tag{14.17}$$

with A_0 as the 'beginning', or creation point, of particle a, and B_0 for particle b, ...

This evidently raises the question of how the 'creation points' A_0, B_0, ..., are to be determined mathematically. For an action of the form of (14.17), the creation points cannot be fixed from a principle of least action, since (14.17) does not have $\delta A = 0$ with respect to variations of A_0, B_0 .... However, a scalar field $C(X)$ can be introduced to overcome this difficulty. A field term

$$-\frac{1}{2}f \int C_i C^i \sqrt{-g}d^4x, \tag{14.18}$$

where $C_i = \partial C/\partial x^i$ was introduced into A, as was the particle interaction term

$$\sum_a \int_{A_0} C_i da^i = -C(A_0) - C(B_0) - \cdots \tag{14.19}$$

Thus the action was taken to be given by

$$A = \frac{1}{16\pi G} \int R\sqrt{-g}d^4x - \sum_a m_a \int_{A_0} dx \tag{14.20}$$

$$-\frac{1}{2}f \int C_i C^i \sqrt{-g}d^4x + \sum_a \int_{A_0} C_i da^i.$$

From this it is easily seen that provided

$$C^i = m_a \frac{da^i}{da}, \tag{14.21}$$

the point A_0 can be varied slightly without changing A in first order. Thus (14.21) was the creation condition given by $\delta A = 0$ for particle a. And similarly for b, c,.... Apart from the coupling constant f, the $C_i C^i$ term in (14.20) is the same as the $\phi_i \phi^i$ term in (14.5).

In the homogeneous isotropic case, $C(X)$ is a function only of t, and varying $\mathcal{A}$ with respect to g_{ik} leads in the case of the metric (13.1) to

$$\frac{2\ddot{S}}{S} + \frac{\dot{S}^2 + k}{S^2} = 4\pi Gf\dot{C}^2, \tag{14.22}$$

$$3\frac{\dot{S}^2 + k}{S^2} = 8\pi G\left(\rho - \frac{1}{2}f\dot{C}^2\right), \tag{14.23}$$

ρ being the proper mass density. When the constant f is taken positive, the term $\dot{C}^2$ in (14.22) acts as a negative pressure, as does the dominant V term in (14.8) for an inflationary scenario. Unlike (14.9), in which the $\dot{\varphi}^2$ term has a positive sign, the $\dot{C}^2$ term for $f > 0$ appears with a negative sign in (14.23). And so it should, because one cannot have creation of matter without a negative energy to compensate it. The difference here is important. In the inflationary case, matter is derived from the positive energy of the ϕ field, which is derived from the positive energy of the potential $V(\varphi)$, which is put in *ad hoc* at the beginning. Here, however, the energies are self-compensating within the universe itself, and therefore in principle open to observational study.

The effect of matter creation is to produce a negative pressure in (14.22) which tends to accelerate the expansion of the universe. Thus the expansion of the universe is a consequence of matter creation. It is not something which is assumed *ad hoc* at some imagined beginning. Each particle of matter has a beginning. If we prefer not to include particle creation in our equations, the price is a singularity, the big bang for the whole universe. But if we do include a beginning for individual particles there need be no singularity for $S(t)$ and no lower limit for time in the field-dependent terms of (14.20). The gain here seems to compensate very adequately for the introduction of $C(X)$.

From a quantum point of view it is also necessary to have $\delta\mathcal{A} = 0$ with respect to the variation $C \rightarrow C + \delta C$, giving

$$\Box_x C = f^{-1}n(X) = F(X), \text{ say}, \tag{14.24}$$

where $n(X)$ is the number of creation events occurring per unit proper four-dimensional volume at the spacetime point X, and $F(X)$ is a 'creation function' defined in (14.24) to be $f^{-1}n(X)$. For $n(X)$ to be non-zero, the creation condition (14.21) is required to be satisfied at X. The value assigned to the constant f^{-1} determines the intensity of the creation which ensues when (14.21) is satisfied. Thus, one has a closed logical loop, similar to that in the electromagnetic theory, with $C(X)$ determined by (14.24) when $F(X)$ is known, and with $n(X)$ determined by (14.21) when $C(X)$ is known. In the electromagnetic case the motion of the charges and the ensuing currents are determined when the field is known, and the field is determined when the charges and their motions are known, forming a closed loop. Although the theory itself can be fully stated, the loop is rarely, if ever, solved in its full generality. Either the charges and currents are specified and the field components calculated, or the field is specified and the motions of the charges are calculated. When both are intermixed, the equations become too complicated to be handled, and much the same can be expected here for $C(X)$. We shall proceed therefore by specifying $n(X)$ and calculate the resulting model, but verifying that (14.21) is satisfied for consistency at spacetime points X where $n(X)$ has been assumed to be non-zero.

References

Chapter 14

1. Guth, A. 1981, *Phys. Rev. D.*, **23**, 347.
2. Kazanas, D. 1980, *Ap. J.*, **241**, L59.
3. Sato, K. 1981, *M.N.R.A.S.*, **195**, 467.
4. Coleman, S. & Weinberg, E. 1973, *Phys. Rev. D.*, **7**, 1888.

This is our view of the conformist approach to the standard (hot big bang) cosmology. We have resisted the temptation to name some of the leading geese.

15

New cosmological models

Basic solutions of the equations

We start from a static empty universe at $t < t_0$ and at $t = t_0$ let

$$f^{-1}n(X) = A\delta(t - t_0), \tag{15.1}$$

where A is here some constant, implying a total creation for unit proper spatial volume of fA. For $t > t_0$ the scale factor $S(t)$ is no longer static and it is the objective here to calculate $S(t)$. With $\dot{C} = 0$ before time t_0, the solution of $\Box C = f^{-1}n(x)$ for $t > t_0$ is

$$\dot{C} = A \frac{S_0^3}{S^3} , \ t > t_0. \tag{15.2}$$

This follows because

$$\Box C = \frac{1}{S^3} \frac{d}{dt}(S^3 \dot{C}) = 0, \ t > t_0,$$

while $\dot{C}$ is the Heaviside function, 0 for $t < t_0$ with a jump to A at $t = t_0$, giving $A\delta(t - t_0)$ for $\ddot{C}$.

Next we notice that the general equation

$$3 \frac{\dot{S}^2 + k}{S^2} = 8\pi G\left(\rho - \frac{1}{2}f\dot{C}^2\right) \tag{15.3}$$

requires $k = 0$ since $\dot{S}$, ρ, $\dot{C}$ are all zero up to $t > t_0$. Moreover, for created particles each of mass m

$$\rho = Afm \frac{S_0^3}{S^3} , \ t > t_0, \tag{15.4}$$

so that (15.2), (15.3), (15.4) give, for $t > t_0$,

$$\frac{\dot{S}^2}{S^2} = \frac{8\pi G}{3} fA \frac{S_0^3}{S^3} \left(m - \frac{1}{2} A \frac{S_0^3}{S^3} \right). \tag{15.5}$$

And the equation for the acceleration $\ddot{S}$,

$$\frac{2\ddot{S}}{S} + \frac{\dot{S}^2}{S^2} = 4\pi G f \dot{C}^2, \qquad t > t_0, \tag{15.6}$$

becomes

$$\frac{2\ddot{S}}{S} + \frac{\dot{S}^2}{S^2} = 4\pi G f A^2 \frac{S_0^6}{S^6}, \qquad t > t_0. \tag{15.7}$$

The dynamical effect of creation is to give $\ddot{S} > 0$, $\dot{S} = 0$ at $t = t_0$. There is now a non-zero expansive acceleration leading to $\dot{S} > 0$ as t increases above t_0, but not an impulsive sudden non-zero value of $\dot{S}$ at $t = t_0$. For $\dot{S} = 0$ at $S = S_0$ in (15.5) we require

$$A = 2m. \tag{15.8}$$

The creation condition (14.21), $C^i = mda^i/da$, applied at $t = t_0$ leads to (15.8). Because of the Heaviside jump of $\dot{C}$ from zero to A, the mean value of $A/2$ has to be used for $\dot{C}$, and with only the time component of (14.21) involved in a homogeneous isotropic situation, (15.8) is consistent with $C^i = mda^i/da$. Whence (15.5) gives

$$\frac{\dot{S}^2}{S^2} = \frac{16\pi G}{3} fm^2 \frac{S_0^3}{S^3} \left(1 - \frac{S_0^3}{S^3} \right). \tag{15.9}$$

This equation was solved by one of us (JVN)[1]

$$S^3 = S_0^3 + \frac{3}{4} B(t + t_0)^2, \quad t_0 = \text{ constant},$$

which has the asymptotic form,

$$S \simeq \left(\frac{3}{4} B \right)^{1/3} t^{2/3}, \tag{15.10}$$

as discussed in Chapter 13, now with

$$B = \frac{16\pi}{3} Gfm^2 S_0^3. \tag{15.11}$$

Since the density ρ_0 immediately after $t = t_0$ is given by $\rho_0 = fAm = 2fm^2$, B can be written as

$$B = \frac{8\pi G}{3} \rho_0 S_0^3 \tag{15.12}$$

which is also the same as the constant B in (13.15). Hence the creation (15.1) *is equivalent asymptotically to the big bang*, but without there being any singular origin for the universe. The circumstance that we get expansion with the same eventual results, but without any singularity or any *ad hoc* initial energy being required, has always made this seem a superior approach to the extraordinary scenarios described in Chapter 14.

Repeated episodes of creation

Consider the more complicated situation with a number of homogeneous creation episodes at times $t_0, t_1, \ldots, t_n$, i.e.

$$f^{-1}n(X) = K_r \delta(t - t_r). \tag{15.13}$$

The solution now for $\dot{C}$ is

$$\dot{C} = 0, \qquad t < t_0$$

$$\dot{C} = \frac{K_0 S_0^3}{S^3} + \frac{K_1 S_1^3}{S^3} + \cdots + \frac{K_r S_r^3}{S^3}, \qquad t_r < t < t_{r+1}, \tag{15.14}$$

for $r = 0, 1, 2, \ldots, n-1$, and

$$\dot{C} = \frac{K_0 S_0^3}{S^3} + \frac{K_1 S_1^3}{S^3} + \cdots + \frac{K_n S_n^3}{S^3}, \qquad t_n < t.$$

The creation coefficients $K_0, K_1, \ldots$ are to be determined by the requirement that the mean value of $\dot{C}$ at each of the episodes is to be m. This gives $K_0 = 2m$, as was seen in the previous section.

Now choose the times $t_0, t_1, \ldots, t_n$ such that

$$\left(\frac{S_n}{S_{n-1}}\right)^3 = \left(\frac{S_{n-1}}{S_{n-2}}\right)^3 = \cdots = \left(\frac{S_1}{S_0}\right)^3 = 1 + \alpha \text{ say}, \quad \alpha > 0. \tag{15.15}$$

Then for r large and

$$K_r = \frac{\alpha m}{1 + \alpha/2} \tag{15.16}$$

$\dot{C}$ jumps at t_r from

$$\frac{\alpha m}{1 + \alpha/2}\left[\frac{1}{1 + \alpha} + \frac{1}{(1 + \alpha)^2} + \cdots + \frac{1}{(1 + \alpha)^r}\right] \simeq \frac{m}{1 + \alpha/2} \tag{15.17}$$

to

$$\frac{\alpha m}{1 + \alpha/2}\left[1 + \frac{1}{1 + \alpha} + \frac{1}{(1 + \alpha)^2} + \cdots + \frac{1}{(1 + \alpha)^r}\right] \simeq \frac{m(1 + \alpha)}{1 + \alpha/2} \qquad (15.18)$$

with the mean value $\dot{C} = m$ required for conservation of energy and momentum density during particle creation.

The mass density ρ tends similarly to a nearly constant value for $\alpha << 1$,

$$\rho \simeq \frac{\alpha m^2}{1 + \alpha/2}\left[1 + \frac{1}{1 + \alpha} + \frac{1}{(1 + \alpha)^2} + \cdots + \frac{1}{(1 + \alpha)^r}\right] \simeq fm^2 \qquad (15.19)$$

and the equation

$$\frac{3\dot{S}^2}{S^2} = 8\pi G(\rho - \tfrac{1}{2}f\dot{C}^2) \qquad (15.20)$$

gives $\quad \dfrac{3\dot{S}^2}{S^2} = 4\pi Gfm^2, \qquad (15.21)$

with the solution $S \propto \exp Ht$ and

$$3H^2 = 4\pi Gfm^2, \qquad (15.22)$$

which is the steady-state model.

The steady-state model is thus seen to be a special case of an otherwise much more general theory, a theory that is easily capable of coping tactically with observational objections that have been raised against the simple steady-state model, objections that are minor, once it is seen that the theory can explain the cosmic microwave temperature, as we shall show in Chapter 16. We also show in Chapter 16 how this theory can equally well explain the counts of radio sources, an objection that was brought against the earlier steady-state theory (Chapter 7).

An oscillatory model

Returning to the first section, repeating the argument there up to (15.7), but not requiring the creation coefficient A in

$$f^{-1}n(X) = A\delta(t - t_0) \qquad (15.23)$$

to be $2m$ as in (15.8), we ask what happens? In this case the right-hand side of (15.5) is not zero at $S = S_0$. Since, however, $\dot{S}$ must be zero – the creation event does not set the universe instantaneously in motion – we must reinstate the k term in (15.5):

$$3\frac{\dot{S}^2 + k}{S^2} = 8\pi G f A \frac{S_0^3}{S^3}\left(m - \frac{1}{2}A\frac{S_0^3}{S^3}\right), \tag{15.24}$$

subject to $\dot{S} = 0$ at $t = 0$. And (15.7) becomes

$$2\frac{\ddot{S}}{S} + \frac{\dot{S}^2 + k}{S^2} = 4\pi G f A^2 \frac{S_0^6}{S^6}, \quad t > t_0, \tag{15.25}$$

so that it is $\ddot{S}$ that is instantaneously non-zero as a result of the creation event.

Eliminating $\dfrac{\dot{S}^2 + k}{S^2}$ gives

$$\frac{\ddot{S}}{S} = -\frac{4\pi G}{3}fmA\frac{S_0^3}{S^3} + \frac{8\pi G}{3}fA^2\frac{S_0^6}{S^6} \tag{15.26}$$

and the universe either expands ($\ddot{S} > 0$) or contracts ($\ddot{S} < 0$) according to whether $A > \frac{1}{2}m$ or $A < \frac{1}{2}m$. The effect of (15.23) is to cause $\dot{C}$ to increase from zero to A. Thus $\dot{C} = m$, the creation condition, is satisfied during the jump provided $A > m$. With this as a necessary requirement for creation to take place, the case $A < \frac{1}{2}m$ is thus excluded, and the universe necessarily expands immediately after the creation event.

Equation (15.24) shows that for $m < A < 2m$ we must have $k > 0$, while for $A > 2m$, $k < 0$. By adjusting the unit of the r-coordinate in (13.1), the constant k is -1 for $A > 2m$, 0 for $A = 2m$, and $+1$ for $m < A < 2m$. The case $k = 1$ is evidently oscillatory, since equation (15.24) then has two roots for S when $\dot{S} = 0$. The smaller of the two is $S = S_0$, the scale of the universe when the creation event occurred. As S increases above S_0, the negative term in S^{-6} on the right-hand side of (15.25) falls away, so that

$$3\frac{\dot{S}^2}{S^2} \simeq 8\pi G f A \frac{S_0^3}{S^3}m - \frac{3}{S^2}, \tag{15.27}$$

and the negative term on the right-hand side of (15.27) decreases less rapidly with increasing S than does the positive S^{-3} term. So then for sufficiently large S, $S = S_1$ say, the right-hand side falls to zero and $\dot{S} = 0$ at S_1. The universe falls back thereafter with $\dot{S} < 0$ until S decreases back to S_0, when the former root is repeated. Also repeated is the previous value of $\dot{C}$, and since the creation conservation condition $\dot{C} = m$ occurred on the first occasion that condition will be repeated, with the consequence that a second creation event becomes possible. Hence we have an oscillating universe with the possibility of a creation event occurring at each of its minima. This is what we have called the *quasi-steady-state model*.

When $k = -1$ there is no such second value for $\dot{S}$, and in this case with

$$3\frac{\dot{S}^2}{S^2} = 8\pi GfA(m - \frac{1}{2}AS_0^3/S^3)\frac{S_0^3}{S^3} + \frac{3}{S^2} \tag{15.28}$$

the universe expands indefinitely.

Inhomogeneity

The implication that a creation event with A in (15.23) not equal to $2m$ can change the topology of the universe, with the k value in the metric

$$ds^2 = dt^2 - S^2(t)\left[\frac{dr^2}{1 - kr^2} + r^2(d\theta^2 + \sin^2\theta d\varphi^2)\right] \tag{15.29}$$

changing from $k = 0$ to $k = \pm 1$ surely cannot be correct. Moreover, $A = 2m$ is required by energy conservation, by the need for the right-hand side of (15.3), $8\pi G(\rho - \frac{1}{2}f\dot{C}^2)$, to be unchanged by the creation process. Whereas with $A \neq 2m$ it is indeed changed, as is seen immediately from (15.5) with $S = S_0$. Does this mean that A must necessarily be $2m$, the interesting possibilities considered in the preceding section not being possible? If we impose the conditions of homogeneity and isotropy for the whole universe the answer is yes. It is homogeneity and isotropy that forces $A = 2m$ and inevitably gives the case studied in Section 15.1. When, however, there is inhomogeneity in $n(X)$, the creation function, the situation is otherwise.

At each spatial position $(\rho - \frac{1}{2}f\dot{C}^2)$ remains unchanged at the moment of creation. The matter increment of ρ remains at the spatial position in question. But the C-field, satisfying the wave-equation, tends to uniformity as t increases. The negative part of the increment flows out of a region when the creation rate is above the average, making $(\rho - \frac{1}{2}f\dot{C}^2)$ positive and so leading to the case $k = +1$; whereas regions in which the creation rate is low have negative C-field energy flowing into them, thus making $\rho - \frac{1}{2}f\dot{C}^2$ negative, lead to $k = -1$. Thus the two cases $k = \pm 1$ studied above arise as consequences of inhomogeneities in the creation rate. A low local creation rate leads to indefinitely large local expansion, i.e. a growing hole, whereas a high local creation rate leads to a region that does not in itself become indefinitely large. It expands to a maximum degree and then falls back, oscillating repeatedly thereafter, except that further creation episodes at oscillatory minima each produce some increase in the scale at the maxima, and so in that sense, lead to a greater and greater measure of expansion. This is an effect we have discussed in detail in the context of the quasi-steady state model (*loc. cit*).

It is often remarked that the actual universe shows a hierarchical structure ranging in scale from galaxies up to clusters of galaxies, superclusters, and perhaps a scale still larger than superclusters, until the whole universe is reached (see Chapter 20). Now it is easily seen how this hierarchical structure may arise. We modify this concept from its usually stated observational form, by extending it both at its upper and lower ends. At its upper end we regard the entire observable universe as a $k = +1$ region within a larger universe. Since the process of formation consists in the propagation of a C-field from one of the $k = +1$ regions into the surrounding universe, the observable universe cannot be strictly closed as it would be if the $k = +1$ case in the metric (13.1) were to apply strictly. For in that case the observable universe would be a strict black hole as seen from outside, with nothing from its interior able to escape into the exterior. Instead, we may speak of the observable universe being a near-black hole as seen from outside. The difference is that cosmological redshift values from objects within our near-black hole do not go strictly to infinity. They have a largest value, reflecting the degree to which our region remains open to the surrounding universe.

Within the observable universe the same story is repeated time and again. There are regions of condensation and there are expanding holes, but never strictly formed in a topological sense. The regions of condensation are only very partially developed, at least until we arrive at the nuclei of galaxies where the nearness to black holes is most advanced. The process of formation of a near-black hole is always the same, whatever its scale; equal creation of positive and negative energy components within the interior, followed by a preferential escape of the negative component. The nearer to a perfect black hole, the harder an escape of the negative component becomes, so that the process is self-limiting. The near-black holes never become perfect.

In this way, everything is made out of nothing, despite the saying attributed to Lucretius that only nothing can be created out of nothing. The positive (matter) stays essentially rooted where it is created, at any rate if it is baryonic, while the negative (C-field) propagates into ever-wider spaces. This is a conceptually simple process, responsible for much of what we observe cosmogonically.

The baryonic matter will not stay stationary, however, since the space near a creation event expands rapidly and some of this matter is ejected[a].

Reference

Chapter 15

1. Narlikar, J.V. 1973, *Nature*, **242**, 35.

[a] In detail the argument goes as follows. When the inner regions where creation takes place is taken to be uniform, the Robertson–Walker metric can be used in the interior, giving

$$3\frac{\dot{S}^2 + k}{S^2} = 8\pi G(\rho - \frac{1}{2}f\dot{C}^2) \tag{1}$$

$$2\frac{\ddot{S}}{S} + \frac{\dot{S}^2 + k}{S^2} = 4\pi Gf\dot{C}^2. \tag{2}$$

Here $S(t)$ is a scale factor for the interior. Eliminating $(\dot{S}^2 + k)/S^2$ we have

$$\frac{\ddot{S}}{S} = \frac{4\pi G}{3}(-\rho + 2f\dot{C}^2). \tag{3}$$

If creation takes place so that the right-hand side of (1) is zero then

$$\frac{\ddot{S}}{S} = 2\pi Gf\dot{C}^2 \text{ instantaneously and the region expands.}$$

16

The observations explained in terms of the quasi-steady-state model

The present chapter is a combined exposition of the so-called quasi-steady-state theory developed by the authors in three papers[1,2,3]. The model has its origin in the oscillatory case just given in mathematical form in Chapter 15, in which the scale function $S(t)$ of a homogeneous isotropic universe takes the form

$$S(t) = \exp\frac{t}{P} \times F\left(\cos\frac{2\pi t}{Q}\right), \qquad (16.1)$$

where P, Q are constants of the model and $F(\cos 2\pi t/Q)$ is oscillatory with respect to $\cos 2\pi t/Q$. The first factor in $S(t)$ is the simple form appearing in the old steady-state theory. It represents the effect of averaging over the cycles arising from the second term in (16.1). Our aim in this chapter is to consider in detail how a theory of this kind relates to the observations we have discussed in earlier chapters.

Figure 16.1 shows the plot of (16.1) above, while Fig. 16.2 gives a typical cycle on enlarged scale. We proceed more accurately, beginning by differentiating (16.1). This gives

$$H \equiv \frac{\dot{S}}{S} = \frac{1}{P} + \frac{\dot{F}}{F} = \frac{1}{P} - \frac{2\pi}{Q}\sin\frac{2\pi t}{Q}\frac{dF}{d(\cos 2\pi t/Q)}. \qquad (16.2)$$

At a particular time the second term here is in general much larger in magnitude than the $1/P$ term, although the second term averages to zero when taken over a complete cycle. The difficulties encountered by the old steady-state theory arose from identifying H_o, the present-day value of H, with the $1/P$ term in (16.2), instead of realizing that the main contribution to H_o comes in general from the second term. This is because the parameters of the model are taken to be such that $P \gg Q$.

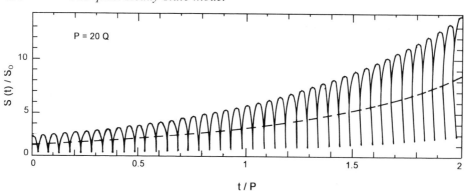

Fig. 16.1. Plot of $S(t)/S_0$ against t/Q. The cycles are scaled by the factor exp t/P for the case $P/Q = 20$, but are otherwise the same.

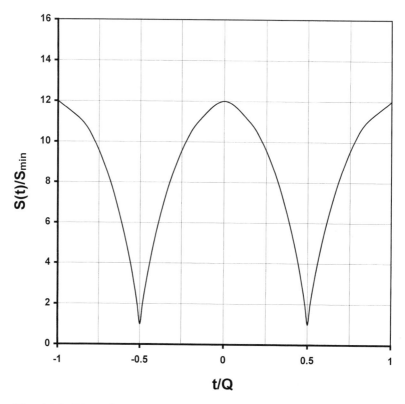

Fig. 16.2. Plot of $S(t)/S_{min}$ against t/Q according to the more accurate values in Table 16.1. This is the form of a single oscillation. Successive oscillations are scaled by the factor exp t/P, as in Fig. 16.1.

The parameters of QSSC

With $P \gg Q$ the factor $\exp t/P$ varies only slightly over a particular cycle, which we consider to begin at maximum phase, i.e. maximum of F. Choosing the zero of t at the maximum of a particular cycle there is first a contraction of $S(t)$ to a minimum at $t = \frac{1}{2}Q$, followed by expansion to the next maximum at $t = Q$. And throughout the oscillation we can put $\exp t/P = 1$ to a sufficient approximation.

Writing S_0 for the scale factor $S(t)$ at minimum and ρ_0 for the mass density also at minimum, and putting

$$x = \frac{S(t)}{S_0}, \quad \alpha = \frac{8\pi G}{3}\rho_0, \quad \beta = \frac{k}{S_0^2}, \tag{16.3}$$

equation (15.24) in the oscillating model obtained in Chapter 15 can be written in the form

$$\frac{\dot{S}^2}{S^2} = \frac{\dot{x}^2}{x^2} = \frac{\alpha}{x^3}\left(1 - \frac{1}{x^3}\right) - \frac{\beta}{x^2}\left(1 - \frac{1}{x^4}\right) + \frac{\gamma}{x^6}.$$

This follows because the term in S_0^6/S^6 in (15.24) can be written for some γ as $(-\alpha + \beta + \gamma)/x^6$.

Then the condition $\dot{S} = 0$ at $x = 1$ gives $\gamma = 0$ and the above equation becomes

$$\frac{\dot{S}^2}{S^2} \equiv \frac{\dot{x}^2}{x^2} = \frac{\alpha}{x^3}\left(1 - \frac{1}{x^3}\right) - \frac{\beta}{x^2}\left(1 - \frac{1}{x^4}\right). \tag{16.4}$$

Putting $\dot{x} = 0$ in (16.4) gives $x = 1$ at minimum phase and $x = x_1$, say, at maximum. With the constants α and β taken such that $x_1 \gg 1$, we have

$$x_1 \simeq \alpha/\beta \tag{16.5}$$

to a close approximation. Using (16.5) to eliminate β we get

$$\frac{1}{\alpha}\dot{x}^2 = \frac{1}{x}\left(1 - \frac{1}{x^3}\right) - \frac{1}{x_1}\left(1 - \frac{1}{x^4}\right) \equiv G(x), \quad \text{say}. \tag{16.6}$$

The time required for expansion from the minimum at $x = 1$ to $1 < x < x_1$ is given by

$$t(x) - t_{min} = \frac{1}{\sqrt{\alpha}}\int_1^x \frac{dx}{\sqrt{G(x)}}. \tag{16.7}$$

Table 16.1. *Minimum to maximum of oscillation for* $x_1 = 12$

x	$\sqrt{G(x)}$	$\frac{1}{x}\sqrt{G(x)}$	$\int_1^x \frac{dx}{\sqrt{G(x)}}$
1	—	—	—
2	0.660	0.330	1.515
3	0.562	0.187	3.295
4	0.487	0.122	5.348
5	0.430	0.086	7.673
6	0.382	0.064	10.291
7	0.339	0.048	13.241
8	0.297	0.037	16.608
9	0.254	0.028	20.545
10	0.206	0.021	25.400
11	0.146	0.013	32.294
12	—	—	46.2

Once values have been specified for $\sqrt{\alpha}$ and x_1 all features of the oscillation are determined. Thus for the particular choice $x_1 = 12$ we have the values shown in Table 16.1.

Oscillation according to (16.6) is symmetrical, minimum to maximum, maximum to minimum. Thus the corresponding values for a contracting half-cycle are the same as for the expanding half-cycle in Table 16.1.

There is still a further parametric choice to be made, however, namely the value of x at the present day. If this is set at $x = 6$, making the expansion factor $S(t)$ at present one-half of its maximum value at $x_1 = 12$, then at the present day

$$\frac{\dot{x}}{x} = 0.064\sqrt{\alpha}. \tag{16.8}$$

But $\dot{x}/x = \dot{S}/S$ and the present value of $\dot{S}/S$ is the Hubble constant H_o, giving

$$0.064\sqrt{\alpha} = H_o. \tag{16.9}$$

Furthermore the oscillatory half-period $\frac{1}{2}Q$ is $46.2/\sqrt{\alpha}$ according to the last entry in the table. Whence

$$Q = 5.91H_o^{-1}, \tag{16.10}$$

so that Q for $H_o = 60$ km s^{-1}Mpc^{-1} is given by

$$Q = 96.3 \times 10^9 \text{ years.} \tag{16.11}$$

By reducing the chosen value for x_1, this could be made less. The form of the oscillation of $S(t)$ with t in units of Q is shown in Fig. 16.2. The difference here in the form of $S(t)$ from a simple sinusoidal curve comes from treating the function $F(\cos 2\pi t/Q)$ in $S(t)$ more accurately.

It is at first sight remarkable that this result should have emerged from a model which appears to contain only unknown parameters, and it is as well to see how this has happened. At the beginning in (16.4) we had three unspecified parameters, α, β and k (or ρ_0, S_0 and k), with $1 \leq x \leq x_1$ and x_1 serving to replace β using (16.5), which is approximate but only very slightly so when $x_1 \gg 1$, i.e. when the oscillations are large. The next step was to specify a value of x_1. The value $x_1 = 12$ was chosen, leading to the numbers in Table 16.1.

The last step was to choose where in the range of x we are at the time of observation, and it was this choice that decided the result for the value of Q in terms of the observed Hubble constant. Had we chosen $x = 9$ instead of $x = 6$ for the present day we would have had

$$0.028\sqrt{\alpha} = H_0 \tag{16.12}$$

instead of (16.9), with the result

$$Q = 42.1 \times 10^9 \text{ years}, \tag{16.13}$$

which is close to the value that we used in an earlier paper[2]. And the redshift of galaxies observed at the last minimum would then be $z = 8$ instead of $z = 5$.

The present value of $Q \sim 100 \times 10^9$ years has many points of interest to be considered in the present chapter.

The origin of the microwave background

If there were no additions from cycle to cycle to the microwave background, its energy density W, taken at our particular phase, would decline as $\exp -4t/P$. Thus an addition is needed cycle by cycle to maintain W. Writing dW as the increment over an element of the cycle we require

$$\oint \frac{dW}{W} = \frac{4Q}{P}. \tag{16.14}$$

This form takes account of W varying as x^{-4} with the parameter x in Table 16.1. For convenience in what follows, we shall give energy densities as they would be at the present-day value of x.

For a background temperature of 2.74K we have $W = 4.26 \times 10^{-13} \text{erg cm}^{-3}$, so that for P/Q in the range 10–20 the addition of new background radiation per cycle is required to be $\sim 5 \times 10^{-14} \text{erg cm}^{-3}$, which we take to come from the thermalization of starlight occurring dominantly near minimum phase when the spatial density of matter is at its highest. If we choose the zero of t at the last maximum, the last minimum is at $t = \frac{1}{2}Q$ with most thermalization of starlight occurring at the latter phase.

According to the choice of parameters in the previous section the present day is at $t = 0.611Q$, and the present-day observed density of starlight, about $10^{-14} \text{erg cm}^{-3}$, would then have had to accumulate between $t = \frac{1}{2}Q$, the last minimum, and $t = 0.611Q$, i.e. over only some ten percent of the present cycle. Clearly by $t = \frac{5}{4}Q$ to $t = \frac{3}{2}Q$ as the next minimum is approached, the energy density of starlight will have risen appreciably above the present-day value, appreciably above $10^{-14} \text{erg cm}^{-3}$. We require it to have risen to $\sim 10^{-13} \text{erg cm}^{-3}$.

It is well-known that stars evolve down the main-sequence to about one solar mass in a time of 10^{10} years. In the extended period $Q \simeq 96.3 \times 10^{9}$ years with which we are involved here, evolution will proceed significantly further to about $0.4 M_{\odot}$. With evolution requiring much the same fractional conversion of hydrogen to helium, the total luminous output obtained by burning the main-sequence down to $0.4 M_{\odot}$ will be determined by the total mass of all the stars going down to this value. Using the usual stellar luminosity function this is about ten times greater than the mass down to $M_{\odot}$. It therefore follows that if burning the main-sequence down to $M_{\odot}$ gives a universal density of starlight that is of order $10^{-14} \text{ erg cm}^{-3}$, burning down to $\sim 0.4 M_{\odot}$ in the time scale Q will deliver a universal energy density of $\sim 10^{-13} \text{ erg cm}^{-3}$.

The spectral energy distribution of the stellar radiation from a typical galaxy varies markedly through each cycle. At minimum, when new gaseous material is acquired and new star formation occurs, most of the energy output is in the blue and ultraviolet, coming from stars of masses greater than the Sun. But as a cycle progresses, especially through the long plateau close to maximum phase when the time scale rises to $\sim 50 \times 10^{9}$ years, the main sequence becomes burnt down to dwarfs of types K to M, with the consequence that the emitted radiation is then mainly in the red and near infrared. However, as minimum is approached during contraction, wavelengths are shortened by a considerable factor. Thus a wavelength of $10\,000$ Å at $x = 6$ would be shortened to only ~ 1600 Å at $x = 1$ if it were not subject to absorption. Moreover, absorption also increases strongly with decreasing x

towards minimum phase, particularly because the intergalactic density of absorbing particles increases as x^{-3}, and also because the most effective absorbing particles – graphite whiskers – absorb most effectively in the ultraviolet.

While iron at cryogenic temperatures and in the form of long thread-like particles (whiskers) is likely to be the dominant source of opacity in the microwave region, carbon particles also in the form of whiskers considerably exceed the effect of iron at shorter wavelengths. This is both because carbon is produced by stars more abundantly than iron, and because the efficiency of carbon is greater at shorter wavelengths.

The value of Q_{abs} for graphite whiskers is essentially constant for all wavelengths longer than $\sim 1\,\mu m$, extending even to long radio wavelengths, and is ~ 0.33 for whiskers of diameter $0.01\,\mu m$ and length $\sim 1\,mm$, equivalent to an absorption coefficient of $10^5 cm^2 g^{-1}$. On the other hand the absorption coefficient is about three times greater than this for ultraviolet radiation, $\sim 3 \times 10^5 cm^2 g^{-1}$, requiring an intergalactic density of $\sim 10^{-34} g\,cm^{-3}$ to produce appreciable absorption of ultraviolet light as the parameter x decreases below 6 as the minimum at $x = 1$ is approached in the contracting phase of each cycle. Thus an intergalactic density of $\sim 10^{-34} g\,cm^{-3}$ at $x = 6$ would rise to $\sim 2 \times 10^{-32} g\,cm^{-3}$ at $x = 1$. Over a cosmological distance of $10^{27} cm$ at $x = 1$ this would give an optical depth of ~ 6 for ultraviolet light and ~ 1 for wavelengths longer than $1\mu m$.

The great bulk of the optical radiation that becomes subject to thermalization in the contracting phase in each cycle has been traveling for 10^{11} years, and even more in the case of microwaves. The radiation incident on a carbon whisker has mostly been in propagation through the last maximum phase of the oscillation. This includes all the microwave radiation existing before the present oscillation, and all the starlight generated by galaxies in the first half of the current cycle, with the consequence that all such radiation is exceedingly uniform in its energy density. For the moment we assume the radiation to be entirely uniform, returning to the slight deviations from uniformity shortly. What we do not assume, however, is that the carbon whiskers responsible for absorbing the starlight and re-emitting it into microwaves are uniformly distributed. The carbon whiskers can be lumpy on the scale of clusters of galaxies. This means that the conversion of starlight to microwaves will occur lumpily as the minimum of the oscillation is approached.

Nevertheless each carbon whisker, wherever it is situated, finds itself in a radiation bath of uniform energy density, a radiation bath of which the major fraction already consists of microwaves from previous cycles and the rest is mostly starlight still to be converted to microwaves. If the whole were microwaves, then having regard to the total flatness of Q_{abs} with respect to wavelength through the range longward of $\sim 1\,\mu\mathrm{m}$, the temperature attained by the particles would be simply the standard microwave temperature, ~ 2.73 K as we know it to be. But because a fraction, say ten percent, of the radiation is ultraviolet, blueshifted from the decreasing values of the parameter x, the stellar component has the higher value of Q_{abs} discussed above. This forces up the temperature of the grains, to a value of order

$$\sim (0.9 + 3^{1/4} \times 0.1)2.73 \simeq 2.82\ \mathrm{K},$$

the second term in the brackets being contributed by the absorption of the starlight. As the starlight is progressively absorbed with optical depth τ, the factor 0.1 in this equation is replaced by $0.1\exp-\tau$ and the factor 0.9 is replaced by $(1 - 0.1\exp-\tau)$, so that the grain temperature varies according to

$$\left[1 + 0.1e^{-\tau}(3^{1/4} - 1)\right]2.73\ \mathrm{K}.$$

Thus as the starlight begins to be absorbed the whisker temperature goes about 0.1 K higher and then lapses back to 2.73 K as the starlight is progressively absorbed.

The effect of τ being lumpily distributed on the scale of clusters of galaxies causes this process of a slight temperature rise followed by a fall back to 2.73 K to be correspondingly lumpy. *But what it does not do is to make the total radiation energy density lumpy at all.* Once the radiation energy density is uniform in total, this essential property is not changed by absorption and re-emission due to particles. Because of course each particle emits just as much energy as it absorbs – the total assembly of particles has itself only a negligible heat content – the particles do not store heat except in a very small amount. Thus, objections to this theory that have been raised by some proponents of the standard model, based on lumpiness of the particles, are not correct.

Since the emissivity of the particles has no wavelength dependence they simply emit a Planck distribution $1/\left[\exp(\frac{h\nu}{kT}) - 1\right]$ at whatever value of T they may happen to have, according to the above temperatures which may range up to about 2.82 K. But in general they do not produce the Planck intensity $\nu^3/[(\exp(\frac{h\nu}{kT}) - 1]$. When T is raised slightly to 2.82 K the intensity distribution is slightly diluted.

So what is the outcome from this first absorption of the starlight? It is a uniform energy density of microwaves with a distribution approximately that of a black body at 2.73 K, but with some fluctuations in the details of the intensity curve with those details having initially a somewhat uneven distribution to the extent that the carbon whiskers are distributed unevenly.

But now, with the starlight absorbed, and with the temperature of the particles everywhere the same, further absorption and re-emission inevitably generates the strict Planck distribution for 2.73 K. Only a few further absorptions at the oscillatory minimum are sufficient for this last step. It can be done with carbon whiskers, as Narlikar, Edmunds and Wickramasinghe[4] suggested as long ago as 1975. However, the addition of even a small quantity of iron whiskers, with very high opacity at the center of the microwave distribution, would make this final step even more decisive.

The essential point is the one already emphasized, that expansion through an oscillatory maximum produces mixing of radiation coming from very large distances, $\sim 10^{29}$cm, and it does so with the intergalactic particle density at a low value, permitting radiation to travel freely.

However, there will always be some radiation emitted by some local source that is capable of producing a small fluctuation in the energy density, just as at a typical spatial point there will be radiation from the nearest galaxy which in general would be about 3 Mpc away. Galaxies in which the light comes mostly from main-sequence dwarfs would have an absolute magnitude of about −21, that is to say an emission of about 10^{44} erg s^{-1}. At a distance of about 3 Mpc, the energy density from such a galaxy would be about 3×10^{-18} erg cm^{-3}, as compared to a cosmological energy density of starlight that is about 10^{-14} erg cm^3, i.e. a local variation of the energy density amounting to one part in 3000. This leads to variability in the temperature of the thermalizing particles by one part in 12 000, an effect that becomes negligible, however, when account is taken of all the thermalizing particles reaching out to the Olbers limit. However, this is not the most extreme case of variability due to local sources. For that we have to consider thermalizing particles that lie near rich clusters of galaxies. Thus, for a particle lying 3 Mpc away from a rich cluster of 1000 galaxies, the local modification of the cosmic field of starlight (10^{-14} erg cm^{-3}) is $1000 \times 3 \times 10^{-18}$ erg cm$^{-3} = 3 \times 10^{-15}$ erg cm^{-3}. That is to say a fluctuation in the energy density of about 30 per cent, capable of changing the temperatures of particles from 2.73 K by an amount that is 30 percent of the change calculated above, which was from 2.73 K to about 2.82 K. The change is thus by about 3×10^{-2} K.

When the sky is examined on an angular scale such that beam widths are compared, one containing a rich cluster of galaxies and the other not, then we shall expect to find a detectable variation in the effective background temperature. However, the amount of the variation will be by no means as large as the value of 3×10^{-2} K just calculated. This is because the particles for which this variation applies must lie within about 3 Mpc of the cluster of galaxies and this purely local fluctuation has to be considered together with all the thermalizing particles along the line of sight out to the Olbers limit. The latter reduce the fluctuation by a factor of the order of 1000, to of the order of 3×10^{-5} K, or about 30 μK, in good general agreement with the observed fluctuations of the microwave temperature plotted against beam width shown in Fig. 16.3.

Thus, we have deduced from very simple considerations the expected fluctuation in the microwave background temperature. In other words, in this theory the fluctuation described as the 'Face of God' (cf. Preface and Chapter 21) is simply a small local fluctuation, showing that the 'Face of God' is not to be seen quite as easily as is usually supposed, a conclusion that is well supported by the whole history of cosmology. Many people and much money has been directed into efforts to explore this 'Face of God' aspect of big-bang cosmology, much as many people entered the abbeys of medieval Europe, abbeys that were supported financially as tithes on the activities of the people. Just as the achievements of those abbeys turned out to be nil, so we expect it to be in the present instance. The fluctuations are nothing but minor

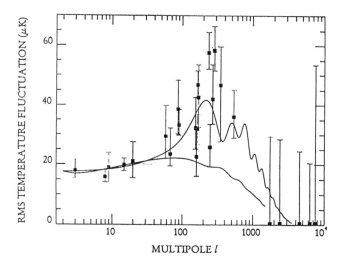

Fig. 16.3. Temperature fluctuations in the cosmic microwave background (Fig. 2 from Bennett *et al.*[10]) are responsible for the curves plotted in this figure. The upper curve, according to them, favored the expectations for inflation plus cold dark matter, while the lower curve is that expected for topological-defect theory. The multipole number ℓ implies a separation on the sky of $100°/\ell$. The linear scale corresponding to $\ell \simeq 200$ when interpreted as a cosmological distance is about that of rich clusters of galaxies. The temperature fluctuations are of the order calculated in the text.

deviations occasioned by spatial variations in the energy density of starlight due to the irregular distribution of galaxies.

We have therefore deduced, first the energy of the microwave background – the temperature already obtained in Chapter 8 – and now also the amplitude of its fluctuations. So how much does this set QSSC ahead of the big bang? By the standards currently operative in big-bang cosmology, in which nothing connected with the background has been proved, but only assumed *ad hoc*, a very long way indeed.

The counting of radio sources

It was discovered in the late 1950s that the count of radiogalaxies against flux F at a given frequency behaved peculiarly. Defining $N(F)$ as the number of sources with flux values greater than F, it was easily proved that for standard sources distributed uniformly in Euclidean space $NF^{3/2}$ should be independent of F. In an expanding cosmological space, $NF^{3/2}$ after remaining approximately constant for a while should eventually decline with decreasing F, i.e. with increasing distance of the sources.

As was discussed in Chapter 7, this was not what the group of radio-astronomers under Martin Ryle at Cambridge claimed to find. Using a unit of $10^{-26} \text{Wm}^{-2}\,\text{Hz}^{-1}$ for F and a frequency of 178 MHz (later 408 MHz) the count was said to be such that $NF^{3/2}$ remained approximately constant for $F > \sim 20$. But thereafter as F decreased $NF^{3/2}$ increased down to $F \simeq 1$. The increase was claimed at first to be very large. It declined year by year from one scientific conference to another until by the early 1960s the increase was claimed to be by a factor of about 4. It will be shown in this subsection how very easily this increase can be understood in the quasi-steady-state theory.

It will be sufficient again to put $\exp t/P \simeq 1$ in the expression for $S(t)$, when the equation governing the oscillations of QSSC is that given in (16.4), with x, α, β defined in (16.3), viz.

$$\frac{\dot{S}^2}{S^2} = \frac{8\pi G}{3}\rho_0 \frac{S_0^3}{S^3}\left(1 - \frac{S_0^3}{S^3}\right) - \frac{k}{S^2}\left(1 - \frac{S_0^4}{S^4}\right), \tag{16.15}$$

the minimum of the oscillation being at $S = S_0$. The k term in (16.15) is important only near maximum phase, and is negligible otherwise. The values of S/S_0 at the present time of observation and at maximum are taken as 6 and 12 respectively.

Except near minimum phase, the S^{-6} terms in (16.15) are negligible. Whence it follows that, during an oscillation, neither close to minimum

nor to maximum, (16.15) can be written with sufficient accuracy in the simple form

$$\frac{\dot{S}^2}{S^2} = \frac{8\pi G}{3}\rho_0\frac{S_0^3}{S^3} = \frac{8\pi G}{3}\rho_*\frac{S_*^3}{S^3},$$

(16.16)

where * in the subscript refers to values at the present day.

Additionally we have

$$\frac{S_*}{S} = 1 + z = \left(\frac{t_*}{t}\right)^{2/3},$$

(16.17)

$$r = \int_t^{t_*}\frac{dt}{S} = \frac{3t_*}{S_*}\left(1 - \frac{1}{\sqrt{1+z}}\right),$$

(16.18)

where z is the observed cosmological redshift and t is a time intermediate between t_{min} and t_*. Thus (16.17) follows from the integration of (16.16), and (16.18) follows from the metric with the k-term neglected,

$$ds^2 = dt^2 - S^2(t)\left[dr^2 + r^2(d\theta^2 + \sin^2\theta d\varphi^2)\right].$$

(16.19)

The flux F from an object of intrinsic luminosity L is given by

$$F = \frac{L}{4\pi r^2 S_*^2(1+z)^2},$$

(16.20)

from which with the choice (say)

$$L = 10^{26}\text{W Hz}^{-1}$$

(16.21)

we obtain in units of $10^{-26}\text{W m}^{-2}\text{Hz}^{-1}$

$$F = \frac{8.51 \times 10^{-3}}{(1+z)^2(1 - 1/\sqrt{1+z})^2}\text{Jy}.$$

(16.22)

Here at $x = 6$ in Table 16.1 the time t_* has been taken as $0.112\,Q$ with $Q = 96.3 \times 10^9$ years.

Assuming the sources to occur uniformly with respect to the r-coordinate, the number with fluxes larger than (16.20) is proportional to r^3, $N \propto r^3$ say. Thus $NF^{3/2}$ is seen to be independent of r, and simply

$$NF^{3/2} \propto \frac{1}{(1+z)^3}.$$

(16.23)

In Table 16.2 we tabulate F given by (16.22) as a function of z. We refer to these sources as Class I. A second class with intrinsic luminosity five times higher at $5 \times 10^{26}\text{ W Hz}^{-1}$, and with one-quarter of the coordinate density is

Table 16.2. *Count of radio sources of intrinsic luminosity* 10^{26} WHz^{-1} *uniformly distributed*

z	$F(z)$	$(1+z)^{-3} \propto NF^{3/2}$
0.02	87.55	0.9423
0.03	38.72	0.9151
0.04	21.68	0.8890
0.05	13.81	0.8638
0.06	9.544	0.8396
0.07	8.980	0.8163
0.08	5.320	0.7938
0.09	4.184	0.7722
0.10	3.374	0.7513
0.12	2.323	0.7118
0.14	1.692	0.6750
0.17	1.133	0.6244
0.20	0.809	0.5787
0.25	0.508	0.5120
0.30	0.346	0.4552
0.40	0.188	0.3644
0.50	0.117	0.2963
0.60	0.0788	0.2441
0.80	0.0421	0.1715
1.00	0.0258	0.1250
1.20	0.0172	0.0939
1.40	0.0122	0.0723
1.60	0.0091	0.0569

Values of F are in flux units (10^{-26} Wm^{-2} Hz^{-1})

also considered. Such Class II sources at the same value of z evidently have fluxes five times higher. While the corresponding value of $NF^{3/2}$ becomes multiplied by $5\sqrt{5}$ for the factor $F^{3/2}$ and divided by four for N, i.e. multiplied in total by $1.25 \times \sqrt{5}$.

However, for the second class we now concentrate their distribution towards the last oscillatory minimum by requiring their smallest possible redshift to be at $z = 0.14$. They occur uniformly from the last oscillatory minimum until $(t_*/t)^{2/3} = 1.14$, i.e. until a time

Table 16.3. *Count of radio sources of intrinsic luminosity* $5 \times 10^{26}\,\mathrm{WHz}^{-1}$ *uniformly distributed up to time* 8.86×10^9 *years from last oscillatory minimum (i.e. $z > 0.14$)*

z	$F(z)$	Count on scale of Table 16.2
0.14	8.460	0.000
0.17	5.665	0.711
0.20	4.045	0.994
0.25	2.540	1.121
0.30	1.730	1.099
0.40	0.940	0.949
0.50	0.585	0.794
0.60	0.394	0.663
0.80	0.211	0.472
1.00	0.129	0.346
1.20	0.086	0.261
1.40	0.061	0.201
1.60	0.046	0.153
2.00	0.027	0.130
2.50	0.016	0.065
3.00	0.011	0.046

$$t = 0.822t_* = 8.86 \times 10^9 \text{ years} \tag{16.24}$$

after the last minimum, again taking $t_* = 0.112Q$ with $Q = 96.3 \times 10^9$ years. In addition to the factor $1.25\sqrt{5}$, the values in the third column of Table 16.2 are multiplied also by

$$1 - \frac{r^3(z = 0.14)}{r^3(z)} = \frac{1 - \left(1 - \frac{1}{\sqrt{1+z}}\right)^3}{1 - \left(1 - \frac{1}{\sqrt{1.14}}\right)^3} \text{ for } z \geq 0.14 \tag{16.25}$$

in order to obtain the values in Table 16.3.

It remains to sum the third columns of the two tables, but at the same value of F, not the same value of z. When this has been done, the last step is to normalize so that the largest value of the count is 1.000, with the result given in Table 16.4. These values agree extremely well with the observations[1,2,3], as can be seen in Fig. 16.4 where the observations at 0.4 GHz and 1.4 GHz are given for comparison with the theoretical values.

Table 16.4. *Sum of counts in Tables 16.2 and 16.3 made at the same value of F*

F	log F	Normalized Count	log Normalized Count
87.55	1.942	0.513	−0.290
38.72	1.588	0.493	−0.307
21.68	1.336	0.484	−0.315
13.81	1.140	0.471	−0.327
9.54	0.979	0.445	−0.352
8.98	0.953	0.432	−0.365
5.665	0.753	0.821	−0.0857
4.045	0.607	0.962	−0.0170
2.540	0.405	1.000	0.000
1.730	0.238	0.969	−0.0139
0.940	−0.0269	0.844	−0.0737
0.585	−0.233	0.716	−0.145
0.394	−0.405	0.614	−0.212
0.211	−0.676	0.459	−0.338
0.129	−0.889	0.352	−0.454
0.086	−1.066	0.278	−0.556
0.061	−1.215	0.224	−0.650
0.046	−1.337	0.176	−0.755
0.027	−1.569	0.127	−0.896
0.016	−1.796	0.0844	−1.074
0.011	−1.959	0.0577	−1.239

We emphasize how very simple are the assumptions leading to the excellent agreement between theory and observation shown in Fig. 16.4. Just two types of source are used, L_I and L_{II}, with $L_I = 10^{26}\,\mathrm{W\,Hz^{-1}}$ and $L_{II} = 5 \times 10^{26}\,\mathrm{W\,Hz^{-1}}$, with the density of L_I four times higher than L_{II}, together with a concentration of L_{II} to $t < 0.822 t_*$. Otherwise the properties of the sources are uniform. Thus using this theory no complicated evolutionary picture is required, as it is in the big-bang model.

The redshift–magnitude relation for galaxies

The redshift–magnitude relation for galaxies has in effect already been given in equations (16.20), (16.21) and (16.22), subject to the small approximation used in obtaining (16.18) for the relation between the radial coordinate r in

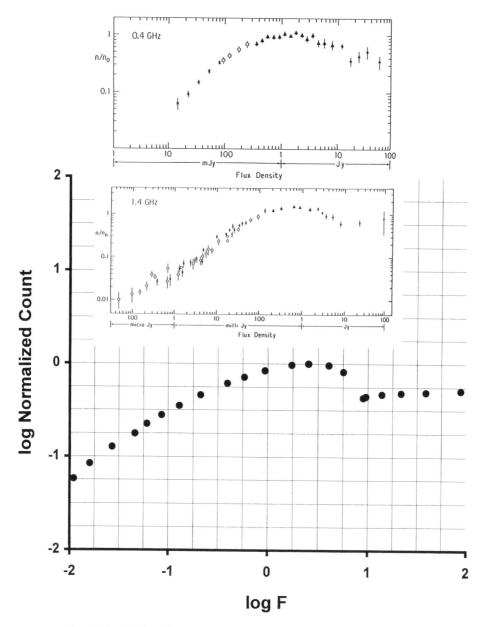

Fig. 16.4. $N(F)$ is the count of the number of sources brighter than F. The plot is of $\log NF^{3/2}$ against $\log F$ for a simple two-component model, as explained in the text. The insets show the experimental counts at (a) 0.4 GHz and (b) 1.4 GHz. The excellent agreement disposes of an early criticism that the experimental counts cannot be explained in QSSC. The details are listed in Tables 16.2, 16.3 and 16.4.

the metric (16.19) and the redshift z. However, (16.22) is in terms of the Jansky as the unit of flux, whereas tradition in optical astronomy has imposed the burden of using a magnitude scale. This requires a reworking of the equations.

Write m for the magnitude of a galaxy of intrinsic magnitude M and using seconds and centimeters (i.e. bringing in the speed of light c) we have

$$m = M + 5\log\left(\frac{crS_*}{3.08 \times 10^{19}}\right) + 5\log(1 + z) + \Delta. \tag{16.26}$$

Then with (16.18) for rS_* and again putting $t_* = 0.112Q$, (16.26) becomes

$$m - \Delta = M + 5\log\left(\frac{0.336cQ}{3.08 \times 10^{19}}\right) + 5\log(1 + z) + 5\log\left(1 - \frac{1}{\sqrt{1+z}}\right), \tag{16.27}$$

where Δ is the extinction produced by intergalactic material in the passage of light from the galaxy to the observer.

The oscillatory period $Q = 96.3 \times 10^9$ years $= 3.03 \times 10^{18}$ s, according to (16.11). Whence with $M = -22.44$ (as in ref. 1), we have

$$m - \Delta = 22.6 + 5\log(1 + z) + 5\log\left(1 - \frac{1}{\sqrt{1+z}}\right) \tag{16.28}$$

and (16.28) gives the plot shown in Fig. 16.5 for z up to ~ 6. This is for galaxies observed at the last oscillatory minimum. Proceeding backwards further with respect to time in QSSC, the value of r in (16.26) continues to increase as z declines, in a symmetrical relationship to the minimum. This causes the curve in Fig. 16.5 to turn back on itself as shown. By choosing the constants t_*, Q and M differently the curve in Fig. 16.5 could be varied numerically, but its form is essentially an invariant of the theory.

It will be recalled that Hubble believed the redshift–magnitude relation to be a straight line, and that Sandage held the same view in the 1950s, at any rate for values of z up to about 0.5 (see Chapter 6). This form is followed in Fig. 16.5. Then at moderately large z the curve turns gently to the right, as it does in the so-called standard big-bang model. However, at the largest values of z, as $z \simeq 5$ is approached, a dramatic change from any big-bang model takes place, with the curve ultimately turning over and returning on itself. Unfortunately even for an intrinsically very bright galaxy this behavior occurs at $m - \Delta \simeq +27$. Since appreciable values for the absorption Δ are to be expected this is at a very faint level indeed, where spectroscopic investigations are not feasible at the present time.

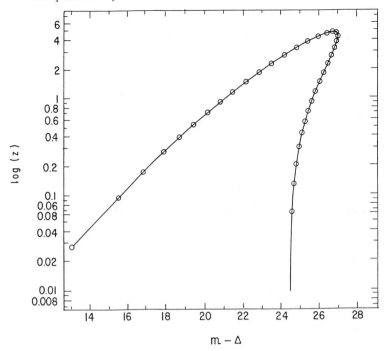

Fig. 16.5. The Hubble diagram for galaxies of intrinsic absolute magnitude −22.44.

The second-order term in the redshift–magnitude relation

Earlier, in Chapters 6 and 7, we gave a brief historical account of the attempts that were made to determine the so-called deceleration parameter q_0. It is well known that for the classical steady-state model $q_0 = -1$, and this can be interpreted as the effect of a repulsive force – a creation term, in the framework of the quasi-steady-state model.

Using Type Ia supernovae as standard candles, Perlmutter *et al.*[5] and Riess *et al.*[6] have recently claimed that the redshift–apparent magnitude relation out to $z = 0.8$ is already showing signs of this type of negative curvature, and since they have only analyzed their results in terms of the standard model they have claimed that this shows that the cosmological constant is non-zero and positive. This has been reported[7] to be the most important scientific discovery of 1998[a], but already in 1999 doubts are being raised.

In what follows we show analytically that this is just what is naturally expected in a quasi-steady-state universe.

---

[a] This is in the modern tradition of maximum hype for big-bang cosmology.

Equation (16.26) for the apparent magnitude of an object of intrinsic magnitude M continues to hold, with the coordinate r given as in (16.18) by

$$r = \int_t^{t_x} \frac{dt}{S}.$$

And with the notation as in (16.3),

$$r = \int \frac{dx}{S\dot{x}} = \frac{1}{S_0} \int_x^{x_x} \frac{dx}{x\dot{x}}, \tag{16.29}$$

where x_* is the present-day value of $x = S/S_0$. Choosing $x_* = 6$, as in previous sections, we have for an object observed at redshift z

$$r = \frac{1}{\sqrt{\alpha}S_0} \int_{6/(1+z)}^{6} \frac{dx}{x\sqrt{G(x)}}, \tag{16.30}$$

the function $G(x)$ being defined in (16.6).

Applying (16.26) with $\Delta = 0$ both to an object of redshift z and to a similar object of redshift z' we have

$$m(z) = m(z') + 5 \log \left[\frac{r(z)(1+z)}{r(z')(1+z')} \right]. \tag{16.31}$$

The first-order theory studied above arises when only the first term in (16.6) for $G(x)$, viz $G(x) = 1/x$, is included. So long as z in the lower limit of the integral in (16.30) is ~ 1 or less, this is an adequate approximation, for then x is greater than ~ 3 in $G(x)$. With the maximum value, x_1, of x taken as 12 we then have $G(x)$ effectively equal to $1/x$. What happens in QSSC is that the expansion of the universe accelerates away from oscillatory minima under the influence of the C-field term in (15.7) in such a way that the first-order redshift–magnitude relation is effectively the same as it is in the standard model of big-bang cosmology. This is so long as we are concerned with redshift values from zero to ~ 1.

As was stated earlier, Perlmutter et al.[5] and Riess et al.[6] have claimed that this first-order theory does not give satisfactory results in the case of about 60 supernovae with redshift values up to about 0.8. For the low-redshift supernovae, Hamuy et al.[8] find that $m(z') = +14.0$ at $z' = 0.01$ to adequate accuracy, when (16.31) becomes

$$m(z) = 14.0 + 5 \log \left[\frac{r(z)(1+z)}{1.01 r(z = 1.01)} \right], \tag{16.32}$$

and with $G = 1/x$ in (16.30) we have

$$m(z) = 14.0 + 5\log\left[\frac{\frac{(1+z)}{1.01}\int_{6/(1+z)}^{6}\frac{dx}{x^{1/2}}}{\int_{6/1.01}^{6}\frac{dx}{x^{1/2}}}\right].$$ (16.33)

While it is true that a visual display of results conveys the quickest form of accessing results (see Fig. 16.6), we think the tabular form given here in Table 16.5 prevents the eye being biased by particular cases. The overwhelming impression one has from Table 16.5 is the persistence of negative differences between the $m(z)$ values calculated from (16.33) and the observed magnitudes. Either the observed magnitudes must contain a systematic error of about a third of a magnitude significance, or a difference from $G(x) \simeq 1/x$ in

$$m(z) = 14.0 + 5\log\left[\frac{\frac{(1+z)}{1.01}\int_{6/(1+z)}^{6}\frac{dx}{x\sqrt{G(x)}}}{\int_{6/1.01}^{6}\frac{dx}{x\sqrt{G(x)}}}\right].$$ (16.34)

is required.

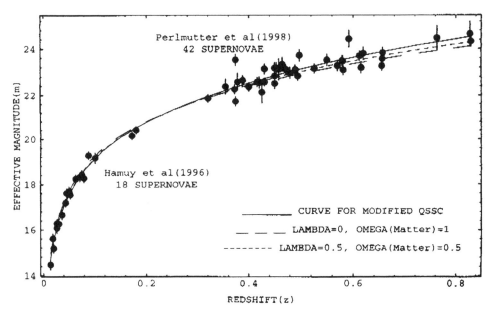

Fig. 16.6. The second-order term in the redshift–magnitude relation. Here we compare three cosmological models: the Friedmann model with $k = 0$, $\Lambda = 0$ shown by long dashed line; the best fit for a QSSC with contemporary creation and $p = 2$, continuous line, see also Table 16.5; the best fit for a Friedmann model with $\Lambda = 0.5$, $k = 0$, dotted curve.

Table 16.5. *Differences between predicted and observed magnitudes for 60 supernovae of Type Ia (after Perlmutter et al.[5]). Column 1 $\equiv$ standard model (k = 0 and Λ = 0), column 2 $\equiv$ QSSC as modified in text, column 3 $\equiv$ λ-dominant cosmology (k = 0 and Λ = 0.5)*

SN	z	Observed magnitude m_{obs}	1	2	3
1990O	0.030	16.26	0.364	0.158	0.215
1990af	0.050	17.63	0.114	−0.072	−0.021
1992P	0.026	16.08	0.232	0.020	0.079
1992ae	0.075	18.43	0.207	0.047	0.091
1992ag	0.026	16.28	0.032	−0.180	−0.121
1992al	0.014	14.47	0.491	0.267	0.329
1992aq	0.101	19.16	0.136	0.002	0.037
1992bc	0.020	15.18	0.559	0.341	0.401
1992bg	0.036	16.66	0.363	0.163	0.219
1992bh	0.045	17.61	−0.097	−0.289	−0.236
1992bl	0.043	17.19	0.223	0.030	0.083
1992bo	0.018	15.61	−0.101	−0.321	−0.260
1992bp*	0.079	18.27	–	–	–
1992br*	0.088	19.28	–	–	–
1992bs	0.063	18.24	0.012	−0.160	−0.113
1993B	0.071	18.33	0.186	0.022	0.067
1993O	0.052	17.54	0.290	0.106	0.157
1993ag	0.050	17.69	0.054	−0.132	−0.081
1992bi	0.458	23.11	−0.388	−0.217	−0.313
1994F	0.354	22.38	−0.254	−0.164	−0.219
1994G	0.425	22.13	0.419	0.564	0.481
1994H	0.374	21.72	0.533	0.639	0.576
1994al	0.420	22.55	−0.029	0.113	0.032
1994am	0.372	22.26	−0.020	0.085	0.023
1994an	0.378	22.58	−0.303	−0.194	−0.258
1995aq	0.453	23.17	−0.473	−0.307	−0.401
1995ar	0.465	23.33	−0.573	−0.397	−0.496
1995as	0.498	23.71	−0.793	−0.593	−0.705
1995at	0.655	23.27	0.290	0.598	0.424
1995aw	0.400	22.36	0.048	0.174	0.101
1995ax	0.615	23.19	0.222	0.503	0.345
1995ay	0.480	22.96	−0.129	0.058	−0.047
1995az	0.450	22.51	0.171	0.336	0.243
1995ba	0.388	22.65	−0.312	−0.195	−0.264

Table 16.5 (*contd*)

SN	z	Observed magnitude m_{obs}	1	2	3
1996cf	0.570	23.27	−0.037	0.214	0.074
1996cg	0.490	23.10	−0.221	−0.027	−0.135
1996ci	0.495	22.83	0.073	0.271	0.160
1996ck	0.656	23.57	−0.006	0.303	0.128
1996cl	0.828	24.65	−0.534	−0.118	−0.362
1996cm	0.450	23.17	−0.489	−0.324	−0.417
1996cn	0.430	23.13	−0.554	−0.405	−0.490
1997F	0.580	23.46	−0.186	0.072	−0.073
1997G	0.763	24.47	−0.549	−0.172	−0.390
1997H	0.526	23.15	−0.105	0.115	−0.008
1997I	0.172	20.17	0.315	0.248	0.260
1997J	0.619	23.80	−0.373	−0.089	−0.249
1997K*	0.592	24.42	–	–	–
1997L	0.550	23.51	−0.361	−0.124	−0.256
1997N	0.180	20.43	0.157	0.098	0.107
1997O*	0.374	23.52	–	–	–
1997P	0.472	23.11	−0.318	−0.137	−0.238
1997Q	0.430	22.57	0.006	0.155	0.070
1997R	0.657	23.83	−0.263	0.047	−0.129
1997S	0.612	23.69	−0.290	−0.010	−0.168
1997ai	0.450	22.83	−0.149	0.016	−0.077
1997ac	0.320	21.86	0.034	0.097	0.054
1997af	0.579	23.48	−0.210	0.047	−0.097
1997aj	0.581	23.09	0.188	0.447	0.302
1997am	0.416	22.57	−0.071	0.068	−0.012
1997ap	0.830	24.32	−0.198	0.219	−0.026
	Average error		−0.048	0.036	−0.029

*For 1992bp, 1992bf, 1997K and 1997) the errors are so large as to suggest that m_{obs} for these cases is too large. We have omitted them from the calculation of the average errors for the various models.

Although we have regarded 'creation events' as occurring during the expansion phases of QSSC in our published work, we have taken the total creation rate to be so low, except at oscillatory minima, that no dynamic effect on the universe occurs through $\sqrt{G(x)}$ in (16.30). But these new results on supernovae suggests that it may be otherwise. The continuity equation for the density N of non-relativistic particles can be written in the presence of creation as

$$\frac{1}{S^3}\frac{d}{dt}(S^3 N) = pN\frac{\dot{S}}{S} \tag{16.35}$$

where p is a measure of the creation rate. The term on the right-hand side of (16.35) is a measure of the creation rate, with the proportionality of N arising from the circumstance that creation events arise from already-existing matter – by processes that we shall discuss in detail in later chapters. Taking p to be a constant, equation (16.35) integrates to give $N \propto S^{p-3}$, where the main term in $G(x)$ is changed from $1/x$ to $1/x^{1-p}$ and (16.30) becomes

$$r = \frac{1}{\sqrt{\alpha}}\frac{1}{S_0}\int_{6/(1+z)}^{6}\frac{dx}{x^{(1+p)/2}}, \tag{16.36}$$

and predicted magnitudes for supernovae of Type Ia are changed from (16.33) to

$$m(z) = 14.0 + 5\log\left[\frac{\frac{(1+z)}{1.01}\int_{6/(1+z)}^{6}\frac{dx}{x^{(1+p)/2}}}{\int_{6/1.01}^{6}\frac{dx}{x^{(1+p)/2}}}\right]. \tag{16.37}$$

Differences between these predicted magnitudes and the observed magnitudes are shown in Table 16.5 for the case $p = 2$. The errors now are much smaller, with an average difference between $m(z)$ and the observed magnitudes of only about 0.05 mag, well within the observational scatter, one might think.

Returning to (16.6) for $G(x)$ the curvature term $-1/x_1[1 - (1/x^4)]$ has been omitted because of the large value of x_1, taken as 12 in the earlier discussion and the second term, $-x^{-4}$, is also small for values of x between 3 and 6 such as we have in the above integrals. The term $-x^{-4}$ arises from the C-field and it takes a different form when 'creation events' have the effect of changing the matter term from $1/x$ to x^{p-1}. Writing W for the energy density of the C-field we have

$$\frac{1}{S^6}\frac{d}{dt}(WS^6) = -pN\frac{\dot{S}}{S} \tag{16.38}$$

where from energy considerations the creation term on the right-hand side must balance the matter term on the right-hand side of (16.35). According to this equation, the magnitude of the C-field energy density falls to $-\frac{2}{5}$ of the particle energy density and is maintained thereafter at this level. Thus, only the numerical coefficient in (16.36) is changed through the inclusion of the C-field energy density – the form of the integral is unchanged and (16.37) with $p = 2$ is unchanged. Hence the inclusion of the C-field energy density in $G(x)$ does not disturb the agreement between predicted and observed magnitudes in the case where creation events contribute sufficient new material to give $p = 2$ in QSSC.

What now is to be done in big-bang cosmology? With the standard model leading to the unsatisfactory discrepancies shown in Table 16.5 a fundamental change is required, which Perlmutter *et al.* attribute to a cosmological constant λ. This leads to a term λx^2 appearing in $G(x)$. If the constant λ is large enough for the new term to be dominant at the values of x in question, say $3 < x < 6$, the integral in the expression (16.36) for the radial coordinate r takes the form

$$r = \frac{1}{\sqrt{\alpha \lambda} S_0} \int_{6/(1+z)}^{6} \frac{dx}{x^2}, \tag{16.39}$$

and predicted magnitudes are to be calculated from

$$m(z) = 14.0 + 5 \log \left[\frac{\frac{(1+z)}{1.01} \int_{6/(1+z)}^{6} \frac{dx}{x^2}}{\int_{6/1.01}^{6} \frac{dx}{x^2}} \right], \tag{16.40}$$

instead of from (16.33).

The differences between (16.40) and the observed magnitudes are also given in Table 16.5. The steady-state theory in its original form also leads to (16.40). In the fourth, fifth and sixth columns of Table 16.5 we therefore have the following cases:

   (i) standard model and QSSC without including contemporary creation events in the fourth column;
   (ii) QSSC with contemporary creation events giving $p = 2$ in the fifth column;
   (iii) a big-bang model dominated by a cosmological constant or the original steady-state model in the sixth column.

Of these possibilities (ii) gives the best agreement with the observations. But in the big-bang case a further resource is possible, namely to combine (i) and (iii). This is done by making the λ term in $G(x)$ dominant at the larger values of $x \sim 6$ but to have the matter term dominant at smaller values, $x \sim 3$. This leads to error values intermediate between (i) and (iii), as in (ii). This tactic implies giving a just-so value to λ, as it does in all other aspects of big-bang cosmology.

The remaining possibility is that the claimed supernova magnitudes contain a systematic error. It will be recalled that early attempts to measure second-order effects in the redshift–magnitude diagram lost their impact on astronomers as possible sources of error began to emerge and it has been a general experience in cosmology that difficulties tend to persist into later work. Hubble's attempt to determine second-order effects through the counting of galaxies is an example. In this respect it may be noted that both the standard model and QSSC in its published form predict supernova magnitudes that are a fraction of a magnitude too bright to agree with the claims of Perlmutter *et al.*[5], which suggest the possibility that the supernovae were not quite caught at their maximum luminosities. Had the observations given brighter magnitudes, the other way round, there would not have been this possibility.

Asymmetric oscillations

The expression for the time dependence of the scale factor $S(t)$, viz.

$$S(t) = F_Q \cdot \exp(t/P), \tag{16.41}$$

when F_Q is an oscillatory function of period Q is not strictly a symmetrical function of t even when $Q \ll P$, because of the increasing value of the factor $\exp t/P$ with increasing t. In QSSC as so far developed with $Q \ll P$, the asymmetric effect is small, however. Its amount depends on the importance of creation. As this become larger the asymmetry becomes more pronounced. Indeed, for an intensity of creation that is sufficient to explain the new SNIa data of Perlmutter *et al.* the effect is pronounced. Write N, W for the energy densities of matter (non-relativistic) and the C-field respectively. Then we have from energy considerations

$$\frac{1}{S^3}\frac{d}{dt}(S^3 N) = 2N\frac{\dot{S}}{S} \tag{16.42}$$

(SN data) and

$$\frac{1}{S^4}\frac{d}{dt}(S^4 W) = -2N\frac{\dot{S}}{S} \tag{16.43}$$

(energy conservation).

Here we take the adiabatic relation for the C-field to be S^{-4} rather than S^{-6}, as in the work of Sachs, Narlikar and Hoyle[9]. According to (16.42) the mass density N falls proportionally to $1/S$, as is required by the SN data, and W also falls as $1/S$ when the magnitude of W is 2/3 of N.

A negative value of the cosmological constant λ causes this solution to attain a maximum of S. In the subsequent reversal from $\dot{S}$ positive to $\dot{S}$

negative the contraction can only be by a factor $3/2$, when $-W$ comes into balance with N. Asymmetry arises between expanding and contracting episodes because of the one-sidedness of the creation process. Matter and C-field are not destroyed. Thus the matter density rises between a maximum and a minimum by the factor $(3/2)^3 = 27/8$, and this is the factor by which S can rise between minima and maxima.

Writing S_0 for S at the last minimum and S_* for the scale at present, the redshift at the last minimum is given by $1 + z = S_*/S_0$, which can be taken as the largest SN redshift, viz. $z \simeq 0.8$. At the maxima and minima the redshift values are given by

$$1 + z = \frac{S_*}{S_0}\left[1, \frac{2}{3}, \frac{9}{4}, \frac{3}{2}, \frac{81}{16}, \frac{27}{8}, \frac{729}{64}, \frac{243}{32}, \cdots\right]$$

where the minima are at 1, $(9/4)$, $(81/16)$, $(729/64)$, and the maxima are at $(2/3)$, $(3/2)$, $(27/8)$, $(243/32)$ etc. At the last maximum the redshift is given by $1 + z = (2/3) \times (S_*/S_0) \simeq (2/3) \times 1.8 = 1.2$, or $z = 0.2$. Galaxies observed at the last maximum will be faint because of their distance, but will be comparatively blue, with a small redshift of about 0.2. These might be the so-called 'blue' galaxies of Tyson, with small redshifts as found by Ellis. The time step between alternating maxima and minima is about H_o^{-1}.

As the amount of creation increases up to $3N\dot{S}/S$ on the right-hand side of (16.42) the model tends to the old steady-state theory, with a scale factor then increasing exponentially with time.

Decisive observational differences between QSSC and the big bang

The present value of $x = S/S_0$, where S_0 is the universal scale factor at the oscillatory minima of QSSC, determines the maximum possible redshift of cosmological origin, $z_{max} = x - 1$. Thus the choice made above of $x = 6$ in Table 16.1 gives $z_{max} = 5$, close to the currently maximum observed redshift of 5.3, which may or may not be completely of cosmological origin. In this theory observation to greater distances would not yield a redshift value significantly larger than this, since for larger distances we enter the previous contracting phase of the universe. In particular, observations to $x = 6$ in the contracting phase would give practically zero redshift. There would only be a very small redshift due to the overall expansion factor $\exp t/P$ which gives a redshift of order $(\exp Q/P - 1)$.

Observations back to still larger x, towards the last maximum of the oscillatory cycle, would even yield a blueshift given by $1 + z = 6/x$. With the values in Table 16.1, observation back to the last maximum at $x = 12$

would therefore give a blueshift effect of $z = -0.5$. At a distance then of $\sim 3 \times 10^{28}$ cm, a galaxy of absolute magnitude -21 would have an apparent magnitude easily calculated from $L/4\pi r^2 (1+z)^2$ of about $+22.5$. This is without taking into account the absorption due to carbon whiskers occurring near the last oscillatory minimum. It was estimated earlier, that the absorption could be as high as six magnitudes, pushing the apparent magnitude down to $\sim +28.5$, beyond the range of current spectroscopy. Such an absorption would seem to imply an observational cut-off at the last oscillatory minimum, except perhaps for galaxies observed not much beyond the last oscillatory minimum, and these would not show blueshifts. The best hope of finding galaxies from the previous cycle would be those with galaxies from $x \simeq 3$ in the contracting phase of the previous cycle, i.e. galaxies with a redshift of $z = 1$. With their light emitted by stars of types K and M, passage through the last oscillatory minimum would shorten the wavelength of the light from the range ~ 8000 Å to $\sim 1\mu m$ to the range ~ 2700 Å to ~ 3300 Å, which is not sufficiently blue to experience the worst of the absorption due to carbon whiskers. Thus if we suppose that there are two magnitudes of absorption instead of six magnitudes, a galaxy of intrinsic magnitude -20 observed at $r = 2 \times 10^{28}$cm and redshift $z = 1$ would have apparent magnitude $+25.5$ without absorption, and $+27.5$ with two magnitudes of absorption. We might thus expect a profusion of galaxies to appear at this magnitude level, their light being very red, in part because the light comes from stars of small mass, and in part from the reddening due to redshift.

Thus we predict that the larger modern telescopes are *already seeing into the last oscillatory minimum of QSSC*. But the degree of absorption which provides the augmentation of the microwave background largely prevents more distant observations being made at a level currently possible even with the 10-meter telescopes.

We saw in Chapter 11 that there is overwhelming evidence which shows that highly luminous objects, and QSOs in particular, have an intrinsic component in their redshifts, i.e that

$$1 + z_0 = (1 + z_i)(1 + z_c),$$

where only z_c is a distance effect.

Thus our choice of x based on the upper limits of observed redshifts (of QSOs) may be too high, since the cosmological component of the highest redshift QSO is likely to be less than 4.9. Suppose we put $x = 4$, rather than $x = 6$. This then requires some adjustment to be made to the numbers given above, but no change of principle. Instead of (16.8) and (16.9) we would then have

$$\frac{\dot{x}}{x} = 0.122\sqrt{\alpha} = H_{\mathrm{o}}$$

at the present epoch. The time back to the last minimum given by the last column of Table 16.1 is then $0.652H_{\mathrm{o}}^{-1}$ which for $H_{\mathrm{o}} = 60\,\mathrm{km\,s^{-1}Mpc^{-1}}$ is 11.0×10^9 years, a good value for the ages of Type I stars in the disk of our Galaxy.

There is no difficulty about the greater ages of Type II stars. These are simply stars that were born in advance of the exact moment of the oscillatory minimum, at about $x = 3$ in the preceding contracting phase, which adds $\sim 7 \times 10^9$ years to the ages of stars born at that epoch for a total of 18×10^9 years. The standard model in big-bang cosmology gives a maximum age of $0.667H_{\mathrm{o}}^{-1}$, little different from $0.652H_{\mathrm{o}}^{-1}$, the time back to the last minimum.

A decade ago it was often argued by big-bang advocates that galaxies were formed at $z \simeq 100$, implying the possibility of observing galaxies at redshifts significantly higher than 5. Observational spectroscopic techniques extend in the red to at least 1 μm, and Lyα could be detected out to a redshift $z \approx 7$. Thus detectable redshifts of 6 or 7 are nowadays within the range of the possible, and some observers have not given up hope that such redshifts will indeed be found and possibly they may be. However, what can already be said is that they are likely to be few and far between, judging from the sharp decline in numbers of QSOs beyond $z \simeq 2.5$ out to about 4. This decline is attributed by the conventional big-bang theorists and observers to a rather mysterious evolutionary effect. Our view is that two factors are involved, the appreciable contribution of the intrinsic redshift components on the one hand, and the cosmological effect in the QSSC just discussed.

It is fair, but very unpopular, to say that faced with observational necessity, and some changes in theoretical belief, $z = 100$ as the epoch of galaxy formation in big-bang cosmology has become replaced by $z = 5$ as the popular theoretical choice.

Of course in that model the entire observed modern universe came from the expansion of a minute fraction of the post-inflation universe, no more than a few centimeters in size.

Expansion from 10 cm to the present scale of 10^{28} cm for the observable universe implies an expansion of $\sim 10^{27}$ in $S(t)$. Since galaxies are not forming at a wholesale rate at the present moment, and since they were apparently not doing so much before a redshift value of $z = 4$, we have to suppose that from the scale of 10 cm it all happened rather precisely at an expansion factor given by

$$S(t) \sim 10^{27}/5 = 2 \times 10^{26}.$$

The asymptotic state of expansion of the universe in big-bang cosmology is no different from the expansion of a uniform cloud in Newtonian gravitation. There one can have three possibilities: the cloud has insufficient kinetic energy to expand to infinity against its self-gravitation, it has presently sufficient energy to arrive at infinity with zero speed, and it has an excess kinetic energy, arriving at infinity with finite speed. Those are the three cases $k = +1, 0, -1$ of big-bang cosmology. In the first case the cloud expands to a maximum extension and then falls back.

Now imagine that in a particular subregion of the cloud there is a uniform perturbation of density. The subregion behaves like a cloud in itself. With $k = +1$ for the main cloud, an increase of density in the subregion causes the subregion to stop expanding and to fall back ahead of the main cloud. With $k = -1$ a *small* perturbation has no effect, regardless of whether the perturbation is an increase or a decrease of density. Because the extra gravitating power of a positive perturbation is overcome by the excess kinetic energy of the cloud generally. The case $k = 0$, however, is razor-edged since a positive perturbation, even though small, will cause cessation of expansion *at some stage during the expansion of the main cloud,* thereby forming a localized condensation. It is this razor-edged property that makes the case $k = 0$ attractive to big-bang theoreticians.

If perturbations of random amounts occurred, condensations in the $k = 0$ case would occur at correspondingly random moments of t, which of course would not do. Because if we are seeking to explain the formation of galaxies in this way, it is necessary to understand how the entire distribution of galaxies is very uniform from place to place. So the perturbations have to possess a corresponding uniformity, leading to a complicated story of how the perturbations follow a law decided by quantum fluctuations in the early universe.

In other branches of science, observational or experimental confirmation of what was being said would be asked at this stage. Show me exactly how the fluctuations are generated, not by theory but by experiment, would be the request. For as Dirac used to tell his students in the early 1930s:

> That which is not observable does not exist.

But science has run a long way downhill, at least in this field, since those days. What is said nowadays is:

> I know my theory is right. Therefore anything required to make it work must also be right whether observable or not.

This, of course, was exactly the methodology of the medieval theologians.

In the days when galaxy formation was thought to have happened at $z \simeq 100$, the perturbations were chosen so that condensing subregions would evolve into clusters of galaxies, rather than into individual galaxies. This led to a smear of smaller condensations forming at $z < 100$, and not to a sudden appearance of galaxies everywhere at $z \simeq 4$ to 5. This was because the gravitational development of subregions of excess density $\Delta\rho$ inevitably involves a smear in time of the order of $(G\Delta\rho)^{-1/2}$. Setting $\Delta\rho = 10^{-27}\mathrm{g\,cm}^{-3}$, appropriate to clusters of galaxies, the smear is $\sim 5 \times 10^9$ years, much too long for the galaxies generally to be able to condense all at once at $z = 5$.

So the next step was to complicate the model, as it always is when a wrong path is being followed. Instead of a subregion with a constant density fluctuation we have subregions within subregions. The sub-subregions are given high enough density fluctuations $\Delta\rho$ for $(G\Delta\rho)^{-1/2}$ to be shortened adequately. Then as the main subregion aggregates on a longer time scale, larger structures are supposed to be built from mergers of the now condensed sub-subregions. Provided such models are made complicated enough, nobody can be sure of whether they work or not. And of course extensive numerical modelling has led to the idea that it is possible, or it will be possible to get agreement with observations. In our view while consistency may be obtained nothing will really have been established. It is simply a case of adjusting a number of free parameters in order to obtain an apparent agreement which is very likely nothing but an outcome of the arbitrary freedoms of the model.

Summary

The quasi-steady-state theory is a development of the general theory discussed in Chapter 15. The latter arises because of a different view of the physical structure of the universe taken in this book from the usual sequence of positive energy transformations that lead inevitably into the big bang, in which, as one goes backwards in time, energy is squeezed into smaller and smaller proper volumes, with spacetime inevitably headed towards a singularity, the singularity known as the big bang. Still thinking backwards in time, the singularity can be postponed through the energy eventually being converted into a scalar field ϕ that itself becomes soaked up into a potential $V(\varphi)$, which is given semantic status, through its being referred to as a 'false vacuum'. Thereafter, the only recourse is to a passage to some mystical state, wholly unattested either by observation or experiment. Leaving the unfortunate astrophysicist at the mercy of a breed of new theologians, who like the old theologians claim to be privy to revelations of which the writings of the compulsively numerate St. John the Divine is an example:

> The number of the beast is the number of man. It is six hundred three scores and six.

Rather than deliver ourselves into the ranks of the multitude that has believed in the sayings of St. John the Divine over very many generations, we have taken the commonsense alternative of supposing that energy appears in the universe in compensating positive and negative forms, and we have worked from an action principle formulation of this point of view, a formulation that we described in Chapters 14 and 15. Out of the possibilities offered by this approach we chose an oscillatory model, to which in the present chapter we have applied explicit parametric choices. Particularly of the oscillatory period Q in relation to the Hubble constant H_0, and of the phase at which we ourselves live at present. These choices could be varied if the relation of the theory to observation should require it. However, we consider the choices made in the first section of the present chapter to be well-motivated in relation to the observations, which are discussed severally in subsequent sections, especially with respect to the microwave background, its energy density, and fluctuations. Then at the end of this chapter we returned to the old problem of the counting of radio sources, showing how very easily the observations can be explained by the model.

Negative energy fields are inherently explosive. Concentrated locally they produce violent events, offering a possibility of explaining a wide range of observations – radiogalaxies, QSOs and active galactic nuclei. When uniformly distributed, a negative energy field exerts a negative pressure that shows itself in our view in the expansion of the universe. So far our discussion has been concerned with this last feature, namely the possibility of explaining the observed expansion of the universe without recourse to a primordial explosion, the expansion of the universe being necessarily concomitant on the appearance of compensating positive and negative forms of energy as an ongoing phenomenon.

From here on, we move from observations that are fully attested to a class of phenomena that are so far from explanation in the big-bang theory that entirely clear-cut observations are being widely ignored by the astronomical community, apparently on the basis that what one cannot understand in terms of a previously known theory can safely be ignored. From here on we shall attempt to extend the theoretical ideas so far developed so as to offer some perceptions that may permit us to grapple with these problems, beginning in the next chapter with a formulation of the theory of relativity that, while yielding the usual theory if one wants it that way, can also be turned in new directions. It offers some chance of dealing with problems that lie outside the range of the usual theory.

References

Chapter 16

1. Hoyle, F., Burbidge, G. & Narlikar, J.V. 1993, *Ap. J.*, **410**, 437.
2. Hoyle, F., Burbidge, G. & Narlikar, J.V. 1994, *M.N.R.A.S.*, **267**, 1007.
3. Hoyle, F., Burbidge, G. & Narlikar, J.V. 1994, *A&A*, **289**, 729.
4. Narlikar, J.V., Edmunds, M.G. & Wickramasinghe, N.C. 1976, in *Far Infrared Astronomy, Proc. 1975 Conf. of Roy. Astron. Soc.* (ed. M. Rowan-Robinson, Pergamon), p. 131.
5. Perlmutter, S. *et al.*, 1999, *Ap. J.*, **517**, 565.
6. Riess, A. *et al.* 1998, *A. J.*, **116**, 1009.
7. Glanz, J. 1998, *Science*, **282**, 1249.
8. Hamuy, M., Phillips, M.M., Maza, J., Suntzeff, N., Schommer, R. & Avilés, R. 1995, *A. J.*, **109**, 1.
9. Sachs, R., Narlikar, J. & Hoyle, F. 1996, *A & A*, **313**, 703.
10. Bennett, T.L., Turner, M.S. & White, M. 1997, *Phys. Today*, **50**, 32.

17

The intrinsic redshift problem

Scale invariant gravitational equations

We are well-used to physical results being independent of the units in which quantities are expressed. This is because results are always dimensionless numbers. This usual situation is for units that stay the same at every space-time point X. But should anything in physics be altered, if units were changed differently at different spacetime points? Using the Compton wavelength of some specified particle as the length unit, and noting that $c = 1$ requires the time and space units to be the same, this question can be discussed by a general change of the length scale achieved by transforming from $ds^2 = g_{ik}dx^i dx^k$ to

$$ds^{*2} = \Omega^2(X)g_{ik}dx^i dx^k. \tag{17.1}$$

This transformation is called a *conformal* or *scale* tranformation. In such a transformation the spacetime coordinates of points on the path of a particle stay fixed. It is the proper distance between adjacent points that changes according to the choice of the twice differentiable scalar function $\Omega(X)$. Experiments confined to a locality over which Ω does not change appreciably will evidently be unaffected. It is possible, however, for events in one locality to be related to distant localities through the propagation of some field, for example the electromagnetic field. The possibility that physics will be unaffected by $\Omega(X)$ even for observations related to widely separated localities is suggested by the circumstance that light cones are not affected by the transformation (17.1), provided Ω is restricted to the so-called conformal condition $\Omega \neq 0$. And indeed Maxwell's equations are invariant with respect to (17.1), with the electromagnetic field tensor unchanged,

$$F_{ik}^* = F_{ik}, \tag{17.2}$$

and with $A_i^* = A_i$ followed by a suitable gauge transformation leaving the wave-equation for the four-potential also invariant.

In fact, it can be shown quite generally that light cones remain invariant under a change of spacetime geometry if and only if the corresponding line elements are conformal transforms of each other. Since normal physical influences propagate along light cones, it is natural to expect that their equations are scale invariant.

Since in a typical scale transformation the coordinate positions of particles remain unchanged, the number of particles counted in a specified three-dimensional coordinate volume must be unaltered, despite the proper three-dimensional volume being changed by Ω^3. Because $|\psi|^2$ measures particle probabilities per unit proper three-volume, the wave-function ψ in quantum mechanics is therefore required to transform from ψ to ψ^* according to

$$\psi^* = \Omega^{-3/2}\psi. \tag{17.3}$$

Moreover, the spatial coordinate distance between two particles remains unchanged whereas the proper distance is altered by Ω. Since the Compton wavelength m^{-1} of some standard particle measures the latter, it is necessary for m to transform to m^* according to

$$m^* = \Omega^{-1}m. \tag{17.4}$$

The number of Compton wavelengths separating two particles then remains the same. This inference can be tested strictly by considering the behavior of the Dirac equation

$$i\gamma^k \frac{\partial \psi}{\partial x^k} + m\psi = 0, \tag{17.5}$$

under the transformations (17.1), (17.3) and (17.4). To this end it is necessary firstly to generalize (17.5) to Riemannian space, writing

$$i\gamma^k \psi_{;k} + m\psi = 0 \tag{17.6}$$

in place of (17.5), with non-Euclidean terms entering into the covariant derivative of ψ (cf. ref. 1, p. 149). A demonstration of the invariance of the Dirac equation, i.e. the transformation to

$$i\gamma^{*k} \psi_{;k}^* + m^*\psi^* = 0 \tag{17.7}$$

can then be given (ref. 1, p. 254).

With quantum mechanics invariant to (17.1), subject to (17.3) and (17.4), and with the electromagnetic field also invariant, quantum electrodynamics is invariant. More recent developments in physics concerned with abstract particle spaces are also considered to be scale invariant. So why should gravitation be the only aspect of physics that is not? We think an important step towards a better understanding of cosmology is to remedy this deficiency, which we shall now proceed to do.

It is necessary to begin by finding an action $\mathcal{A}$ that is unaffected in its value by a scale transformation. The second term on the right-hand side of (14.2) satisfies this requirement. For a set of particles $a, b, \ldots$ of masses $m_a, m_b, \ldots$ the form of the Lagrangian density, $\mathcal{L}_{\text{phys}}$, usually considered in gravitational theory is

$$\sum_{a,b,\ldots} \int \frac{\delta_4(X, A)}{\sqrt{-g(A)}} m_a(A) da, \tag{17.8}$$

where the possibility of the particle masses varying with the spacetime position requires the mass $m_a(A)$ of particle a be able to vary with the point A on its path, and similarly for the other particles. Hence the second term on the right-hand side of (14.2) is

$$-\sum_a \int m_a(A) da. \tag{17.9}$$

With $da^* = \Omega da$ and $m_a^* = \Omega^{-1} m_a$ it is clear that (17.9) is invariant with respect to a conformal (scale) transformation. On the other hand it can be shown that (ref. 1, p. 28)

$$R^* = \Omega^{-2}(R + 6\Omega^{-1} \square \Omega) \tag{17.10}$$

and it is equally clear that the first term on the right-hand side of (14.2) is not invariant, the lack of invariance in general relativity coming from this first term. Since one can already remark on the artificial appearance of (14.2), in its strange combination of physical and geometrical quantities, it is surely here that a change must be made. Investigation shows that an attempt to replace the geometrical integral in (14.2) with some other geometrical quantity does not succeed, leaving

$$\mathcal{A} = -\sum_a \int m_a(A) da \tag{17.11}$$

as seemingly the only possibility, a possibility that is startling in its simplicity provided we define $m_a(A)$ as a direct particle field entirely related to its source points, just as an electromagnetic field can be looked upon as entirely related to its sources, i.e. charged particles. We will shortly elaborate on this concept further.

The gravitational equations are to be obtained in the usual way, by making a slight change of the metric tensor, $g_{ik} \to g_{ik} + \delta g_{ik}$, in a general four-dimensional volume with $\delta g_{ik} = 0$ on the boundary of the volume. The outcome can be written in the form

$$\delta \mathcal{A} = -\frac{1}{2} \int [T^{ik} + (?)] \delta g_{ik} \sqrt{-g} d^4 x, \tag{17.12}$$

where the energy-momentum tensor T^{ik} has its usual form, viz.

$$T^{ik}(X) = \sum_a \int m_a(A) \frac{\delta_4(X, A)}{\sqrt{-g(A)}} \frac{da^i}{da} \frac{da^k}{da} da, \tag{17.13}$$

and (?) is a tensor whose form is determined by the properties of the masses of the particles, by the variations of $m_a(X), m_b(X), \dots$ with respect to space-time position. The gravitational equations following from the principle of stationary action, $\delta \mathcal{A} = 0$ for all δg_{ik}, are given by

$$T^{ik} + (?) = 0, \tag{17.14}$$

in which the quantity denoted by (?) has still to be determined explicitly. To this end we now consider the direct particle concept in detail. Choose a 'mass field' $M(X)$ to satisfy

$$\Box_X M(X) + \frac{1}{6} R M(X) = \sum_a \int \frac{\delta_4(X, A)}{\sqrt{-g(A)}} da. \tag{17.15}$$

The reason for this combination on the left-hand side will shortly become clear. Equation (17.15) has both advanced and retarded solutions to which can be added any solution of the homogeneous equation, which has the right-hand side of (17.15) equal to zero. We particularize an advanced solution $M^{\text{adv}}(X)$ and a retarded solution $M^{\text{ret}}(X)$ in the following way. $M^{\text{ret}}(X)$ is to be the so-called fundamental solution in the flat spacetime limit[2]. This removes for $M^{\text{ret}}(X)$ the ambiguity that would obviously arise from the homogeneous wave equation. The corresponding ambiguity for $M^{\text{adv}}(X)$ is removed by the physical requirement that fields without sources are to be zero. Since

$$\Box[M^{\text{adv}} - M^{\text{ret}}] + \frac{1}{6} R[M^{\text{adv}} - M^{\text{ret}}] = 0, \tag{17.16}$$

the immediate consequences of this boundary condition is that $M^{\text{adv}} - M^{\text{ret}}$, being without sources, must be zero, so that

$$M^{\text{adv}}(X) = M^{\text{ret}}(X) = M(X), \text{ say}. \tag{17.17}$$

We call such 'fields' direct particle fields, as they do not have any extra degrees of freedom beyond those 'vested' in their source particles.

The gravitational equations are now obtained by putting

$$m_a(A) = M(A), m_b(B) = M(B), \ldots.$$ (17.18)

The variation of spacetime geometry alters the fundamental solution of (17.15) and leads to a variation of $M(A)$ and hence of the action (17.11). The tensor (?) can then be determined and it can also be shown that in a conformal transformation the mass field $M(X)$ transforms as in (17.4),

$$M^*(X) = \Omega^{-1}(X)M(X),$$ (17.19)

because it satisfies the wave equation (17.15) (cf. ref. 1, p. 111). The outcome (loc. cit., 112 *et seq.*) is

$$K\left(R_{ik} - \frac{1}{2}g_{ik}R\right) = -T_{ik} + M_iM_k - \frac{1}{2}g_{ik}g^{pq}M_pM_q + g_{ik}\Box K - K_{;ik},$$ (17.20)

where

$$K = \frac{1}{6}M^2.$$ (17.21)

These gravitational equations are scale invariant thanks to the combination of terms in the left-hand side of (17.15). It may seem curious that from a simpler beginning, using (17.9) for the action rather than (14.2), the outcome is more complicated, but this seems to be a characteristic of the physical laws. As the laws are improved they become simpler and more elegant in their initial statement but more complicated in their consequences.

Now make the scale change

$$\Omega(X) = M(X)/\tilde{m}_0,$$ (17.22)

where $\tilde{m}_0$ is a constant with the dimensionality of $M(X)$. After the scale change the particle masses simply become $\tilde{m}_0$ everywhere and in terms of transformed masses the derivative terms drop out of the gravitational equations. And defining the gravitational constant G by

$$G = \frac{3}{4\pi\tilde{m}_0^2},$$ (17.23)

the equations (17.20) take the form of general relativity

$$R_{ik} - \frac{1}{2}g_{ik}R = -8\pi GT_{ik}.$$ (17.24)

It now becomes clear why the equations of general relativity are not scale invariant. They are the special form to which the scale invariant equations (17.20) reduce with respect to a particular scale, namely that in which particle masses are everywhere the same.

It is also clear that the transition from (17.20) to (17.24) is justified provided $\Omega(X) \neq 0$ and $\Omega(X) \neq \infty$. For example, if $M(X) = 0$ on a space-like hypersurface the above conformal transformation breaks down. It is because of the existence of such time-sections that the use of (17.24) leads to the (unphysical) conclusion of a spacetime singularity. It has been shown[1,3] that the various spacetime singularities like that in the big bang or in a black hole collapse arise because of the illegitimate use of (17.24) in place of (17.20).

Planck particles

It is easily seen from the wave equation (17.15) that $M(X)$ has dimensionality $(\text{length})^{-1}$, and so has $\tilde{m}_0$. Units are frequently used in particle physics for which both the speed of light c and Planck's constant $\hbar$ are unity, and in these units mass has dimensionality $(\text{length})^{-1}$. If we suppose these units apply to the above discussion then from (17.23)

$$\tilde{m}_0 = (3/4\pi G)^{1/2}, \tag{17.25}$$

which with $c = \hbar = 1$ is the mass of the Planck particle. This suggests that in a gravitational theory without other physical interactions the particles must be of mass as determined in (17.25), which in ordinary practical units is about 10^{-5} g, the empirically determined value of G being used. This conclusion can be supported at greater length.

The action (17.9) from which the theory is derived is a dimensionless number obtained by counting the number of units $\tilde{m}_0^{-1}$ there are along portions of the paths of all the particles within some chosen four-dimensional volume. Since this count is the same in all conformal frames, it does not matter that the frame used is the special one in (17.24). For such a count to have the basic significance that the theory requires it to have, it is necessary that the length unit $\tilde{m}_0^{-1}$ have basic significance. The significance is required to relate to the nature of the particles and so to quantum mechanics, as it does if we require $\tilde{m}_0^{-1}$ to be the Compton wavelength of the particles. There seems to be no other plausible way of developing the theory.

Let m_0 be the mass of the particles with respect to some practical unit and consider practical units also for time and length, so that neither c nor $\hbar$ is unity. Then the relation between m_0 and the Planck mass $\tilde{m}_0$ is

$$\tilde{m}_0 = m_0 c / \hbar. \tag{17.26}$$

The gravitational equations (17.20) are

$$\frac{1}{6} \tilde{m}_0^2 (R^{ik} - \frac{1}{2} g^{ik} R) = - \sum_a \tilde{m}_0 \int \frac{\delta_4(X, A)}{\sqrt{-g(A)}} \frac{da^i}{da} \frac{da^k}{da} da. \tag{17.27}$$

Using (17.26) to replace $\tilde{m}_0$ by m_0 we get

$$R^{ik} - \frac{1}{2} g^{ik} R = - \frac{6 \hbar}{c^3 m_0^2} \sum_a m_0 c^2 \int \frac{\delta_4(X, A)}{\sqrt{-g(A)}} \frac{da^i}{da} \frac{da^k}{da} da. \tag{17.28}$$

Identifying $8 \pi G / c^4$ with $6 \hbar / m_0^2 c^3$ these are the equations of general relativity, i.e. for

$$m_0 = \left(\frac{3 \hbar c}{4 \pi G} \right)^{1/2}. \tag{17.29}$$

When the empirically determined value of G is used in (17.29) this is indeed the mass of the Planck particle, the value 1.06×10^{-5} g quoted above.

Anomalous redshifts

One of the reasons for believing that these ideas are generally correct is that after some reformulation in accordance with QSSC they lead to a possible explanation of the cause of intrinsic redshifts. This potential for understanding the intrinsic redshift phenomenon is very satisfying, considering the formidable observational case for QSSC made in Chapter 16. There we showed that QSSC could account for the genesis of the microwave background, the form of the radio source counts, and the marked reduction in the number of objects, galaxies and QSOs beyond $z \sim 2.5$–3.

QSSC has a functional form

$$S(t) = \exp \frac{t}{P} \cdot F(\cos \frac{2 \pi t}{Q}), \quad P \gg Q, \tag{17.30}$$

for the universal scale factor $S(t)$, with $F(\cos \frac{2 \pi t}{Q})$ a periodic function of the time t and $\exp t / P$ being an expansion factor arising from the effect of matter creation at minimum phases of the oscillation. The zero of t can be taken so that the minimum phases occur at

$$t_s = \frac{1}{2} Q - s Q, \quad s = 0, +1, +2, \ldots. \tag{17.31}$$

The model is therefore characterized by two time parameters, the period Q of the oscillation and the longer time P of overall expansion, and a metric that for small k can be approximated, as in Chapter 16, by

$$ds^2 = dt^2 - S^2(t)\big[dr^2 + r^2(d\theta^2 + \sin\theta d\varphi^2)\big]. \tag{17.32}$$

Because the theory is conformally invariant the metric can be changed with physical invariance not only with respect to a coordinate change from r, θ, φ, t but also by electing to work with $\Omega^2(X)ds^2$ instead of ds^2. The mass scale is then multiplied by Ω^{-1}. A useful conformal frame is obtained from $\Omega = S^{-1}$,

$$\Omega^2 ds^2 = d\tau^2 - \big[dr^2 + r^2(d\theta^2 + \sin^2\theta d\varphi^2)\big], \tag{17.33}$$

in which $d\tau$ is defined by

$$d\tau = dt/S(t). \tag{17.34}$$

In this τ-frame, the mass scale varies with time as $S(t)$.

Nothing new is involved to this point. But if the hypothesis is made that the only terms to be included in $\sum_b$ in the calculation of the mass field $M(X)$,

$$\Box M(X) + \frac{1}{6}RM(X) = \sum_b \int \frac{\delta_4(x, B)}{\sqrt{-g(B)}}\, db \tag{17.35}$$

acting on particle a,

$$m_a(A) = M(A), \tag{17.36}$$

in (17.15) and (17.18) are those terms that existed at the birth of particle a (i.e. were then non-zero on the right-hand side of (17.35)), a radically new situation arises.

Using the τ-frame defined above, the contribution of particle b to $M(X)$ is simply $1/r$, and the times τ_s in the τ-frame corresponding to $t_s = (\frac{1}{2} - s)Q$ are given by

$$\tau_s = \int_{\frac{1}{2}Q}^{(\frac{1}{2}-s)Q} \frac{dt}{\exp(t/P)\, F(\cos 2\pi t/Q)} \tag{17.37}$$

$$= (\text{constant}) \times \sum_{n=0}^{s-1} \exp\frac{nQ}{P} = (\text{constant}) \times \left(1 - \exp\frac{sQ}{P}\right).$$

To calculate the mass m_s of particles created at τ_s, consider the past light cone from $\tau = 0$ to $\tau = \tau_s$. The structure of the wave-equation (17.35) with $R = 0$ in the τ-frame requires the contribution to m_s from particle b to propagate on this cone, which is given by

$$\tau = \tau_s - r, \ \tau < \tau_s. \tag{17.38}$$

Now write $\triangle_u m_s$ for the contribution to the mass scale at τ_s coming from particles $b, \ldots,$ that were created at τ_u. Then the mass scale m_s for particles created at τ_s is given by

$$m_s = \sum_{u>s}^{\infty} \triangle_u m_s. \tag{17.39}$$

Next we note that the particles created at τ_u in a homogeneous isotropic model form a homogeneous spatial distribution at $\tau > \tau_u$, with a constant density proportional to $\exp -3uQ/P$. Hence with a spatial volume factor proportional to $(\tau_s - \tau_u)^3$, and with a solution of the wave-equation proportional to $1/(\tau_s - \tau_u)$ we obtain

$$\triangle_u m_s = (\text{constant})(\tau_s - \tau_u)^2 \ \exp -3u\frac{Q}{P}. \tag{17.40}$$

From (17.37), (17.39) and (17.40) we therefore get

$$m_s = (\text{constant}) \sum_{u>s} \exp -3u\frac{Q}{P}\left(\exp\frac{uQ}{P} - \exp\frac{sQ}{P}\right)^2. \tag{17.41}$$

Whence writing $n = u - s$

$$m_s = (\text{constant}) \exp\frac{-sQ}{P} \sum_{n=1}^{\infty}\left(\exp\frac{nQ}{P} - 1\right)^2 \exp\frac{-3nQ}{P}, \tag{17.42}$$

which depends on s as $\exp -sQ/P$.

Now let samples of matter created at the minima $s = 1, 2, \ldots,$ be brought forward in time and compared in their mass scales to that of matter created at $s = 0$. Since masses remain constant in the τ-frame, the mass scales of such samples will evidently be in the ratios

$$1 : \exp\frac{-Q}{P} : \exp\frac{-2Q}{P} : \exp\frac{-3Q}{P} \ldots \tag{17.43}$$

These ratios compared at the most recent oscillatory minimum of the cosmological model, at $t = \frac{1}{2}Q$, $r = 0$, are unaffected by transformation back from the τ-frame to the t-frame. Accordingly it follows that the older the samples in such a comparison the smaller will the mass scale be, and the redshift of light emitted by that matter will be larger when compared to emissions from similar atoms in material created at $t = \frac{1}{2}Q$. The anomalous redshift $1 + z_i$, for matter created at t_s is given by

$$1 + z_i = \exp\frac{sQ}{P} \tag{17.44}$$

showing that $\triangle \log(1 + z) = $ constant. The scale invariant theory developed above thus has the potential to lead to an understanding of intrinsic redshifts, a phenomenon that is otherwise outside the range of gravitational theory. We do not claim here to have covered all possibilities. We have simply shown the direction in which progress can be made in a problem that is otherwise completely outside the range of conventional theory.

References

Chapter 17

1. Hoyle, F. & Narlikar, J.V. 1974, *Action at a Distance in Physics and Cosmology* (San Francisco, W.H. Freeman).
2. Courant, R. & Hilbert , D. 1962, *Methods of Mathematical Physics* (New York, Interscience).
3. Khembavi, A.K. 1978, *M.N.R.A.S.*, **185**, 807.

18

Creation centers and black holes

Preliminaries

Setting aside the question of intrinsic redshifts examined in the previous chapter, we return now to the line of reasoning started in Chapter 14 and continued to the end of Chapter 16. By the mid-1950s it had already been realized that by introducing a scalar field into cosmology, new models, outside those previously studied, could be obtained. The distinction lay between the rules of general relativity contained in the action formula (14.16) and the action (14.20) containing the scalar field $C(X)$. Otherwise the two forms of theory were handled mathematically in similar ways, with equations (14.22) and (14.23) instead of (13.12) and (13.13). Whereas the latter lead to the Friedmann models of big-bang cosmology, the former leads to a new class of models through the equation (14.24),

$$\Box_x C = f^{-1} n(X) = F(X), \text{ say.} \tag{18.1}$$

The constant f was a coupling constant appearing in the new action formula (14.20), and $F(X)$ was a so-called creation function whose form decided the model. As with (14.22), (14.23) for the time dependence of the scale factor $S(t)$ in the Robertson–Walker metric (13.1), so (18.1) was obtained by applying the principle of least action to (14.20).

The simplest case to study was that in which $F(X)$ was a constant, and it was this which led to the so-called steady-state model. However, quite generally, the creation of matter implied by non-zero $F(X)$ did not involve creation of energy. Thus the matter created in a small time interval added equally in magnitude to ρ and to the $-1/2 f \dot{C}^2$ in (14.23). There was no impulsive reaction on $\dot{S}^2$. But matter, once created, varied in density as in the big-bang models, i.e. proportionally to S^{-3}. Whereas it could be shown that the C-field dependent term in (14.23) varied as S^{-6}. Hence equal and opposite initial additions to the two terms on the right-hand side of (14.23), viz.

$$3.\frac{\dot{S}^2 + k}{S^2} = 8\pi G\left(\rho - \frac{1}{2}f\dot{C}^2\right) \qquad (18.2)$$

did not stay equal and opposite. The addition to the ρ term became larger in magnitude than the addition to the $\dot{C}^2$ term, yielding a new positive contribution to the right-hand side of (18.2), even though locally no energy had been created at all.

Using (18.2) it was possible to eliminate $(\dot{S}^2 + k)/S^2$ from (14.22), viz.

$$\frac{2\ddot{S}}{S} + \frac{\dot{S}^2 + k}{S^2} = 4\pi Gf\dot{C}^2, \qquad (18.3)$$

giving

$$\frac{3\ddot{S}}{S} = 4\pi G(-\rho + 2f\dot{C}^2). \qquad (18.4)$$

The negative density term on the right-hand side here was just the usual deceleration of the universe due to ordinary gravitation. But the positive $\dot{C}^2$ term gave an outward acceleration of the universe. It offered the possibility of explaining the expansion of the universe without the need for a big-bang origin.

So long as F continued as a constant in (18.1) the universe continued to expand, with the consequence that the k/S^2 term in (18.2) eventually became negligible, giving

$$\frac{3\dot{S}^2}{S^2} = 8\pi G\left(\rho - \frac{1}{2}f\dot{C}^2\right). \qquad (18.5)$$

This was the same resolution of the so-called flatness problem that in the 1980s became acclaimed in the inflationary models, except that it was already known by the early 1960s.

It was also well-known that, if creation were to stop, the universe could not fall back into a singularity. This was because (18.5) took the form

$$\frac{3\dot{S}^2}{S^2} = \frac{A}{S^3} - \frac{B}{S^6} \qquad (18.6)$$

with A,B both positive constants. Thus a model, such as was studied in Chapter 13 with $\dot{S}$ initially negative, could not fall into a singularity. The model would bounce with $\dot{S} = 0$ at some finite value of S, and then expand back out again.

We have described in Chapter 7 how the model was attacked, essentially from its inception in 1948, with a vigor that trespassed on the irrational. The claimed observational disproofs were all weak or wrong – the Stebbins–Whitford effect, the value of q_0, and the counting of radio sources. The most influential astronomers of the day took this hostile view, and their effect on the international community was so great that the emotional opposition to the theory effectively killed it. This has meant that further possibilities such as those studied in Chapter 15 have remained uninvestigated for many years. Thus progress has been slow, leaving us in a position as we come to the end of this book where decisive further steps, clearly required by the observations, can only be taken in a generally qualitative fashion. These are the steps underlying the origin of the function $F(X)$. The oscillatory model of Chapter 15 leads to what we have called the quasi-steady-state model. Here the driving agents are the delta functions on the right-hand sides of (15.1), (15.13) and (15.23). It is the sources of these delta functions that require a more penetrating understanding. Here we shall indicate the direction in which observation strongly suggests we should go. But since 30 years have been wasted we shall be into the twenty-first century before this can be done.

It is appropriate to add that the scalar field introduced into cosmology by the inflationists of the 1980s had none of the attractive features of the earlier scalar field. It only solved geometrical problems, for example the flatness problem, which had already been solved decades earlier. It was a field without sources and it had only a positive energy component. Thus it had to be put in arbitrarily at the beginning.

Creation events

A creation event consists of the appearance of a C-field contribution together with a positive energy contribution totaling to zero, at a spacetime point, or in a particular small spacetime volume. That is to say a term $A\delta(X - X_0)$, where $\delta(X - X_0)$ is the four-dimension delta function in the wave equation (18.1) for C, i.e.

$$\delta(X - X_0) = \delta(t - t_0)\delta^{(3)}(\mathbf{x} - \mathbf{x}_0). \tag{18.7}$$

If at a particular moment t_0 there are such events occurring approximately uniformly throughout three-dimensional space, then we can average their effects with respect to $\mathbf{x}_0$, leaving a delta function with respect to the time, $\delta(t - t_0)$ multiplying some constant, thereby arriving at the source terms for the models studied in Chapter 15, the terms in (15.1), (15.13) or (15.23). In our opinion, astrophysical observations point strongly to the points $\mathbf{x}_0$ at which such events occur as being located at the centers of galaxies. And spatial integration with respect to x_0 consists in adding the effects of a large number of galaxies, ultimately with homogeneity depending on the local scale at which averages are taken, as in the standard cosmological principle.

We now consider the line element

$$ds^2 = dt^2 - S^2(t)\left[\frac{dr^2}{1 - kr^2} + r^2(d\theta^2 + \sin^2 d\varphi^2)\right] \qquad (18.8)$$

to apply to the region of the creation event, with the scale factor $S(t)$ applying locally, and satisfying equations similar to (18.2), (18.3) and (18.4). With the same expansionary tendency due to the $f\dot{C}^2$ in (18.4) as before, the locality of the creation region tends to expand.

Following creation events, the C-field particles obey the wave equation $\Box C = 0$, and are consequently massless, whereas the positive component consists of radiation and matter, produced by decays that are similar to those invoked in the early universe of big-bang cosmology. So that with the exception of neutrinos, which escape comparatively freely, the positive component is a hot mass of material possessing inertia and not able therefore to escape freely as the C-field particles do. Thus towards the outer regions of an expanding creation event there is a strong preferential tendency for the negative energy component to run ahead of the positive component, leaving an interior with an excess of positive energy and therefore building towards a black hole condition. *It is this preferential escape of the negative energy component that in our view produces the near-black holes found at the centers of galaxies.*

Because the center of mass of a uniform spherical cloud of gas, or of an axially symmetric uniform cloud, is at its geometrical center, the view is commonly held that gravitation will cause a dense central core to be formed. But actually the one place in such clouds where there is no centrally directed gravitational force is the center.

It is the same old story which repeatedly causes havoc throughout science. Because observation shows near-black holes to be present at the center of galaxies, the community argues inferentially that the concentration of mass at the center must happen 'somehow', with 'somehow' taken to mean a trivial falling together of the interior regions of the cloud, rather than by a process at present outside the range of current thinking. The inability to include what is so far unknown within the range of the possible is responsible repeatedly for wrong turnings being taken in cosmology, as well of course as in all other walks of life. Instead of new concepts being permitted to form gradually over the years, the 'happen somehow' syndrome blocks off the way to progress, leaving the community with no way forward except through senescence and replacement.

There is good evidence to show that mass concentrations from $\sim 10^6 M_\odot$ up to $\sim 10^{10} M_\odot$ actually exist at the centers of galaxies (cf. Chapter 20). We can see now how from creation events occurring, these concentrations can have been formed, without having to argue that extended clouds of gas or stars somehow manage to fall together. The concentrations occur because of the preferential escape of the negative energy component from such events.

Communication through the interior of such an object requires a proper time in conventional units of $2GM/c^3 \simeq 10^{-5} M$ s, where M is in solar units. If signaled outwards to the exterior of an object, this time is magnified by gravitational dilatation, by a factor

$$\gamma \simeq \left(1 - \frac{2GM}{c^2 R}\right)^{-1/2}, \tag{18.9}$$

magnified to $\sim 10^{-5} M\gamma$ s, which can be written as

$$\sim 10^{-5} M \left(1 - \frac{3M}{R}\right)^{-1/2}, \tag{18.10}$$

where R is now the scale of the object in kilometers, M is the positive mass in solar units, and (18.10) is measured in seconds. Strict black hole closure would require $\gamma \to \infty$, ultimately making a sequence of events taking place inside the object impossible for an external observer to detect. Sequential detection of events occurring in near-black holes remains possible, however, until the external time spacing given by (18.10) increases to a cosmological value, $\sim 10^{17}$ to 10^{18} s.

Events implying violence in the interiors of near-black holes are observed with apparent time separations of 10^{12}–10^{14} s. Radio maps of the nearby galaxy NGC 5128, for example, show evidence of repeated nuclear explosions which have expanded around the galaxy on a scale of a few hundred kiloparsecs. The scale of separation of the X-ray QSOs which appear to have been ejected from their parent galaxies is 10–100 kpc. Since expansion seems to occur at speeds of $\sim 0.1c$, the external time scale is typically $\sim 10^{14}$ s, permitting $\gamma \approx 10^9$–10^{10} when $M \approx 10^{10} M_\odot$.

A γ-value as high as this implies a strong gravitational inhibition on the escape of the C-field from the interior of a near-black hole, leading to high negative pressure values becoming established in the interior, with the consequence that break-up modes occur – referred to observationally as violent events – of which the ejection of QSOs is by now a well-observed phenomenon, as we have already discussed in Chapter 11. On this view, QSOs are simply fragments of the original near-black hole forced out of galaxies into surrounding space, where they themselves can become the centers of new galaxies.

Thus the mystery of how a near-black hole comes to be present at the center of a galaxy is a mystery only because the question has been posed the wrong way round. The central near-black hole is not secondary to the galaxy. It is the galaxy that is secondary to the central near-black hole. In the remainder of this chapter we shall examine the consequences of this idea, ending the present section with a brief recapitulation.

(i) In creation events positive energy forms, in particular radiation and matter, are in energy balance with the negative energy C-field.

(ii) The C-field escapes more readily from the creation region than the positive forms, causing an interior excess of the positive forms.

(iii) A gravitational barrier becomes established due to the preferential escape of the negative C-field, which with increasing time approaches more closely to a black-hole condition. However, from the point of view of an exterior observer the strict black-hole condition is never reached.

(iv) The gravitational barrier eventually impedes the escape of the negative C-field, since the latter obeys the wave equation.

(v) For near-black holes with masses of galactic order, events associated with them can be observed sequentially on a cosmological time scale until gravitational barriers γ become of order 10^{10}.

(vi) The mass of a near-black hole is determined by the excess of positive energy components over negative. The total amount in magnitude of either positive or negative taken separately can much exceed their difference.

The external condition on near-black holes

Dynamic disturbances to near-black holes can lead to the emergence of positive energy components as well as negative ones, but we think only in terms of violent events, and of comparatively minor emissions. By no means is this sufficient to account for the observed baryonic content of the universe. However, the baryonic content present in stars and galaxies is no more than a fraction, a few times 10^{-2} of the so-called closure density. The magnitude of the universal C-field energy density is required to be of the order of the closure density, as is the positive contribution of the near-black holes. Thus we have a new candidate for 'missing mass', near-black holes, which we therefore require to be a common phenomenon with a smoothed universal mass density a hundred times larger than the smoothed density of galaxies. This is certainly a major change of perception, but it readily explains why attempts in other directions to trace the origin of missing mass have to this point met with little success (cf. discussion in Chapter 20).

So how does the few percent of the positive energy, present as the baryonic content in galaxies, come free in a remarkably quiescent manner from the large γ barriers present in near-black holes? Turn back now to Fig. 11.15 where we showed the position angles of the jet of the massive galaxy M87, directed exactly towards the major galaxy M84 in the Virgo cluster. If we are to believe our eyes when we look at this figure, and not believe that this is an accidental projection, the implication is that an object was forced out of M87 in the direction of the present jet. The object traveled some distance and then disgorged the material which became M84. How did it do this? Not with immense violence apparently, but in the comparatively quiescent condition needed to permit galaxy formation. Or do we follow the widespread practice in cosmology of saying: 'Because this is beyond comprehension, forget about it.' ?

It is likely that near-black holes exist in a nested condition, one inside another, with the largest scale of all on a cosmological scale, $\sim 10^{28}$ cm at the present time, but at other scales through the oscillatory phases of QSSC. Thus at maximum phase the scale is $\sim 3 \times 10^{28}$ cm and at minimum phase it is $\sim 10^{27}$ cm. Since the mass M associated with a scale R cm is $R/3 \times 10^5$ in solar units, the near-black-hole mass associated with R varies from $\sim 10^{23} M_\odot$ at maximum to $\sim 3 \times 10^{21} M_\odot$ at minimum.

Now the average positive mass density inside a black hole of mass M in solar units is $\sim 2 \times 10^{16} M^{-2} \mathrm{g\ cm}^{-3}$, higher than the universal C-field density by $(3 \times 10^{21}/M)^2$ when the latter is considered at a cosmological oscillatory minimum. The C-field energy being negative in sign, its incidence on a near-black hole has a disruptive tendency. However, the magnitude of $(3 \times 10^{21}/M)^2$ for M of a galactic order is so large that the tendency appears at first sight to be negligible. But C-field particles of cosmological origin falling into a near-black hole gain in energy by the barrier factor γ. There is also a gain with the time scale R/c for an interior observer extending to $\gamma R/c$ when exterior infall is considered. Thus we have γ^2 to make up for the apparently large factor $(3 \times 10^{21}/M)^2$, which for

$$M > 3 \times 10^{21} \gamma^{-1} \tag{18.11}$$

it will do. Thus for $\gamma \simeq 10^{10}$ near-black holes with masses $> \sim 3 \times 10^{11} M_\odot$ receive enough incident C-field energy to become expanded apart, with their positive baryonic component released into the outside world. This effect is extremely gentle since it requires the contraction of the universe to minimum phase, $\sim 10^{10}$ years or more, to bring it about. Some light is therefore thrown on Fig. 11.15. And the epochs of emergence of much of the new baryonic material into the external world are at, or close to, the minima of QSSC, as we have hitherto taken them to be. Thus while there are comparatively rare examples of this process at the present epoch it was much more common at an earlier epoch.

The physical interpretation of the constant f

The creation of equal and opposite positive and negative components does not show immediately in the external world. So long as equal and opposite components remain confined inside the γ barrier of a near-black hole no external effects are seen. But let the two separate, either in violent events or in the more placid process just considered in the previous section, and the consequences cosmologically are those discussed in Chapter 15.

Taking the positive component to be in particles of mass m, in Chapter 15 we had the cosmological effects of creation contained in the equation

$$f^{-1}n(X) = 2m\delta(t - t_0). \tag{18.12}$$

The meaning of (18.12) is that at time t_0 there is a uniform release $2mf$ per unit spatial volume of compensating $\pm$ particles of energy magnitude m. However, we now have the difference that (18.12) is not the primary creation process itself. Rather it is the separation of $\pm$ pairs from near-black holes. But this leads as in Chapter 15 to the cosmological equation

$$\frac{\dot{S}^2}{S^2} = \frac{16\pi G}{3} fm^2 \frac{S_0^3}{S^3} \left(1 - \frac{S_0^3}{S^3}\right). \tag{18.13}$$

And if $3Hfm$ is interpreted as the cosmologically smoothed average rate per unit spatial volume at which near-black holes disgorge such energy-compensating pairs then we would obtain the steady-state case of Chapter 15, viz.

$$S \propto \exp Ht, \quad 3H^2 = 4\pi Gfm^2. \tag{18.14}$$

Hence the constant f of Chapter 15 is seen to be related to, and determined by, the rate at which $\pm$ energy-compensating pairs are expelled by near-black holes into the wider universe.

Suppose a particular value of H has become widely established and that in a particular region (say of scale ~ 1000 Mpc) the production rate increases for a while above $3Hfm = 9H^3/4\pi Gm$. Then such a region behaves like the $k = +1$ oscillatory case of Chapter 15, while a deficit of production leads to a cosmological hole with $k = -1$, thus producing both structure and a straightforward interpretation of the constant f.

Interior and exterior time scales

Where small rapidly varying emissions are concerned, an exterior phenomenon is to be preferred to an interior one, agreeing with the popular accretion disk–black hole picture. The above time scale, $10^{-5}M\gamma$ s with M in solar units, is far too long however, because of the γ factor to be associated with such rapidly varying emissions. Exterior emissions lack the γ factor. Thus for $M = 10^8 M_\odot$ as an example, exterior emission can be only a fraction of an hour, agreeing very well with the most rapidly varying observed emissions.

Characteristically, the magnitudes of rapid energy emission are only a few hundred solar luminosities, not remotely in the range of major outbursts. We saw in Chapter 12 that the latter run in the case of radio sources above 10^{60} erg in total, almost surely implying an emission rate far above anything possible for the exterior accretion disk model. Outbursts in the latter case appear to be connected with a dynamical instability of the entire near-black-hole structure.

As we have discussed in an earlier chapter, there is a wide range of astrophysical observations relating to galactic nuclei. What is being said now corresponds very well, we feel, to this observational evidence.

The delta functions by which creation events have been described, as in (18.12), relate ultimately to events in which positive and negative energy pairs, C-field particles for $-$, and a positive matter form of particle for $+$, are considered to appear at a particular time in a particular place. Throughout Chapter 15 and succeeding discussions, we have kept carefully to the constraints of energy and momentum conservation. But this is a necessary requirement, not a sufficient one. In effect we have argued that when there can be conservation there will be creation. More is really required, however. What we have to ask determines the particular time t_0 of creation in the delta functions appearing in the models of Chapter 15, the delta functions appearing explicitly in (15.1), (15.13) and (15.23).

The improved discussion of the relativity theory given in Chapter 17 suggested that particle masses of the order of the Planck mass in (17.29) were involved at the level we are now attempting to discuss, $m_0 \simeq (3\hbar c/4\pi G)^{1/2}$ in conventional units. Stated more meaningfully, a length of the order of $G^{-1/2}$ was involved, both for the positive mass m_0 and for the magnitude of the corresponding C-particle. It is the property of a particle of mass m_0 that its Compton wavelength and its gravitational radius are the same, implying a fundamental relationship to the spacetime structure.

Particle energy magnitudes for the C-field will, however, fall by γ^{-1} when there is escape from a near-black hole of gravitational barrier γ. But the lost energy will be regained if a particle should subsequently be incident on an object of the same barrier factor. Or the energy may be exceeded, perhaps by a large factor, if the particle falls through a larger barrier than the one from which it escaped. In the case of nested black holes, it is possible for a particle even to pick-up a chain of γ-factors, attaining energies above $(3\hbar c/4\pi G)^{1/2}$, considerably above, making for a serious disruption of the spacetime structure. It is this which we take to produce a creation event.

Of course everybody would like to see a detailed physical theory of such an event. In our view this may take some time to develop. But here we have outlined the direction that the observations are indicating.

It is widely agreed that matter cannot be infinitely old. In big-bang cosmology matter has a finite age because the whole universe has a finite age. Conceptually, particles are associated with paths in spacetime, which we can write as $x^i(q)$, $i = 1, 2, 3, 4$, namely as continuously varying functions of some parameter q, x^i being the spacetime coordinates along the path. Although physical quantities such as mass or some abstract group also enter into the description of a 'particle', some modification of spacetime itself is required in order that meaning can be attached to saying a particle of such and such a kind is to be found in the neighborhood of a spacetime point A, but not in the neighborhood of B. Physics generally has been extraordinarily reluctant to specify exactly what it is that decides whether in a particular neighborhood we have a particle or we don't.

The word creation as it is normally used in physics is really a misnomer. It refers only to situations in which the time coordinate $x^4(q)$ does not increase (or decrease) entirely monotonically with respect to q, i.e. the equation $x^4(q) = a$, where a is some constant multivalued with respect to q. Pair creation and its inverse, pair annihilation, are processes of this type. In the classical (big bang) cosmology strict pair creation and annihilation is assumed except for the very early universe, where to avoid a universe entirely of radiation a deviation is allowed in just the amount needed to give the required baryon-to-photon ratio η.

Our approach throughout has been to argue that modern observations in cosmology force a breakout from the strict restraining mold. We have argued that particles are genuinely created in conditions where their Compton wavelengths are comparable to or less than their gravitational radii, at an energy

$\sim 10^{19}$ GeV, which energy is related to the gravitational constant by $(3c\,\hbar/4\pi G)^{1/2}$ in conventional symbols. It is noteworthy that big-bang cosmology never reaches this energy, at any rate in its best-known formulation. Thus the highest energy reached in the inflationary model discussed in Chapter 14 was the quantity σ in the Coleman–Weinberg potential, with $\sigma \simeq 2 \times 10^{15}$ GeV, when the universe disappears without the critical energy threshold for real particle creation ever being reached.

A first step towards a quantitative theory of particle creation was taken in Chapter 14, where an action-principle development was given leading to dynamical equations for the universe, at a level comparable to general relativity, that is to say, at a statistical level representing averages over many particles, not at the level of a field-theoretical treatment of gravity, which development still remains in limbo.

Energy was conserved in this approach, a negative component carried by the scalar field $C(X)$ accompanying the creation of positive energy forms. This field $C(X)$ exerts negative pressure. Generated locally with high intensity, the effects are in accordance with astrophysical observations that have been discussed in the present chapter.

The correspondences between theoretical expectations and the observations appear to us to be good. It is always a question, however, when such correspondences should be given an inferential status, i.e. when we can say that because certain conceptual ideas lead to deductions that agree with what we see in the world, the ideas in question may inferentially be claimed to have some relation to reality. Experience shows that the readiness or otherwise with which scientists make this inversion depends on how they think, and also on the proselytizing zeal of the front-runners. When the leaders become strongly committed, many are likely to follow, as has been the case with the big bang. The same appears to be true for a swarm of bees, when the bees are searching for a new location for their hive. Except that bees dance to order to make a decision, whereas cosmologists write books, publish papers, and sometimes actively court the press.

We turn in the next chapter to the current state of the observations.

19

Modern observations of faint galaxies and related objects

Introduction

So far we have covered four major areas.

1. The history of modern cosmology as we see it from the earliest days up to the 1990s.
2. The key observations and the way they have led to the development of the belief in the hot big-bang model.
3. The observational discoveries stemming from the 1960s which lead us to believe that the simple standard model is not correct.
4. Our development of a modified steady-state model which we believe is a first step towards understanding all of the observed phenomena.

We turn now to the wealth of observations which have started to pour in during the past decade or so. These observations, correctly interpreted, may ultimately determine whether or not the current popular view of the universe involving a hot big-bang beginning and evolution of the galaxy population down to the present can be maintained, or whether a different approach is indicated.

It is our view that at this stage the data do not support either the standard model unambiguously nor clearly indicate that the new direction we have proposed is correct. We shall try to describe what is currently being detected and attempt to separate what is actually observed, from what is often stated to be what is observed. We give some examples of these ambiguities later.

The driver for all of the work has been the increase in the number of survey telescopes, large optical telescopes, and radio telescopes now available. Since the late 1970s many new several-meter class optical telescopes have been put into operation, culminating recently with the two 10-meter Keck telescopes in Hawaii and the Hubble Space Telescope. More very large telescopes with 8-meter (Gemini), 10-meter (Japan), and 16-meter (ESO) apertures are under construction. In the same period very sensitive radio telescopes have been used extensively for extragalactic studies, the sky has been surveyed at infrared wavelengths out to $\sim 10\,\mu$m, and several X-ray telescopes have been used to survey the sky at kiloelectronvolt energies.

These major instruments together with sensitive detectors have led to an explosion in the collection of data on extragalactic objects. In this chapter we discuss the physical properties of these extragalactic objects. In the next chapter we shall discuss the large-scale structure defined by the arrangement of these objects.

Surveys at many wavelengths

A photographic record in the red and the blue of the whole sky reachable from Palomar (down to declination $-33°$) was first made using the 48-inch Palomar Schmidt telescope in the 1950s. The Palomar Sky Survey was augmented in the 1970s by the corresponding Schmidt Survey made in the southern hemisphere by ESO in Chile, and by the UK Schmidt telescope in Australia. The ESO-SERC Sky Survey covers the sky in red and blue light from declination $-17°$ to the south pole. These all-sky surveys are photographic records of all of the objects in the sky that are emitting light in optical wavelengths down to apparent magnitude $19^m\!.5$ (blue) and 21^m (red). While many of the objects in the survey are stars and nebulae in our own Galaxy, the majority are faint galaxies (perhaps 10^{10} objects). Surveys of the sky made at all other wavelengths so far contain far fewer objects than this.

If we restrict ourselves to the extragalactic objects which are bright enough to be detected in current surveys at radio wavelengths (centimeters to meters in general), infrared (10–100 μm) or at X-ray energies ($\sim$ 2–100 keV) and γ-ray energies ($\sim$100 keV–100 MeV), only a comparatively luminous subset of sources in each wavelength or energy range has so far been found and cataloged. The objects that are found in surveys at far infrared wavelengths in the range 10–100 μm are emitting low energy thermal photons coming either from cool stars or from heated dust clouds, in our own Galaxy or external galaxies. We believe that the extragalactic radio sources are mostly non-thermal sources emitting power by the incoherent synchrotron process.

Optical and IR sources

The radiation process that we understand best is the thermal emission of optical radiation from stars in galaxies. The bulk of these sources are simply populations of stars generating energy by hydrogen burning where most of the emergent photons are emitted at energies near 1 eV. The basic constituents of galaxies as seen from their radiation are stars, diffuse atomic and molecular gas, and dust, and we know that where there is plenty of diffuse gas and dust, star formation occurs. If large amounts of star formation occur, large fluxes of ultraviolet and optical photons are generated. If this takes place in regions where there is much dust, large fluxes of infrared radiation from the heated dust will be emitted. The currently fashionable term used for galaxies where such regions of active star formation give rise to high luminosities in both optical and infrared wavelengths, is that they are *starburst galaxies*. Thus many of the very bright infrared sources are thought to be starburst galaxies – systems containing large numbers of hot stars and HII regions and dust. Measurements of the molecular lines show that in many of these galaxies there are very large and massive molecular gas clouds. In such galaxies hot stars are probably the ultimate energy sources. However, in other galaxies much of the IR flux may be coming directly from less massive cool stars.

The most extensive survey in the infrared has been made by the Infrared Astronomical Satellite (IRAS)[1] which surveyed over 96% of the sky with a completeness limit of 0.5 Jy at 12, 25 and 60 μm and 1.5 Jy at 100 μm.

The IRAS survey detected predominantly spiral and irregular galaxies, since ellipticals contain little gas and dust, and the survey is particularly sensitive to galaxies in which star formation is active since much of the far-infrared radiation is emitted by heated dust clouds. There are also a comparatively small number of objects in the IRAS catalogs with very large infrared fluxes whose underlying sources of optical and ultraviolet radiation may not be stars at all, but non-thermal sources. Probably both kinds of source are present.

Radio sources

We discussed in Chapter 7 the identification of the first extragalactic radio sources with optical galaxies. Identification of radio sources with optical objects has been carried on slowly since the 1960s. While by now 10^5–10^6 radio sources have been cataloged, only a small fraction of these have been optically identified with extragalactic objects, and an even smaller fraction of these have had redshifts measured so that we can say something about their physical properties. The only catalog of radio sources in which the optical identifications and redshifts are complete is the 3CR catalog[2] of sources with fluxes $\geq$ 9 Jy at a frequency of 159 MHz. There are about 300 extragalactic sources in this catalog, which covers about two-thirds, or about 25 000 square degrees, of the sky. Among these sources a small number are supernova remnants in our own Galaxy, and there are some comparatively nearby galaxies which are intrinsically weak sources, but the bulk consists of QSOs and comparatively powerful radio galaxies. The identifications made from this catalog and the corresponding catalogs in the southern hemisphere led to the early conclusion that the following objects give rise to the radio sources.

(a) Normal galaxies, which are very weak radio emitters with radio power $\sim 10^{-5}$–10^{-6} of that in the optical wavelengths.
(b) Bright elliptical galaxies, frequently the brightest central galaxies in clusters (cD galaxies), which give rise to double radio sources emitting at power level $\sim 10^{-3}$–10^{-1} the optical luminosities.
(c) Quasi-stellar objects (QSOs).
(d) High z radio galaxies (HZRG).

The sources in the fourth class (d) are similar in their radio structure to (b), but their optical emission is dominated by a blue continuum and strong broad emission lines. Apart from the QSOs, these are the sources with the largest redshifts. The prototype, though it has a small redshift, is Cygnus A. Many years after Cygnus A was identified, Osterbrock[3] managed to identify stellar absorption features in the continuum, showing that a galaxy of stars does underlie the optical source. Whether or not elliptical galaxies are in general present in the high-redshift radio sources is not known though it is generally assumed. Thus it is generally believed that in the HZRG we are seeing elliptical galaxies at an earlier state in their evolution. We have already discussed the present state of our understanding of this problem in Chapter 6.

The optical spectra of the HZRG are in some ways similar to QSO spectra, though the optical and infrared spectra are extended for $z \gtrsim 0.6$. It is almost certain that the energy associated with the event that gives rise to the radio source also makes a major contribution to the optical spectrum. It is from the spectrum associated with this event that we get the redshift.

Active galaxies and QSOs

As was described in Chapter 5 it was through radio astronomy that the QSOs and related objects were discovered. These objects are very rare compared with normal galaxies. Following their discovery as radio sources, the optical surveys first initiated by Sandage showed that the number of 'radio quiet' QSOs is about 100 times the number of 'radio loud' QSOs. Surveys over small areas of the sky have given surface densities up to about 50 per square degree at 21^m. Since their energy distribution and hence their colors do not define the QSOs uniquely, it is necessary to obtain spectra and hence redshifts to confirm the identifications. At present (1999) about 11 000 QSOs with measured redshifts are known. We show in Figs. 19.1, 19.2 and 19.3 the surface distribution over the sky of the QSOs with known redshifts when about 7300 QSOs were known. The very patchy distribution simply indicates the regions over which the identifications have been made.

QSOs are very rare compared with normal galaxies. Over the whole sky[4], the total number down to 21^m is $\sim 2 \times 10^6$, and for magnitudes fainter than 21^m, the surface density flattens off (Chapter 11). The redshift distribution of the QSOs is also remarkable; it is shown in Fig. 19.4. The numbers clearly fall off steeply beyond redshifts of 2.5–3. It is generally agreed that this is a real effect and not a matter of observational selection.

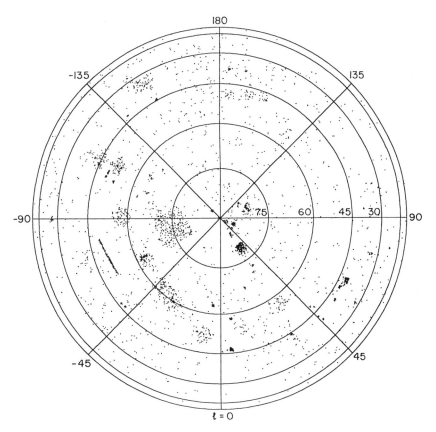

Fig. 19.1. This figure is taken from a paper by Hewitt and Burbidge[33] containing a catalog of QSOs. It shows the distribution of 7315 QSOs with measured redshifts on the sky in Galactic coordinates (Aitoff projection).

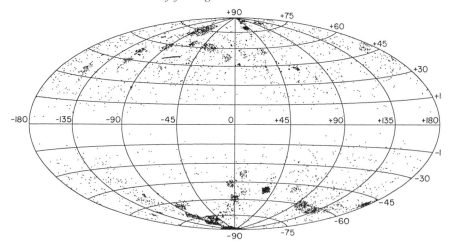

Fig. 19.2. The distribution of 7315 QSOs in Galactic coordinates viewed from the North Galactic Pole (center of plot is $b = 90°$) (ref. 33).

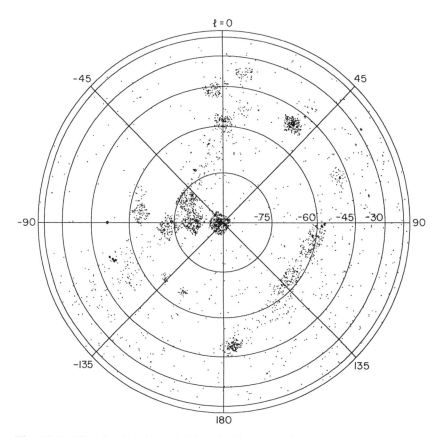

Fig. 19.3. The distribution of 7315 QSOs in Galactic coordinates viewed from the South Galactic Pole (center of plot is $b = 90°$) (ref. 33).

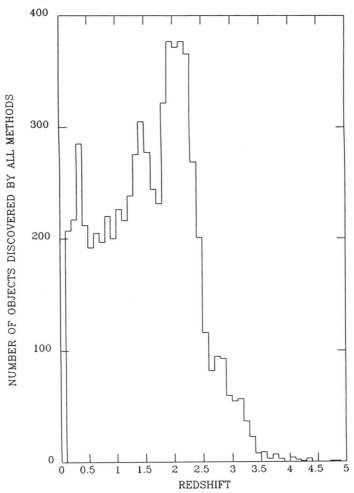

Fig. 19.4. Histogram showing the emission redshift distribution of 7315 QSOs in the catalog of Hewitt and Burbidge[33].

Closely related to QSOs, at least in their optical properties, are active galaxies, often called active galactic nuclei (AGN). What are the real distinctions between QSOs and AGN? Many argue that there is no distinction to be made. We believe that the excitation mechanisms giving rise to the line spectra are similar, and that the underlying continua are basically non-thermal in character, though there is a school of thought that argues that all of the optical characteristics may be due to stellar evolution in a dense star cluster. However, we believe that the term AGN should only be used when it is *known* that a galaxy of stars is present. To establish the presence of a galaxy of stars, it is necessary to (a) detect normal spiral structure or other morphological evidence characteristic of normal galaxies and (b) observe spectra which show the characteristic absorption features due to stars – CaII H and K, the G band etc. In most cases this has not been done. A detailed discussion of the evidence bearing on this question was given in Chapter 11.

There is one very rare class of AGN, known as BL Lac objects after the prototype BL Lac (originally thought to be a variable *star*). These are rapidly variable AGN at radio, optical and sometimes X-ray and γ-ray energies, with optical spectra which show no traces of emission lines, or only very weak emission lines, and absorption features such as the H and K lines of Ca II and the G band which are due to starlight. Studies show that a plot of the redshifts of these objects against their apparent magnitudes at minimum light give a good Hubble relation[5], demonstrating that they are giant elliptical galaxies with absolute magnitudes ~ -23, and a component of variable non-thermal radiation coming from their nuclei. Thus there is strong observational evidence that these AGN *do* lie in the nuclei of elliptical galaxies.

X-ray and γ-ray sources

The extra-galactic X-ray sources with energy emission in the range 1–100 keV fall into several categories.

(a) There are the comparatively close-by normal galaxies in which the X-rays are being emitted from a large number of close binary systems with one star a highly evolved object (neutron star or near-black hole). This is the situation in our own Galaxy. The total X-ray flux from such galaxies is very small, $\lesssim 10^{-4}$ of the optical luminosity of the galaxy.

(b) Investigations at X-ray wavelengths have led to the discovery of a new class of extended clouds of sources which are due to bremsstrahlung and line radiation from hot plasma at temperatures in the range $10^6 - 10^7$K. These sources fill large volumes in the central regions of galaxy clusters, showing that there is a large mass of hot gas in such clusters and groups. In most cases, the X-ray source is centred on a luminous central galaxy which is often a powerful non-thermal radio source. It is often difficult to separate the gas that is physically part of the outer halo of the central galaxy, from what is truly intra-cluster gas. The temperature of the gas is estimated from the X-ray spectrum. On the assumption that the gas is not heavily clumped, an estimate of the mass can be made. It has been concluded that masses comparable to or greater than the mass in visible galaxies are present. The X-ray luminosities of such sources are $\sim 10^{43}$–10^{44} erg s^{-1}.

The cooling time for such a hot low density gas is $\sim 10^8$–10^9 years. It was therefore proposed many years ago that the energy source is gravitational in origin and is due to the accretion and infall into the cluster of inter-galactic gas (for which there is no direct evidence). As the gas falls in, it gives up gravitational energy thus powering the X-ray source. The promotion of this explanation has led to an extensive literature on what are known as 'cooling flows', and practically all of the observers have uncritically accepted that this is the explanation of the X-ray sources[6,7]. Whether or not this is correct is simply not known. This is an attractive band-wagon idea which is easily fitted into the big-bang cosmology. An interesting alternative explanation that is not yet developed but which we find attractive is that the energy source (but not the hot gas) is the galaxy (which is often a powerful radio galaxy) that frequently lies at the center of the cluster. This model follows naturally from the explosive cosmogony which we have described.

(c) The third class of X-ray sources are the compact objects, QSOs and active galaxies. In the recent X-ray surveys many known QSOs have been identified as X-ray sources, while a number of compact X-ray sources have been shown to be previously undetected QSOs. Of particular interest are the discoveries of compact X-ray sources which appear to have been ejected from very closeby active galaxies such as NGC 4258, NGC 2639 and NGC 1068[8,9,10]. These were discussed in Chapter 11. All of these sources have so far turned out to be comparatively low-redshift QSOs which are quite faint at optical wavelengths. They provide further evidence that QSOs with very different redshifts from their parent galaxies are being created and ejected. Since these galaxies are very close to us with distances in the range 10–50 Mpc and the QSOs are faint, 18^m–20^m, this means that they have absolute magnitudes ~ -10 to ~ -14, and are very much fainter than the QSOs found by the normal surveys described earlier. If such faint QSOs are ejected from all active spiral galaxies a very large population is indicated.

Another property of the X-ray QSOs and AGN is that they sometimes show rapid variability in flux. In some of the classical Seyfert nuclei like NGC 4151, very rapid X-ray variability is seen. In that very small class of QSOs which are already known to be rapidly variable in optical and radio frequencies, and in the BL Lac objects — these are together often called 'blazars' — rapid variability is seen up to the highest photon energies measured, i.e. γ-ray energies. In fact the surveys at γ-ray energies show that the only sources so far identified with previously known objects are the very rapidly variable QSOs and BL Lac objects. Fewer than 100 of these have so far been detected[11].

Faint galaxies: time scales

We now turn to a detailed discussion of the population of optically faint galaxies which are the largest visible constituents of the universe. It is reasonable to assume that galaxies of all morphological types are present, with a mix that is similar to the sample that we can study in detail around us.

Since the basic physics of stellar evolution is well understood, the goal for many years has been to understand how large aggregates of stars have evolved since they were first formed. Two time parameters are involved. The first is the age of each galaxy, defined as the time that has elapsed since the first generation of stars formed, and the galaxy began to radiate thermonuclear energy. This is an unknown parameter for nearly all galaxies. Only for our own Galaxy and its companions can we make a hard estimate based on the age dating of star clusters, or the age of the elements determined from radioactivity. As was discussed in Chapter 4 the oldest visible cluster in our Galaxy has an age $\simeq 15$ Gyr with an uncertainty of perhaps ± 2 Gyr.

Only if we suppose that the initial mass function for stars is a universal quantity and that all galaxies have evolved in the same way as our Galaxy are we justified in supposing that other galaxies have the same ages. So already there is a large uncertainty. The second time parameter is the age of a galaxy with a redshift z as we can measure it. Its age will be given by the time since it formed minus the time it has taken for the light to travel to us from redshift z. This parameter depends on the cosmological model.

By studying samples of the large population of galaxies, QSOs, AGN and radio sources we can hope ultimately to find clues to the large scale and small scale cosmogony, i.e. how these systems formed and evolved within the framework of the correct cosmological model. It is reasonable to suppose that the galaxies of stars are the basic building blocks for everything that we can see. This may also be true for the component of dark baryonic matter for which there is some indirect observational evidence which we shall discuss later. Since it is generally believed that most of the visible mass in galaxies consists predominantly of stars with masses close to that of the Sun, the characteristic evolutionary time scale for them is the thermonuclear time scale for $1 M_\odot$ which is about 12 Gyr.

Radio sources, QSOs, and active nuclei all have high luminosities, and characteristic time scales much shorter than this. These latter time scales are probably $\sim 10^8$ years or less. Since many of these high luminosity objects are identified with optical galaxies, it is reasonable to suppose that they are all generated in the galaxies in some way. There are several possibilities. They may be transitory stages through which all galaxies pass. For example,

Seyfert galaxies make up about 1% of spiral galaxies in the same volume. Thus it has been argued that all spiral galaxies go through this phase, so that the time scale for Seyfert activity is $10^{10} \times 10^{-2} = 10^8$ years. Recently, based on the observation that many nearby spirals show evidence of the Seyfert phenomenon (weak broad emission spectra in very tiny nuclei) it has been argued that the Seyfert phenomenon is widespread. From an energetic standpoint this makes little difference to the frequency argument given above, since even over 10^{10} years the non-thermal energy emitted by a tiny Seyfert nucleus will not be important.

Alternatively the activity may be restricted to a very small fraction of spiral galaxies where it goes on all of the time because of some intrinsic property of that galaxy. Similar frequency arguments can be made for the QSOs. Are they the very dominant nuclei of galaxies? If they are, then because they are so rare, they must have lifetimes very short compared with the underlying galaxies, with large energies emitted over short times.

Are many, if not all, of them objects which are ejected from galaxies as some of the evidence discussed earlier suggests? Again, the time scales for ejection and travel away from the present galaxies must be much less than 10^{10} years. But in this case, there is also the likelihood that most of them have very low luminosities and have not been detected. Thus weakly emitting QSOs may be very numerous.

Similarly, the luminous IR galaxies either dominated by non-thermal sources of energy or thermonuclear energy from hot stars must have lifetimes in their luminous phases short compared with 10^{10} years. Do galaxies of old stars underly these systems? In general we simply do not know, though it is usually assumed that they do.

What about stellar systems with ages much greater than 10^{10} years? Stars with masses significantly below that of the sun, say $M \leq 0.5 M_\odot$, have evolutionary times $\gtrsim 50$ Gyr. They will be very faint and extremely difficult to detect but they may well exist. On the other hand, more massive stars with masses $\gg 1 M_\odot$ will evolve in times $\ll 10^{10}$ years, but their remnants will still exist as white dwarfs, neutron stars and near-black holes indefinitely. They will also be extremely difficult to observe. Thus within the framework of what we *really* know, it is perfectly reasonable to argue that much of the dark baryonic matter in the universe is in the form of very low mass stars whose evolutionary times are very long, and evolved stars – stellar remnants which no longer emit appreciable energy. Whole galaxies made up of such objects may well exist. This argument is only constrained if we insist that the universe has a finite age of the order of H_o^{-1}. As we have shown earlier, there is not a strong scientific case for this.

With this overall summary as a backdrop, we turn to the observed properties of faint galaxies, which are the best tool that we have for understanding the basic cosmogony.

Our problem can be likened to that faced by an observer standing in a small clearing in a vast forest of trees of all types, where the trees are widely separated so that they can be distinguished as individual trees as far as the eye can see. Is the forest of finite or infinite extent? We do not know. There are trees of various types, shapes and sizes – how did they begin? We do not know. Did they all start together or do they have a wide span of ages? We do not know. Can one species evolve into another? We do not know. Can one tree give birth to another? We do not know. Some trees appear to stand taller than others. Can these be used as standards to measure distances, i.e. are they all really of the same height? Apparently they are. Some trees are found in clumps. Why? We do not know. Were all the trees born together, or are they of many different ages? We do not know. Some trees spontaneously burn for short periods. Do all trees burn, or only certain kinds? We do not know.

The properties of faint galaxies

As was discussed in Chapter 6, the first driver to the study of the properties of faint galaxies as a function of redshifts was closely tied to the idea that it was necessary to determine the changes in the integrated properties which could be used to make 'evolutionary' corrections. The idea was that after these corrections were taken out it would be possible to determine observationally the deceleration parameter q_0 in the Hubble diagram. The work on the bright galaxies was first based on photographic surveys, and then CCDs were used to carry out fainter (often called deeper) surveys. This work has been summarized by Koo and Kron[12]. Even at the bright end in the range $16^m5 < B < 19^m$ there appeared to be differences between the galaxy counts $A(m)$ between northern and southern hemispheres[12].

The difference may be caused by a variety of factors involving the measuring machines, the software, and the detectors used for calibration, but it already shows the difficulty in obtaining believable results even for the brighter galaxies. Thus the plot of $A(m)$ against Δm shown in Fig. 19.5 must be approached with caution. It has been stressed that the steep slope of $\Delta \log A / \Delta m$ is not compatible with a model which contains no luminosity or color evolution[13,14].

Counts of bright galaxies in the blue have been carried out by many groups. The lines which are drawn on Fig. 19.5 give the predicted counts for the no-evolution model. It can be seen that they fit reasonably well.

b$_j$ magnitude

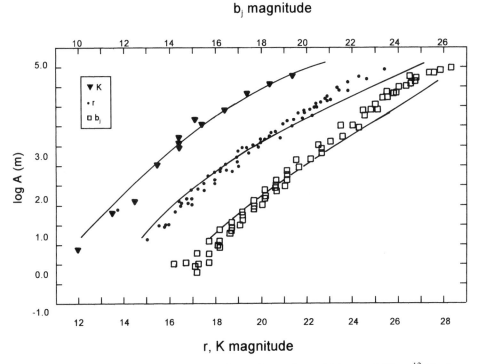

Fig. 19.5. This figure is taken from the review of Koo and Kron[12]. It shows differential galaxy counts $A(m)$ in number per square degree in the blue (b_j), the red (Gunn r) and the near infrared (K) bands. The points are taken from the work of many authors listed in ref. 12. The curves give the predicted counts for a no-evolution model.

However, the fit is due to the fact that the no-evolution model has many parameters which are favourably specified. In Fig. 19.6 we show the red-shift– apparent blue magnitude diagram for all of the data available to Koo and Kron[12] by 1992. It is clear from this diagram that the average under-lying redshift distribution cannot be accurately derived even from samples as large as these. Drawing in the Hubble relations brings out the fact that the galaxies with larger redshifts tend to have lower luminosities. Also intrinsically low luminosity galaxies appear at all distances, but the relative number cannot be determined because of the sampling differences. It is of the greatest importance to realize that because the surface brightness of the galaxies falls off as $(1 + z)^4$ the faintest galaxies in a survey which at a standard luminosity will have the largest redshifts, will be those whose redshifts are most difficult to determine. Yet they are the ones of most interest for cosmology. Koo and Kron[12] pointed out in 1992 that *not one galaxy with z > 1 had so far been measured* in these surveys. This might

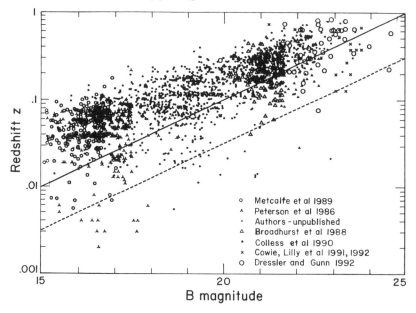

Fig. 19.6. This figure is also taken from the review of Koo and Kron[12] (their Fig. 4). It shows the Hubble diagram: $\log z$ against blue magnitude (B) for the faint galaxies as they were known in 1992.

have been because the observers were not yet looking faint enough, but it might also have been because some of the features used to measure the redshifts, such as the break at $\lambda 4000$, move into the noisier part of the sky background. This speculation – that the absence of higher redshift galaxies in the surveys is not a real effect of nature – is borne out by some of the data; for example among the largest redshift objects ($z > 0.6$) in the sample of Colless *et al.*[15] the spectrum of one object has a very large equivalent width of [OII] $\lambda 3727$, and for another galaxy is much brighter than the magnitude limit.

The frequency with which [OII] $\lambda 3727$ has been detected in the spectra of faint galaxies has been used as a parameter which might bear on galaxy evolution. For example, Broadhurst, Ellis and Shanks[16] argued that the frequency of an appearance of [OII] in galaxies with $z > 0.1$ as compared with its frequency in the spectra of bright nearby galaxies, is evidence in support of the idea that there was a prevalence of bursts of star formation at $z > 0.1$. However, Koo and Kron[12] showed that this evidence is weak since it is not clear that local samples of galaxies which show [OII] have been chosen in the same way as those with larger redshifts, in that the galaxies at larger redshifts are predominantly less luminous.

Since the work of the early 1990s, there has been a steady increase in the number of redshift surveys, including some from the 10-meter Keck telescopes and from the Hubble Deep Field. Also a new and powerful technique has been developed with which it is possible to find galaxies with redshifts in the range $z \simeq 2$ to $z \simeq 4$. A recent review has been given by Ellis[17]. Ellis has given an updated plot of $\log A(m)$ against $\triangle m$ which we show in Fig. 19.7.

The main conclusion appears to be that the Butcher–Oemler effect first found in 1971 in a single cluster[18,19] is a general effect showing that the fraction of galaxies with colors bluer by $0^{m}2$ than the peak of the color distribution increases with increasing z both for galaxies in clusters and in the general field.

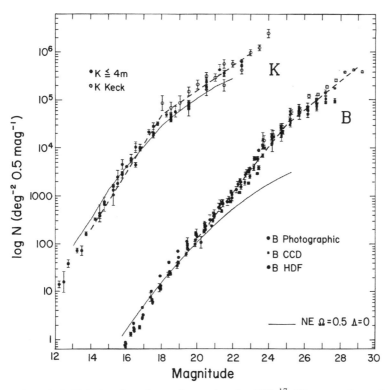

Fig. 19.7. This is taken from the review by Ellis[17]. Here we show the differential galaxy counts in the B and K bands based on ground-based observations and then augmented by Hubble Space Telescope observations out to magnitude 29. The two power law slopes which are fitted have $\gamma = d \log A / dm = 0.47, 0.30$ near $B = 25$ and $0.60, 0.25$ near $K = 18$. The solid curve is based on a no-evolution prediction. More details are given by Ellis[17].

The general idea is then that evolution of the stellar population in the galaxies is responsible, though the details are obscure. As we pointed out earlier, if galaxies all formed at about the same epoch, then looking back to higher redshifts, we should see galaxies identical to those seen locally, but when they are younger. Thus the stellar population on the average will be younger, and the energy distribution and the total luminosity will depend on the redshift. Thus since the number of blue galaxies in high redshift ($z \simeq 0.5$) clusters is similar to the number of S0 galaxies in low redshift ($z \sim 0$) clusters, one suggestion has been that what we are seeing in the blue galaxies in high z clusters are spiral disks with very active star formation, which by the present epoch have exhausted their gas, leading to S0 systems. An alternative suggestion has been that mergers between galaxies at higher redshifts have given rise to more star formation at earlier times. Yet another possibility is that the enhanced star formation at higher redshifts is a result of the interactions involved when the galaxies are first falling in to the clusters.

Some may argue still that the general result is not as well established as many have claimed. One of the reasons for this is that the criteria for the Butcher–Oemler effect in clusters have not always been rigorously applied in the fainter samples.

Serious questions can also be asked about the different explanation(s). For example, if true evolution of whole stellar populations – assuming that all galaxies condensed at the same time – is responsible for the effect, it is very surprising that it can so easily be detected over such a small redshift range – originally, at the low value of $z \simeq 0.2$ corresponding to a look back time $\lesssim 2$ Gyr.

An obvious problem is concerned with the sampling. The universe is quite inhomogeneous. Thus in order to compare similar samples, say in two redshift ranges, we already must survey over two very large volume samples. For example, if we compare samples of galaxies at redshifts near 0.05 with samples having redshifts near 0.2, how do we determine whether the nearby sample is not representative of the population in general, or simply whether the more distant sample is a result of evolution? Of course if a steady progression of color with redshift is found in all samples, i.e. if the fraction of blue galaxies increases with redshift over a significant range of z all over the sky this would rule out the possibility of inherent variations.

Also the change in the relative population of different types of galaxies as a function of redshift cannot easily be disentangled from the total numbers observed as a function of z. Fig. 19.7, based on the more recent work, extends down to magnitudes $K \sim 25$ and $B \sim 29$, about five magnitudes fainter than the results shown by Koo and Kron in Fig. 19.6.

There is a significant change in slope $d \log A/dm$ around $B = 25$, from 0.47 to 0.30, and around $K = 18$ from 0.60 to 0.25. The surface density of the B break is some 30 times higher than that at the K break. If the two effects were manifestations of the same phenomenon, i.e. a decline in the volume density beyond some redshift limit, the $B - K$ color would remain constant across the break. But it does not. Rather the change in slope is accompanied by an increase in the numbers with color $B - K < 5$. These are the blue galaxies.

One suggestion is that these fainter galaxies come from intrinsically less luminous galaxies which have burned down to very low luminosities by the present epoch so that at small redshifts they are not included in the counts. Thus the speculation is that these are part of an immense population of dwarf galaxies not detected locally. Alternatively some would argue that they are present at higher redshifts due to more merger activity and star formation driven by interactions.

We conclude this brief survey of what is known about the faint galaxy population with a discussion of a very small area of the sky that has been examined using the Hubble Space Telescope (HST), and the development of a technique for detecting normal galaxies at high redshifts.

The Hubble Deep Field (HDF)

In 1995 the Hubble Space Telescope was used to intensively observe a very small area of the sky at RA 12h 36m 49s, $Dec + 62°$ 12′58″ (JD 2000) (4 square arc minutes, or about 3×10^{-8} of the total area of the sky) to very faint levels[20] – the total exposure time was about 220 h. With this technique it was possible to reach very faint magnitudes as is shown in the last points plotted in Fig. 19.7, i.e. about 28^m in the B band and about 25^m in the K band. More than 3000 images were recorded in the Hubble Deep Field, using four broad band filters between the ultraviolet and the near infrared. Since the data were very quickly released to the astronomical community (a unique occurrence), there has been a flurry of activity and many claims as to what is being seen. A major objective of this program was to determine the nature of the faint blue galaxy population. To do this it is important to determine *both* the morphology and the redshift distribution of this sample.

Except for the brightest galaxies at the low-redshift end, the galaxies are too faint for ground-based redshift determinations of the usual kind. However, based on a possible interpretation of what is seen in the absorption spectra of QSOs, it has been argued that galaxy formation was at its peak $\lesssim 8$Gyr ago. If this were the case, and the normal (meaning local) luminosity distribution were followed, then it could be argued that among the galaxies observed in the Hubble Deep Field, we should be seeing galaxies all the way through the maximum of galaxy formation, from $z \simeq 4$, down to $z \sim 1.5$ where the peak is well past and the active phase of galaxy formation is decreasing.

The technique of picking out faint candidates by looking for the Lyman limit break (912 Å) as it is shifted into the optical spectrum means that it is likely that galaxies with redshifts in the range $2 < z < 4$ can be identified. This procedure, devised by Steidel *et al.*[21], has been applied to faint galaxies in a number of fields[22-27]. To confirm the redshifts and get exact values of z it is necessary to get slit spectra and this requires the largest ground-based telescopes. The spectra are interpreted as being those of galaxies early in their evolution and thus dominated by star-forming activity, and hence absorption and emission-line spectra of the type seen in nearby starburst galaxies are used for comparison. A prototype which has been used is the starburst region in NGC 4214. Several hundred cases of this type have now been spectroscopically confirmed[24].

This then shows that the fraction of blue galaxies – young in evolutionary terms – increases as z increases, confirming the proposal made a few years ago. These galaxies are much more luminous than galaxies undergoing some star formation at the present epoch.

Studies of the morphology of these galaxies[23,25] suggest that the number of peculiar systems which cannot easily be classified in the normal Hubble sequence is a steeply rising function of redshift amounting to 50% or more as we go from $B = 22$ to $B = 25$.

The relation of the QSOs and radio galaxies to the 'normal' faint galaxies

If it is assumed that all QSOs and all radio galaxies have underlying 'host' galaxies, and redshifts of cosmological origin, it is reasonable to argue that these objects are manifestations of very active phases in the evolution of galaxies in which the galaxies become much more luminous for short periods in their lives. Since the distribution of QSOs with z gives a peak near $z \simeq 2.5$ and falls off rapidly for $z > 3$, this is interpreted as evidence that a major period of galaxy activity took place at the epoch determined by this redshift range. It is supposed that this was a phase when the systems were very energetic in the first parts of their lives.

The look back time to $z \simeq 2.5$ depends on the cosmological model, but for the popular 'flat' model ($\Omega = 1$), at this redshift ($z = 2.5$) it is only about 2 Gyr after the beginning which was $\frac{2}{3}H_o^{-1}(\sim 10 \text{ Gyr})$ ago, so that the galaxies cannot be older than 2 Gyr at this epoch. This then suggests that young galaxies were extremely energetic in their youth. Since it is also argued in the conventional picture that the very large non-thermal luminosities are due to gravitational energy released by matter falling into massive black holes, these mass concentrations must have been formed even earlier in the galaxies' history.

At the same time if this picture is to be connected to the large numbers of faint blue galaxies found at high redshifts, it must be argued that not only are massive black holes formed early in the history of galaxy formation, but also large amounts of star formation are going on in the same period. Since the excess number of blue galaxies extends down to redshifts less than one, the active phase of galaxy formation extends in these schemes over a wide range of z, approximately from $z \simeq 5$ to $z \simeq 0.5$.

Absorption in the spectra of QSOs

Another observational phenomenon found in the spectra of QSOs is also being related to this cosmological period of activity. This is the presence of strong absorption lines in the spectra of QSOs. We discussed the absorption spectra of QSOs in Chapter 11 and described the different possible interpretation of these spectra. Here we only consider them on the assumption that the absorbing systems are probes of the inter-galactic medium at large redshifts.

We show in Figs. 19.8 and 19.9 typical spectra of high-redshift QSOs with many absorption features. Absorbing clouds with a wide range of size, all the way from very large optical depths where the strongest line – Lyα – is saturated, to very weakly absorbing clouds, must be present. As was first mentioned in Chapter 11, strong absorption features which are comparatively rare and are known in the trade as damped Lyα systems (DLA systems) (cf. Fig. 19.8) were first pointed out by A. Wolfe and his colleagues who realized that they must arise in hydrogen clouds, with about the same amounts of gas as are expected in spiral galaxies before many of the stars have formed. Absorption at the same redshifts as the DLA due to heavier elements are also seen in these systems (optical depths $\leq 10^{21}$ atoms/cm^2 column).

At the other end of the range of sizes and optical depths, very weak hydrogen absorption systems detected only through the Lyα line alone or the Lyα and Lyβ lines are found. They occur in profusion immediately to the blue of the Lyα emission feature in high-redshift QSOs. This is the Lyα 'forest'. In between the two extremes, the DLAs and the Lyα forest, there are many narrow line absorption line systems with intermediate optical depths.

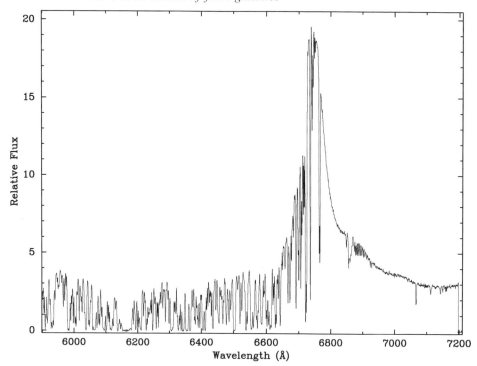

Fig. 19.8. A high-resolution spectrum of Q 2237-607 which has an emission redshift $z_{em} = 4.558$ and an r magnitude of 18.3. This spectrum was taken by Lisa Storrie-Lombardi and A. Wolfe with the low-resolution imaging spectrograph on the Keck II 10-meter telescope. The strong damped Lyα absorption between 6100 and 6200 Å is clearly seen, to the blue of the strong Lyα emission line.

If all of this absorption is due to intervening clouds distributed at random, these spectra provide a wealth of information on discrete absorbing objects at redshifts in the range $1.5 \leq z \leq 4.5$.

Wolfe *et al.* have argued that the DLAs are representative of the thick disks of protogalaxies in the process of formation. Since the QSOs in which the spectra are observed are much brighter than the star-forming galaxies (observed in the same redshift range) described earlier in this chapter, much better high-resolution spectra have been obtained of the DLAs than are available for most of the faint galaxies.

Thus it has been possible to measure abundances of some elements in these systems and also get some idea of the dynamics, always assuming, of course, that we are observing a galaxy in which some stars have evolved. Wolfe *et al.* have argued that the data suggest that the DLAs are disk galaxies in the process of formation at high redshifts. The evidence in support of a disk

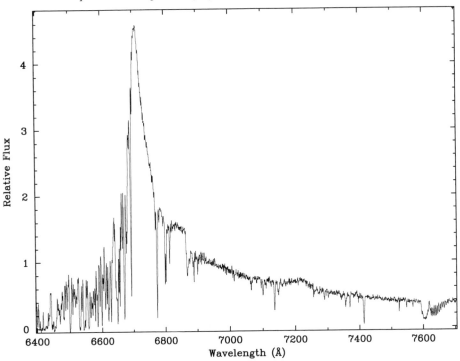

Fig. 19.9. High-resolution spectrum of Q 0019-1522 which has $z_{em} = 4.528$ and an r magnitude of 19.0. This was also taken by the same workers as Fig. 19.8 using the same equipment. In this case there is no damped Lyα feature.

comes from evidence for rotation associated with the asymmetry of the metal lines[28,29,30]. However, in studies of two individual cases[31] kinematical evidence for rotation only occurs for one of them. On the other hand, Pettini *et al.*[32] have argued that the pattern of metallicity over a range of redshifts in the DLAs *when compared with what is believed to have occurred in our own Galaxy over a similar time span* does not point to a rotating disk interpretation. Rather, they believe that the DLAs are associated with a range of morphological types of forming galaxies, spheroidal and irregulars as well as spirals. However, the most recent work of Wolfe strongly supports the thick rotating disk models.

What do the abundances show? They show that the DLAs are mostly metal poor at all redshifts sampled ($z \simeq 0.7$ to 3.4) and from $z > 3$ to $z \simeq 2$ there is an increase in the mean metallicity. The mean under-abundance is $\lesssim 1/10$ that of the Sun. Also there is a large spread of up to a factor of ~ 100 in metallicities measured in the same redshift range.

What can we conclude from these results?

If the QSOs have cosmological redshifts, and if the absorption is due to intervening matter, and if we are looking at galaxies when we observe the DLA absorption, the data can be fitted into the evolutionary picture described earlier. Either this is the 'cutting edge of research' as is claimed by many, or the height of speculation. Readers must investigate and reach their own conclusions.

We have devoted this chapter to faint galaxies and we end with a word of warning. To nearly all astronomers, a galaxy is a well organized distribution of stars, gas and dust in which all of the components can be identified and studied. The DLA systems are already being called 'galaxies' by some, but to call them 'galaxies' is indeed a gross extrapolation. What are actually seen are spectra indicating the presence of some absorbing gas. In one or two cases, weak Lyα emission at the redshift of the DLA has also been seen. There is no direct evidence for stars.

Theoretical remarks

Although the exact implications of what is being observed for faint galaxies and related objects is often imprecise, we can say with a measure of certainty that the redshift range back to $z = 5$ has been an episode of great astrophysical activity. This is as it should be according to the quasi-steady-state cosmology, in which we are looking back towards the last minimum in the oscillation of the scale factor $S(t)$, occurring at $z \simeq 5$ (cf. Table 16.1).

In the standard model, on the other hand, the intergalactic medium is like a cloud in Newtonian physics that just possesses sufficient energy to expand against gravitation to infinity, a condition expressed by the condition $k = 0$. Then initial perturbations are arranged to produce galactic condensations. This has to be done rather precisely at a closely prescribed value of $S(t)$, which has to be $\sim 10^{26}$ times that at the temperature $\sim 10^{14}$ GeV in (14.15), an extremely finely tuned initial condition. It is the artificiality of this requirement that in our minds casts doubt on its validity.

References

Chapter 19

1. Soifer, B.T., Houck, J. & Neugebauer, G. 1987, *ARAA*, **25**, 187.
2. Bennett, A.S. *et al.* 1962, *Mem. Roy. Astr. Soc.*, **68**, 163.
3. Osterbrock, D.E. 1983, *PASP*, **95**, 12.
4. Boyle, B.J., Shanks, T. & Peterson, B.A. 1990, *M.N.R.A.S.*, **243**, 1.
5. Burbidge, G. & Hewitt, A. 1987, *A. J.*, **92**, 1.

6. Fabian, A. & Barcons, X. 1992, *ARAA*, **30**, 429.

7. Fabian, A. 1994, *ARAA*, **32**, 277.

8. Burbidge, E.M. 1995, *A&A*, **298**, L1.

9. Burbidge, E.M. 1997, *Ap. J.*, **484**, L99.

10. Burbidge, E.M., private communication.

11. Thompson, D.J. *et al.* 1995, *Ap. J. S.*, **101**, 259; 1996, *Ap. J. S.*, **107**, 227.

12. Koo, D.C. & Kron, R.G. 1992, *ARAA*, **30**, 613.

13. Maddox, S.J., Sutherland, W.J., Efstathiou, G., Loveday, J. & Peterson, B.A. 1990, *M.N.R.A.S.*, **247**, 1P.

14. Ellis, R. 1987, in *Observational Cosmology*, IAU Symp. 124, eds. A. Hewitt, G. Burbidge & L.Z. Fang (Dordrecht, Reidel), p. 367.

15. Colless, M., Ellis, R., Taylor, K. & Hook, R.N. 1990, *M.N.R.A.S.*, **244**, 408.

16. Broadhurst, T.J., Ellis, R.S. & Shanks, T. 1988, *M.N.R.A.S.*, **235**, 827.

17. Ellis, R.S. 1997, *ARAA*, **35**, 389.

18. Butcher, H. & Oemler, A. 1978, *Ap. J.*, **219**, 18.

19. Butcher, H. & Oemler, A. 1984, *Ap. J.*, **226**, 559.

20. Williams, R.E. (& 14 co-authors) 1996, *A. J.*, **112**, 1335.

21. Steidel, C.C., Pettini, M. & Hamilton, D. 1995, *A. J.*, **110**, 2519.

22. Giavalisco, M. 1997, *The Universe at Low and High Redshifts* (Woodbury, AIP Press), in press.

23. Giavalisco, M., Livio, M., Bohlin, R.C., Macchetto, F.D. & Stecher, T.P. 1996, *A. J.*, **112**, 369.

24. Pettini, M., Steidel, C.C., Dickinson, M., Kellogg, M., Giavalisco, M. & Adelberger, K. 1997, *The Universe at Low and High Redshifts* (Woodbury, AIP Press), in press.

25. Abraham, R.G., Tanvir, N.R., Santigo, B.X., Ellis, R.S., Glazebrook, K. & van den Bergh, S. 1996, *M.N.R.A.S.*, **279**, L47.

26. Giavalisco, M., Steidel, C.C. & Macchetto, F.D. 1996, *Ap. J.*, **470**, 189.

27. Madau, P., Ferguson, H.C., Dickinson, M.E., Giavalisco, M., Steidel, C.C. & Fruchter, A. 1996, *M.N.R.A.S.*, **283**, 1388.

28. Wolfe, A.M., Turnshek, D., Smith, H.E. & Cohen, R.D. 1986, *Ap. J. S.*, **61**, 249.

29. Wolfe, A.M., Lanzetta, K.M. & Bowen, D.V. 1992, *Ap. J.*, **391**, 48.

30. Prochaska, J. & Wolfe, A.M. 1997, *Ap. J.*, **487**, 73.

31. Lu, L., Sargent, W.L.W. & Barlow, T. 1997, *Ap. J.*, **484**, 131.

32. Pettini, M., Smith, L., King, D. & Hunstead, R. 1997, *Ap. J.*, **486**, 665.

33. Hewitt, A. & Burbridge, G. 1993, *Ap. J. S.*, **87**, 451.

20

The large-scale distribution of matter

Luminous matter

Despite the fact that from a theoretical standpoint it is always assumed that the universe is homogeneous and isotropic, the universe as it is represented by the bright galaxies comparatively nearby appears to be highly inhomogeneous. There is a pronounced asymmetry between northern and southern galactic hemispheres with the major concentration of galaxies in the northern hemisphere – the Virgo cluster. This has been known since the time of Charlier.

As was pointed out by de Vaucouleurs[1] more than 40 years ago, we lie on the outskirts of a flattened supercluster of galaxies of which the Virgo cluster is an important central part. Not surprisingly, when de Vaucouleurs first proposed that the local structure was all part of a single flattened physical system with a diameter ~ 50 Mpc, his work was not taken seriously. However, as it became clear that more distant galaxies also are distributed in similarly large structures, the concept of superclusters was accepted.

Historically, in looking for galaxies in order to extend the redshift–apparent magnitude diagram, Hubble and Humason[2] first measured the redshifts of a few galaxies in a number of the brighter clusters of galaxies beyond the Virgo cluster. They included galaxies in the Coma, Centaurus, Ursa Major, Pegasus, Pisces, Perseus and Leo clusters.

Early surveys of galaxies were directed by Shapley[3] using the 16-inch refractor at Harvard and the 24-inch reflector at Bloemfontein in South Africa. More than half of the sky was surveyed[4] and that survey was reasonably complete down to about 17^m6. At Mount Wilson Hubble[5] used the 100-inch and 60-inch telescopes, with a standard exposure of one hour with the 100-inch to carry out surveys. With both the 60-inch and the 100-inch telescopes, more than 1200 fields were examined. From this work Hubble[6] estimated the surface density of galaxies in the range 18^m5–21^m.

The most extensive work done after this was the Lick survey by Shane and Wirtanen[7] with the 20-inch Carnegie astrographic telescope. They covered the entire sky north of declination $-23°$ and counted more than 800 000 galaxies down to a limiting photographic magnitude of 18.8. A plot of these galaxies made by the Princeton group, Peebles and colleagues[8] is shown in Fig 20.1.

Following the survey of the northern sky published in the Palomar Sky Survey, Abell[9] published a catalog of clusters of galaxies containing more than 2700 of the richest clusters, and Zwicky and his associates published a very extensive catalog of galaxies and clusters of galaxies[10]. More recently with the completion of the UK Schmidt in Australia, and the ESO Schmidt in Chile, the southern sky survey has been completed and we have catalogs of clusters in the southern sky also.

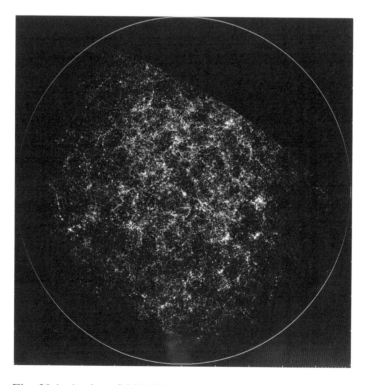

Fig. 20.1. A plot of 800 000 galaxies from the Lick Survey originally published by the Princeton Group[8]. This is the entire northern sky north of declination $-23°$ down to a limiting photographic magnitude of 18.8.

All of these surveys began to show in detail what had already been clear in the early work on the brightest and nearest galaxies – that superimposed on the statistical regularities associated with the homogeneity and isotropy, and the Hubble relation between apparent magnitude and redshift, there is structure on all scales that can be measured.

Clustering appears to be a fundamental property of the spatial distribution of galaxies. Aggregates of galaxies often consist of smaller aggregates, and are members of larger aggregates. Both Zwicky[11] and Hubble[6] concluded that the scale of clustering has a finite upper limit, the maximum scale being 40–100 Mpc. Based on his catalog Abell[9] reached the same conclusion, estimating the typical scale to be about 90 Mpc.

While large numbers of galaxies and groups and clusters had been cataloged, less was known about the redshifts. Most of the redshifts measured up to about 1960 had been obtained to help define the Hubble relation[12] by Hubble, Humason and Mayall or were contained in the bright galaxy catalog of de Vaucouleurs[13]. Only a few thousand redshifts had been measured. In 1983 Palumbo *et al.*[14] listed redshifts for more than 8000 galaxies.

Since then the measurements of redshifts have gone into high gear using both optical spectroscopy and the 21 cm line. By 1990 more than 30 000 had been measured, still largely with $cz \leq 10\,000$ km s^{-1}. This included work from Harvard-Smithsonian groups[15–18], from Cornell[19], and many others[20,21]. With this material it has been possible to plot out the three-dimensional structure in some comparatively nearby regions (for a summary see ref. 19).

By the turn of the century we should have $\sim 10^6$ galaxy redshifts, and these should give us a fair sample of the comparatively nearby three-dimensional structure. We also may have a few deep surveys (over very limited areas) which will give information on the clustering behavior of galaxies at greater distances. Is the clustering tendency truly hierarchical as was originally envisaged by Charlier? In modern terminology is the distribution a purely scale invariant fractal distribution?

Support for such a model (requiring essentially a static universe) is found in the clustering and superclustering tendencies, as has been argued by Coleman and Pietronero[22]. Such structure should continue to the largest scales which can be present, $c/H_\circ \simeq 3.10^3$ Mpc (the Hubble distance).

Peebles[23] has given a detailed discussion of this situation and has concluded that the fractal model within the Hubble distance is ruled out by

the observational evidence involving clustering and galaxy–galaxy correlations, unless the fractal dimension d is very close to 3^a.

Actually the concept of a completely hierarchical universe as proposed by
Charlier[24] meant that in the early surveys fainter and fainter (more distant)
galaxies would be found, but the mean number density would decrease as the
survey covered larger and larger hierarchical structures. However, it was
clear fairly early on that this was not true. The early work and all of the
more recent work has confirmed that there is a uniform mean galaxy density,
but not a decrease in density.

The redshift surveys have allowed us to investigate for the first time the
true three-dimensional galaxy distribution in comparatively nearby systems.
Typical regions show clustering, superclustering, strange line-like structures,
walls and voids. The general result is that galaxies lie on surfaces surrounding
low density regions or voids. Some examples are shown in Figs. 20.2–20.5.

None of this structure so far delineated in these nearby regions has given
any signs of regular patterns. In fact all of the theoretical models that have so
far been devised to relate the large-scale structure of the galaxies to the
evolution of initial fluctuations predict that the structures will be random.

However, the first results from the studies of very faint galaxies in the
direction of the Galactic poles suggested that there is a periodicity in the
structure on a scale[25] of about 120 Mpc. A later study[26] appeared to confirm
this result. Most recently a study has been made involving a large number of
the rich clusters with $z \gtrsim 0.12$. Out of a total of 1304 clusters redshifts are now
available[27] for 869. This cluster sample has been used to construct a catalog
of 220 superclusters where the distances between nearest neighbors among
member clusters are ≤ 50 Mpc. Einasto *et al.*[28] conclude that there is a

---

a The fractal dimension is defined as follows. We suppose that $N(< r)$ is defined as the number of galaxies
found within a distance r of a galaxy, averaged over counts in a fair sample. In a scale invariant fractal
with all galaxies in the same hierarchy,

$$N(< r) = Ar^d.$$

If the distribution of the galaxies is uniform, $N(< r)$ is proportional to the volume so that $d = 3$. To get
d for a construction illustrated in Figs. 20.9–20.12, note that if $r \rightarrow \lambda r$, N is increased by two because on
average we have doubled the number of levels. Thus

$$N(< \lambda r) = 2Ar^d$$
$$\text{so that } \lambda^d = 2.$$

In this model the mean number density

$$n(r) \propto \frac{N(< r)}{r^3} \propto r^{-\gamma}, \ \gamma = 3 - d.$$

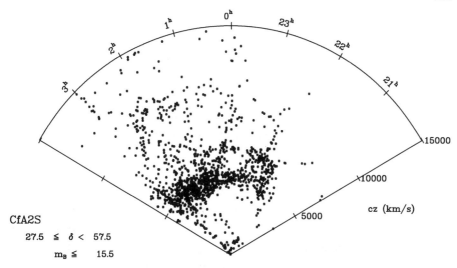

CfA2S

$27.5 \leq \delta < 57.5$

$m_B \leq 15.5$

Fig. 20.2. This shows the distribution of comparatively nearby galaxies for the declination range 27.5° to 57.5° and RA 4^h to 22^h and with B magnitudes ≤ 15.5 ($cz \leq 15\ 000$ km s^{-1}) (from the Center for Astrophysics, Smithsonian Institution Survey).

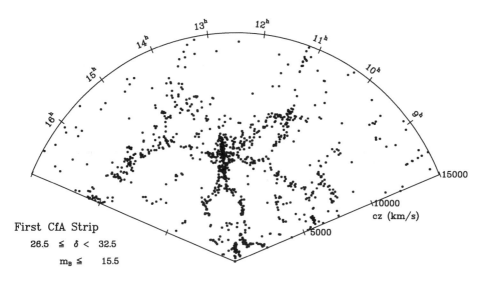

First CfA Strip

$26.5 \leq \delta < 32.5$

$m_B \leq 15.5$

Fig. 20.3. This shows the distribution for the declination range 26.5° to 32.5° and RA 17^h to 8^h ($cz \leq 15\ 000$ km s^{-1}) (from the Center for Astrophysics, Smithsonian Institution Survey).

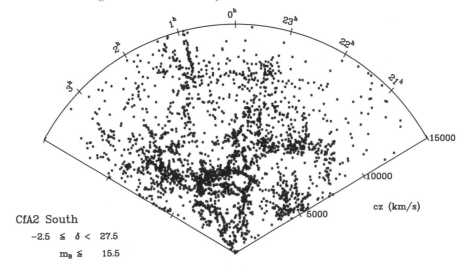

CfA2 South

$-2.5 \leq \delta < 27.5$

$m_B \leq 15.5$

Fig. 20.4. This shows the distribution for the declination range $-2.5°$ to $+27.5°$ and RA 4^h to 22^h ($cz \leq 15\,000$ km s^{-1}) (from the Center for Astrophysics, Smithsonian Institution Survey).

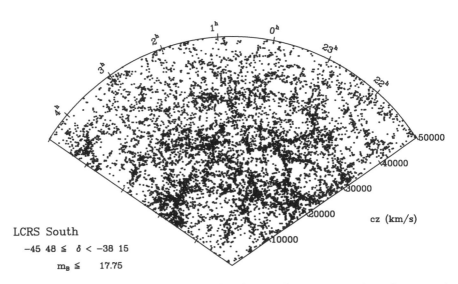

LCRS South

$-45\ 48 \leq \delta < -38\ 15$

$m_B \leq 17.75$

Fig. 20.5. This shows the distribution from a deeper survey (Las Campanas) in the declination range $-45°$ to $-38°$ and RA 5^h to 23^h for $B \leq 17^m 25$ ($cz \leq 50\,000$ km s^{-1}).

network which is moderately *regular* with a unit size $\sim 240 \pm 40$ Mpc. We show their results in Fig. 20.6. This is compatible with the results of Broadhurst *et al.* at higher redshifts[25,26].

This regularity is completely unexpected. No theory at present can accommodate it. However, it appears to us to be real, and it is another pointer suggesting that current theory is inadequate. The popular view at present is, of course, that these data are not to be believed.

Dark matter

If there is matter in the universe which does not emit photons at a detectable rate, we can only deduce its presence indirectly through its gravitational field.

The general scientific case for the existence of dark matter is as follows. Our understanding of stellar evolution leads clearly to the conclusion that in general matter will only be tied up in luminous stars for a comparatively small part of its life, the fraction depending on the star's mass. When the matter is in the form of cold gas, or in planets and other cold condensed objects, it will not be luminous. Only when it is condensed into stars will it radiate, and at the end of the stars' thermonuclear-burning lives, and depending on their masses, much of the matter will be ejected as diffuse gas, while the rest will remain as dead stellar remnants – slowly cooling white dwarfs, neutron stars and near-black holes. The amount of dark baryonic matter must increase steadily as the age of the universe increases. Thus, while there is a natural tendency to concentrate on the properties of visible galaxies, there is every reason to believe that a significant fraction of all of the matter in the universe is dark, and can only be detected through the gravitational fields that it exerts. How can it be detected in various configurations?

Single galaxies

More than 60 years ago, Oort[29] analyzed the dynamics and the numbers of stars in our own galaxy in the vicinity of the Sun, and concluded that he could not account for the gravitational forces which are present without supposing that there must be more matter present than that contained in the visible stars. He concluded that perhaps half of the matter must be in non-luminous form. Oort's analysis of 1932 and later in 1960[30] has generally been accepted up to modern times (cf. Bahcall and Soneira[31]), but Gilmore and his colleagues[32] have concluded that while the local volume mass density near the Sun (the Oort limit) shows that perhaps 50% of the mass measured

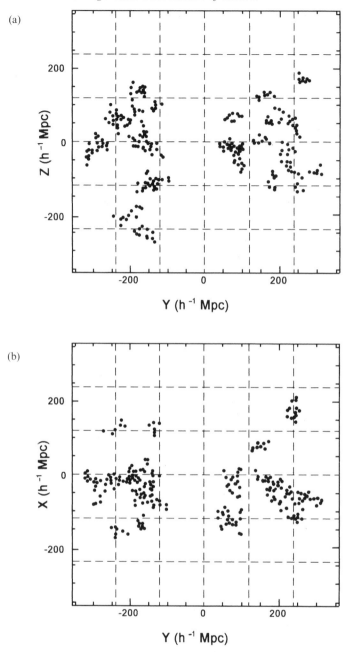

Fig. 20.6. (a, b, c, d). This shows the distribution of nearby clusters of galaxies redrawn from the work of Einasto *et al*[28]. The plot contains the distribution of 319 clusters in 25 very rich superclusters with at least eight members (including 58 clusters with photometric distance estimates), illustrating the network in the cluster distribution in supergalactic coordinates. In (c) and (d), clusters in the northern and southern Galactic hemispheres

(c)

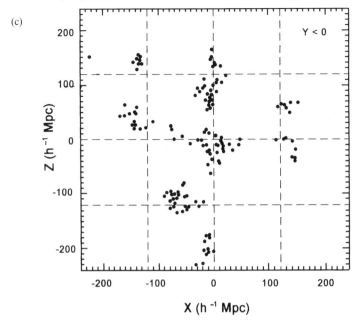

(d)

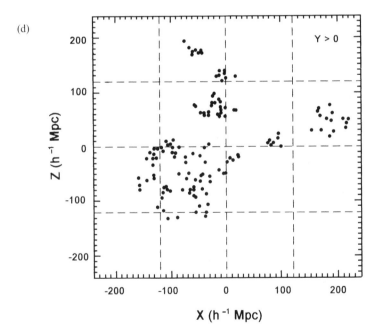

are plotted separately. The supergalactic $Y = 0$ plane coincides almost exactly with the Galactic equatorial plane, that is, with the zone of avoidance due to galactic absorption. The grid with step size 120 h^{-1} Mpc corresponds approximately to distances between high-density regions across voids. In the two panels (a) and (b), several superclusters overlap due to projection but are actually well-separated in space.

dynamically remains unidentified (the so-called missing mass), the *surface* mass density near the Sun shows no significant difference between the mass measured dynamically and the mass identified in stars that shine. Thus, either the missing mass in the Oort limit has a very small scale height, i.e. we are dealing with a very thin layer, in which case the amount of dark matter is very small, or there may be serious uncertainties in the determination of the Oort limit. Thus it must be concluded that the evidence for much dark matter present locally is not very strong.

More convincing evidence for the existence of dark matter comes from the studies of the rotation of spiral galaxies. The visible mass in a typical spiral is concentrated in a rotating disk with a central bulge. This bulge can vary from being very dominant relative to the disk (Hubble type Sa) to being very small (Sc–Sd). If the luminous mass was all that was present, then according to Newton's law the rotation curve of the galaxy would rise steeply from the center at $r = 0$, reach some maximum value v_{max}, and then fall off slowly, approaching the Keplerian velocity v_K asymptotically.

Rotation was first detected in the nearest spiral galaxy, M31, by Slipher[33]. Opik[34] first considered the possibility of determining the mass of M31 from its rotation, and Babcock[35] first measured a rotation curve for M31, which he found to be still rising out to a distance of about 18 kpc from the center.

In the late 1950s a program of spectroscopic measurement of rotation of a large number of spirals was undertaken at the prime focus of the 82-inch telescope of the McDonald Observatory using a long slit technique ($290''$ long), where the slit is placed across the nucleus and along the major axis of the galaxy, and then at many other position angles, in order to establish that only rotation is being measured. Using the Hα and [NII]$\lambda\lambda$ 6548, 6583, emission lines associated with HII regions in the spiral arms the program was very successful, and rotations and hence masses of the inner parts ($\lesssim 15$ kpc) of more than 30 spiral galaxies were obtained[36]. However, in none of these galaxies was it possible to measure rotation significantly further out than the outermost HII regions. On the other hand, radio astronomers[37] began to measure rotation using the 21 cm line of hydrogen. They were able to show that neutral atomic hydrogen is present in the disks of spirals much farther out than we can detect any stars, and were able to show that the rotation curves after reaching maxima may drop a little but tend to remain flat out to very great distances (up to ~ 40 kpc from the centers). A representative curve[37] is shown in Fig. 20.7 for the spiral galaxy NGC 3198. Also with the improvement in optical spectrographs and telescopes it became possible to make optical measurements of rotation curves much further

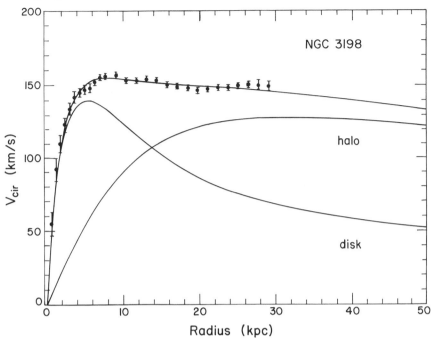

Fig. 20.7. 21 cm rotation curve for the spiral galaxy NGC 3198 taken from the work of van Albada *et al.*[37]. It shows how a flat curve can be made up of the sum of two components – rotation due to a normal disk and a halo of dark matter.

out, and Rubin and her colleagues[38] showed that the rotation measured from optical spectrum lines show the same effect. Some of these results are shown in Fig. 20.8.

The existence of such flat rotation curves was immediately interpreted as meaning that large amounts of mass must be present in spherical or spheroidal components with dimensions as large as the extended disks. It is also necessary to have mass present in the halos of spiral galaxies in order that the disks be stable against bar-mode instabilities. Of course, halos containing Population II stars, globular clusters and some gas are typically present in spiral galaxies, but the mass of these components is not large. The general conclusion from the analysis is that massive halos may be present in most spiral galaxies with masses up to $\lesssim 10$ times the mass in the visible disks.

For our own Galaxy, the total mass out to the distance of the Sun from the Galactic center (8 kpc) is $\sim 1.5 \times 10^{11} M_\odot$. A variety of estimates using globular clusters, halo stars, the tidal radii of satellite galaxies, etc., give a total

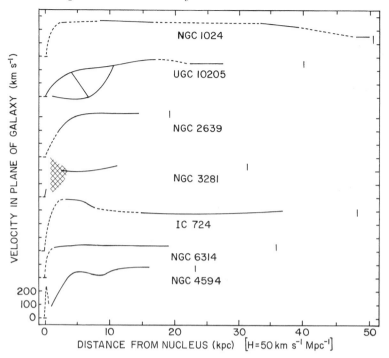

Fig. 20.8. Rotation curves of several spiral galaxies from Rubin and her colleagues[38] based on optical data.

mass $\sim 10^{12} M_\odot$ out to $R \approx 100$ kpc. Similar masses are obtained by analyses of the flat rotation curves which extend out to 50–100 kpc from the centers of other spiral galaxies. Since $M \approx v^2 R/G$, and v is roughly constant, for a large range of R, $M \propto R$. This is the simple result that leads to the conclusion that there is much dark matter in spiral galaxies.

Where could it be in error? The most fundamental and radical suggestion bearing on this has been made by Milgrom[39–42] who has proposed that Newton's second law should be modified.

However, while Milgrom has made a radical modification of the physical laws, it is not a modification that we are very comfortable with. In our own work (Chapter 13) we have kept to the generally accepted procedure of formulating the physical laws through an action principle, ensuring that, although the quantities to which the action principle was applied were changed from the usual forms, they possessed the usual invariance properties with respect to all spacetime transformations. Milgrom's proposal, on the other hand, is an *ad hoc* modification of Newton's law designed to suit the particular phenomenon under consideration.

What about examining critically other methods which are used to measure the mass? All of the methods (including the rotation curve method) essentially use the virial theorem which tells us that

$$2 \times (\text{Kinetic energy}) + (\text{Potential energy}) = 0. \qquad (20.1)$$

This assumes of course that we are always dealing with relaxed systems in equilibrium. For spheroidal galaxies, and the bulges of spiral galaxies, the kinetic energy is determined from the random velocities of the stars. The line of sight components of the stellar random velocities are obtained by determining the line broadening of the integrated stellar spectra, and then corrections are made to obtain a measure of the three-dimensional random velocity field. The forms and the scales of the potential energy are measured from the light distribution, and the size is determined from the distance of the galaxy. It is then very easy to determine the total mass from the virial equation.

The uncertainties in the method are associated with the determination of the true velocity dispersion, and the extent of the galaxy[43]. In a few cases, and in particular M87, a very bright and massive elliptical galaxy in the Virgo cluster, observations using X-ray telescopes have shown that there is an extensive halo of very hot gas with a temperature of 10^6–10^7 degrees surrounding the optical galaxy. If we assume that the extended halo of gas is part of the bound system, the very large radius means that the total mass of the galaxy is very large. The total mass assuming a radius of about 300 kpc for the X-ray source in M87 lies in the range $3 \times 10^{12} \sim 10^{13} M_\odot$. Most of this must be dark matter. A few other elliptical galaxies centred on extended X-ray sources give similar results.

Groups and clusters of galaxies

Studies of pairs of galaxies and small groups have consistently shown when the virial condition is assumed that there is much more kinetic energy in the visible matter than potential energy, i.e. for the visible matter $2 \times (KE) > (PE)$.

In fact as we pointed out in Chapter 11, Ambartsumian[44] argued in the 1950s that many of these systems have total positive energy, i.e.

$$(KE) + (PE) > 0 \qquad (20.2)$$

so that they are expanding associations.

Very much earlier, Zwicky[45] and Smith[46], having only fragmentary information on the velocity dispersion of galaxies in a few of the rich clusters, had used the virial theorem to show that 10–100 times more mass was required to bind the clusters than is present in the form of visible galaxies. The general result has turned out to be as follows.

In almost all cases, dark matter must be present in pairs, groups and clusters if it is supposed that they are bound systems. Apart from the evidence from single galaxies, the *existence of such systems is the primary evidence that large amounts of dark matter are present in the universe.*

In 1974, Einasto, Krasnik and Saar[47] and Ostriker, Yahil and Peebles[48] collected many results together and pointed out that the total mass M both in galaxies and in groups and clusters increases roughly linearly with the scale R of the system. Those papers had a considerable impact and a wide audience began to appreciate the case for the existence of dark matter.

However, it had always appeared to one of us (GB) that the result $M \propto R$ really stemmed completely from the belief that the virial theorem holds for all of these systems[49]. This is because in all situations

$$M(R) = \frac{k}{G} v^2 R \qquad (20.3)$$

where k is a constant of order unity which is different for the different methods used. Here v is a rotational velocity in a spiral galaxy, or it is some measure of the random motions of stars $(\bar{v}_s^2)^{1/2}$ in an elliptical galaxy, or it is the random motion $(\bar{v}_g^2)^{1/2}$ of galaxies in a group or cluster, or it is a measure of the average velocity difference of pairs of galaxies in binary orbits. R is always the size of the system.

The remarkable point is that the numerical values of v in these different situations do not vary very much. We show this in Table 20.1. It is then *this* that leads us from equation (20.3) to the conclusion that $M(R) \propto R$. Most of the values in Table 20.1 are close to 300 km s^{-1}, and only when we get to more populous clusters do the random velocities increase by factors of two or three.

Thus much of the evidence for dark matter rests on the assumptions that all of the systems are bound, relaxed and stable. This point has simply been ignored in the literature. It is important to ask what are the alternatives?

Two arguments can be made:

1. in some cases we are not looking at a single physical system at all, or
2. the system is not stable and is disintegrating.

Let us briefly consider these ideas. A good example of the first may be the so-called Local Group which consists of our Galaxy, M31, M33 and a small number of nearby dwarf galaxies with much smaller masses and luminosities, including the satellites of our own Galaxy and M31.

Table 20.1. *Relative velocities and random velocities associated with galaxies**

Type of system	$(v_r^2)^{1/2}$ (km s^{-1})
Single galaxies:	
Spirals (maximum rotational velocity)	200–300
Ellipticals (velocity dispersion of stars)	200–300
Double galaxies:	
Close doubles:	
5 elliptical pairs (570)	
13 elliptical-spiral pairs (300)	290
13 spiral pairs (150)	
6 irregular-spiral pairs (80)	
Wide doubles	240
Small groups	300
Clusters:	
NGC 541 cluster	410
Virgo	420
Hercules	630
Coma	900
Perseus	1420
Field galaxies	200–300

*Dwarf systems are excluded

Use of the virial theorem directly, together with the fact that M31 and our galaxy are approaching each other in an orbit, and the effect of the Local Group on the velocity field of nearby galaxies, have all been used to estimate a total mass for the Local Group $\sim 3 \times 10^{12} M_\odot$ (with considerable uncertainties) so that it is argued that much dark matter must be present. However, the belief in the physical existence of the Local Group may simply be a result of the fact that we live here. Seen by an astrophysicist living in a galaxy in the Virgo cluster, our Galaxy and M31 would be a very wide pair with a separation of 0.6 Mpc, each galaxy with its own satellites. This separation is large enough so that the system might well be rejected as a physical pair, i.e. it might well be assumed that these two galaxies have no more relationship to each other than two ships passing each other in a large ocean.

What about the other possibility, that many groups and clusters are physical systems which are disintegrating? As was discussed in Chapter 11, Ambartsumian suggested in the 1950s that this was occurring. The apparent difficulty with the hypothesis is that the time scale for such systems to disintegrate is $\lesssim 10^9$ years, a value much less than H_o^{-1}, and much less than the time scale associated with galaxy formation and evolution in the big-bang cosmology.

In 1961 at the IAU Symposium on extragalactic astronomy in Santa Barbara, one of us (GB) in private discussions with Jan Oort was shown very clearly how the argument went, as far as Oort was concerned. During that meeting, Ambartsumian had again discussed his cosmogony, involving expanding associations of galaxies etc., with a number of us, but he was very unclear about the details. Oort had by this time been totally converted to evolutionary cosmology by Ryle's claimed disproof of the steady-state cosmology – an incorrect result – (cf. the discussion in Chapter 7). Thus Oort argued that the galaxies *must* have been formed in the early universe, and must, therefore, be $\sim 10^{10}$ years old. Thus they could not be young, as would be required in expanding associations. Also where must all of the galaxies from previous expanding associations have gone? He concluded that the groups and the clusters *must* be bound, and the extra matter required, which could not be seen, must be present. Thus Oort was, early on, a very strong and influential figure in the attempts to detect dark matter since he *believed* that it must be there.

We still have no conclusive evidence that the bulk of the galaxies are very old, since we can only obtain good ages from color-magnitude diagrams involving star clusters which all formed at the same time. Thus only for our Galaxy and very nearby systems can we be sure that the ages are $\sim 10^{10}$ years. In the 1960s, to talk about young galaxies was dangerous, since this meant to most people that you were invoking the steady-state cosmology. Nowadays, it is accepted that there are extensive regions in galaxies and perhaps whole galaxies which are young in evolutionary terms, but whether there is always an old underlying stellar population is not known. There is also evidence that coherent objects – even galaxies – are being ejected from galaxies (cf. Chapter 11). Perhaps the field is populated by galaxies ejected from older systems.

Gravitational lensing as an indicator of dark matter

Following a very early suggestion by Zwicky[50] and more specific proposals concerning the QSOs by the Barnothys[51], the idea that gravitational lensing by foreground dark matter of background QSOs and faint galaxies might explain some of the anomalous redshift phenomena, has been explored. If the observational evidence is interpreted in this way it provides evidence for dark matter in clusters of galaxies, etc. There are several possibilities.

(1) Multiple QSOs

Starting in 1979 with the discovery by Walsh *et al.*[52] that the QSO 0957 + 561 consisted of two images with identical spectra with a separation of $6''$, it has been argued that a number (about 40 at present) of the known QSOs are gravitationally lensed (cf. Chapter 11). If more than one optical or radio image is detected and if they appear to be identical and have separations of $\lesssim 5''$ it is assumed that they must be gravitationally lensed objects, i.e. there must be a mass, either part of a galaxy or a cluster of galaxies, which is responsible for the multiple images and which lies very close to the line of sight and at a distance much less than the distance to the QSO derived from its redshift.

In a few cases, in particular 0957 + 561 and the famous Einstein Cross $(2237 + 0305)$[53] where four images are seen within $0''.3$ of the center of a 15^m galaxy, the galaxy acting as the lens can be studied. However, in the majority of the cases the lensing mass has not been directly detected. In principle if the interpretation is correct the method will allow us to detect very faint galaxies or genuinely dark matter.

This topic was discussed in a different context in Chapter 11.

(2) Gravitational arcs

It is also possible to detect the presence of dark matter by studying the images of faint galaxies whose line of sight happens to intercept concentrations of matter along the light path. The first well established arcs due to this process were identified independently by Lynds and Petrosian[54] and Soucail *et al.*[55], and redshifts were measured showing that these features are indeed likely to be gravitationally distorted images of background galaxies. It has also been suggested that many small features in the form of arclets which can be detected in very deep exposures of faint galaxies are in fact parts of gravitational arcs associated with the images of very distant galaxies which lie behind dark mass distributions in clusters of galaxies. How far these methods can be taken to detect dark matter is not yet clear.

(3) Low-mass compact dark objects which may be detected through gravitational microlensing

Pacynski[56] suggested that through the process of gravitational microlensing by dark objects with the dimensions of evolved stars it might be possible to detect dark matter in this form in the halo of our own Galaxy, since distant stars will occasionally be occulted or partially occulted by such a dark object crossing the line of sight.

Using this method, several groups[57,58,59] have investigated many lines of sight through the bulge and halo of our Galaxy and the Large Magellanic Cloud. They have detected many events due to gravitational microlensing, most of which are believed to be caused by faint stars of various types. Indeed, the method is beginning to give important evidence on the distribution of faint and possibly dark stars.

This concludes our very brief summary of gravitational lensing as far as it bears on the problem of detecting dark matter. We shall return to it in the context of its applications in cosmology in Chapter 21.

In summary, how strong is the observational evidence for the existence of dark matter in appreciable amounts?

We have discussed the various observational attempts which have been made to detect dark matter. What has been found is that there are many situations in which dark matter must be present, *unless* a less conventional and/or a less popular explanation is involved.

(a) Is there strong evidence for dark matter locally in our Galaxy? According to Gilmore *et al.*[32] this may not be as certain as Oort and other authors have claimed.

(b) Do spiral galaxies have large massive halos of dark matter? Yes, unless Milgrom is correct. So far his arguments have not been refuted, only ignored.

(c) Do physical systems of galaxies, from binaries through small groups of galaxies up to rich clusters, contain dark matter in quantities large enough to stabilize them? Here the obvious answer is that some clusters are bound, and some are unbound. It is clear from the forms of the clusters that some – like the Coma cluster – are relaxed and stable, and others – like the Hercules cluster – are unstable, and may well be systems of positive total energy. The assumption that $M \propto R$ for all of the systems is simply a crude restatement of the virial theorem, and *not* a directly observed result.

Our conclusion is that some dark matter is present in the universe, but there is no evidence from the studies of galaxies to determine how much, and certainly not enough to conclude that $\Omega = 1$.

Even if Milgrom's ideas are ignored, and the virial is assumed to hold for all multiple systems, the mass when measured in terms of mass-to-light ratios where $M_\odot/L_\odot = 1$, gives contributions to Ω as follows.

> For smaller stellar systems with $M/L \approx 1$, the contribution to Ω is only about 0.001.
>
> For individual galaxies with flat rotation curves, etc., $M/L \approx 10$, giving a contribution to Ω of 0.01.
>
> For small groups, $M/L \approx 50$, giving a contribution to Ω of 0.05
>
> For rich clusters and superclusters, $M/L \sim 200 \pm 50$, giving a contribution to Ω of 0.02.

There is no good astronomical reason to suppose that any of the dark matter is non-baryonic. As far as cosmology is concerned, the existence of non-baryonic matter rests on belief and not on any hard evidence. As far as particle physics is concerned, the only theory that predicts the existence of a new weakly interacting massive particle (WIMP), and provides estimates of its range of mass and interaction rates, is supersymmetry. The attraction for the particle physicist is that the discovery of such WIMPs in the cosmos would do a great deal for theory. It is probably for this reason that nearly all physicists who have recently come into astrophysics tend to believe that its existence is a real possibility, but we have found no reputable astronomer who believes in it for reasons based on astronomical data.

Models for large-scale structure

Many attempts are currently being made to explain the large-scale structure delineated by the visible galaxies in terms of the conventional big-bang model.

All of the attempts concentrate on the growth of primordial fluctuations in the post-inflationary phase, through gravitational interaction. In the standard cosmology the structure formation recipe includes the following steps.

1. Assume primordial fluctuations as arising from quantum fluctuations in the Planck era.
2. Use some suitable brand of the inflationary universe as a means to generate a scale invariant (Harrison–Zeldovich) spectrum in the fluctuations.
3. Study the growth of fluctuations first as a linear problem through gravitational equations.

4. Use a dark matter component in varying amounts to induce the desired growth of fluctuations of visible matter.

5. Select the dark matter principally to be non-baryonic, so as not to produce too large an effect on the growth of fluctuations of temperature of the radiation background.

6. Adjust the biasing factor, i.e. fluctuations of dark matter *vis-à-vis* visible matter, to arrive at the desired final result.

7. Adjust the composition of dark matter to achieve the desired final result.

8. When the density fluctuations $\delta\rho$ grow comparable to the average density ρ, switch over to non-linear techniques which involve techniques like the Zeldovich approximation, percolation, genus, N-body simulations, the minimum spanning tree etc.

9. Check the final product with the present large-scale distribution of galaxies, clusters, superclusters, etc., the present large scale peculiar motions, and the results of microwave background fluctuations $\triangle T/T$ as observed by COBE and other experiments.

10. If the answers do not match, go back to 2 above and change the parameters in steps 2–7 and redo the calculations.

So far, it appears that a satisfactory structure has not been achieved despite the existence of a large industry trying to cover as much as possible of parameter space. The infinite patience, unbounded human hours and grant money spent on gigantic computer simulations are matched by the firm faith that the approach is correct.

It is against this background that we must consider structure formation in the QSSC. That purely gravitational effects will prove inadequate, was shown by S. Banerjee[60] and one of us (JVN) by studying the growth of density fluctuations in the QSSC through one oscillatory phase. The quantity $\delta\rho/\rho$ decreases during expansion and increases during contraction. However, unless some non-gravitational effect is playing a role, it is hard to see how this quantity will show a secular increase from one oscillation to the next. In short, purely gravitational effects *will not* deliver the kind of inhomogeneous and filamentary large scale structure that the redshift surveys are pointing to.

The QSSC, however, can rely on the creation process in the so-called minicreation events (MCE) to give shape to large scale structure. The process works in the following sequence.

1. Creation takes place near the minimum of the scale factor during each oscillation.

2. Each MCE is centred around a spinning compact object simulating a near-Kerr black-hole-type situation. The created matter can emerge more natu-

rally along the polar axis (any equatorial direction tending to drag the created matter within the ergosphere).

3. The emerging jet may start with high speed ($v \sim c$) but slows down as it encounters the intergalactic medium, finally condensing into a coherent mass which may also have a spin aligned with the spin of the original mass. Alternatively, the central mass in the MCE may break up and eject a coherent piece as discussed in Chapter 18.

4. The ejected mass may eventually act as the center of another creation event, probably in the next oscillation.

5. The process goes on, maintaining a steady state from one oscillation to the next.

This last point may be seen as follows. Consider the number density of collapsed massive objects at one oscillatory minimum to be f. Then the number density to the next oscillatory minimum would fall to $f \exp(-3Q/P)$, if no new massive objects were added. To restore a steady state therefore,

$$\alpha f \equiv [1 - \exp(-3Q/P)]f \sim 3f \, Q/P$$

masses must be created anew. In other words, a fraction $3Q/P$ of the total number of massive objects must duplicate themselves in the above fashion. To study the cumulative effect of this activity the following toy model was tried.

What will happen when a system of points on a plane are duplicated in this fashion? A clue is provided by an experiment of computer simulations.

Take a unit area and generate N random points on it. Then create one random point as a neighbor to each of the N points, the 'neighborhood' being defined as a circle of radius x times the average separation between two typical neighbors in the first set. The number x is taken to be a fraction between 0 and 1.

Next scale the area and the point distribution homologously to twice the size, and take the inner unit square. On an average it should have N points. Repeat the process a number of times n, say, each time duplicating the original set of points, doubling the area and taking the inner unit square. Figure 20.9 gives the result of this process in a typical case.

Notice that points tend to cluster, with large voids in between, very similar to the cluster/voids seen in the distribution of galaxies.

A variation of the process is shown in Fig. 20.10, in which at each subsequent doubling after the first one, memory is retained of the direction in which the first neighbor was created. Thus if the first point was A and its created neighbor was at B, B's neighbor C is randomly produced in the

N = 100,000, n = 10, x = 0.75

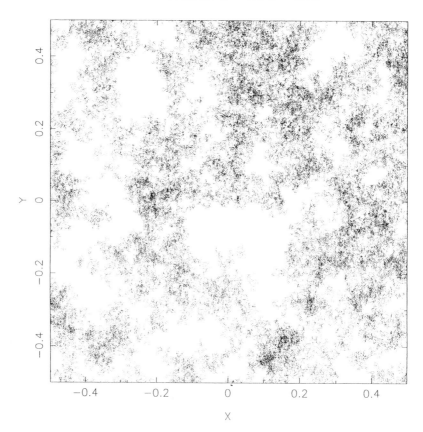

Fig. 20.9. A cluster/void distribution generated in the toy model for $N = 10^5$, $x = 0.75$ and $n = 10$ (computations for this and Figs. 20.11 and 20.12 made by Ali Nayeri).

'forward' half plane of the line perpendicular to AB, that is, pointing away from A. Thus the creation center B retains the memory of the direction A to B and preferentially creates the next point C in a direction BC making an acute angle with respect to line AB produced. The resulting plot of points shows voids with filamentary boundaries. These filaments are expected to be sharper if the direction BC is constrained to make a small angle with the forward direction AB of previous ejection.

The exercise can be repeated in three dimensions also, with a cube of unit volume. Figure 20.11 shows a thin slice of the cube perpendicular to one of its axes. The distribution of points in it looks remarkably similar to the observed 'slices' of large scale structure produced by redshift surveys.

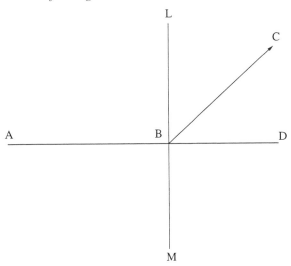

Fig. 20.10. LM is perpendicular to AB which is continued to D. The point C is the right half plane of LM, i.e. BC makes an acute angle with BD.

To bring the toy model closer to reality of the QSSC, the following may be done. Instead of creating a new neighbor around each of the original set of N points, do so only around αN of these points, where the fraction α is defined above. Then expand the volume homologously by the factor $\exp 3Q/P$ only instead of by factor 2, and choose the inner unit cube. Figure 20.12 shows one such simulation.

The agreement could be improved by taking into consideration gravitational clustering of the points during a typical oscillatory cycle. However, the driving mechanism in the process, as illustrated above, is not gravitation but the creation process. This is the 'non-gravitational' feature missing from the standard structure formation scenario in big-bang cosmology.

It is also possible to devise computer codes to carry out a correlation analysis on the point distribution. Ali Nayeri and Sunu Engineer have run several simulation cases of the above model for calculating the two-point correlation function $E(r)$, which is defined as follows. Taking any point in the distribution, define the probability of finding another point in a small volume δV at a distance r from it as

$$\delta P = n\delta V[1 + \xi(r)].$$

Here n = number density of galaxies on the average. If the distribution were uniform $\xi(r) = 0$. In a clustered distribution, however, $\xi(r)$ is positive for

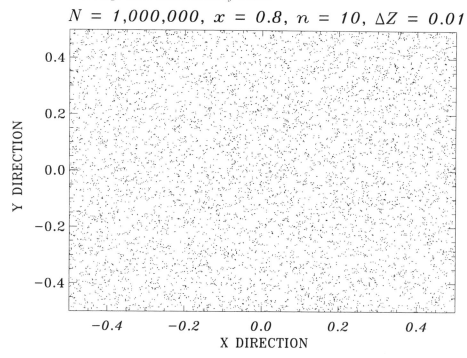

Fig. 20.11. A three-dimensional version of Fig. 20.10 with $N = 10^6$, $x = 0.8$, $n = 10$ and slice thickness $= 0.01$.

small r, dropping down to zero as r increases. In a void, $\xi(r) < 0$. Typically the galaxy distribution shows $\xi(r) \propto (r_0/r)^\gamma$, with $\gamma \sim 1.8$.

It is interesting to find that preliminary studies of the above QSSC toy model give $\gamma \cong 1.8$ for a few iterations. Figure 20.13 demonstrates a typical case. For details see Nayari *et al.*[61].

Compared to the enormous effects put in by big-bang structure formation enthusiasts, a modest effort here seems to lead to a structure distribution remarkably close to what is observed – perhaps pointing to a new direction!

$P/Q = 20, \ N = 100{,}000, \ x = 0.3, \ n = 5, \ \Delta Z = 0.01$

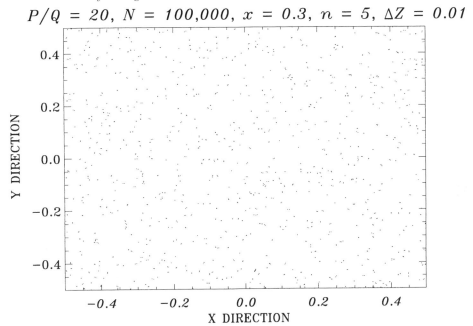

Fig. 20.12. A three-dimensional version of the rule of Fig. 20.10 adapted for the QSSC with $N = 10^5$, $x = 0.3$, $n = 5$, and $P/Q = 20$ the slice thickness being 0.01.

n=100000,s=0.2

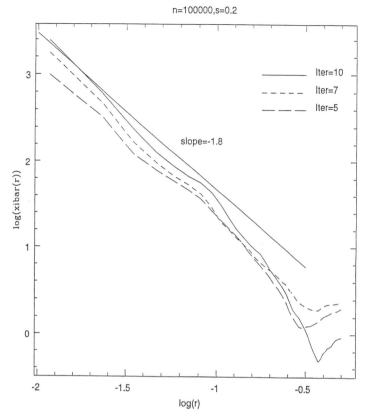

Fig. 20.13. Two-point correlation function $\xi(r)$ is plotted in the $\log \xi - \log r$ diagram for the case of QSSC with $N = 10^5$, $x = 0.2$, $n = 5$, 7, 10, $P/Q = 20$. The straight line has slope $\xi(r) \propto r^{-1.8}$, i.e., $\gamma = 1.8 =$ constant. The actual distributions for iterations 5–10 do run parallel to this slope before dropping down steeply, where the cluster ends and a void begins.

References

Chapter 20

1. de Vaucouleurs, G. 1953, *A. J.*, **58**, 30; 1956, in *Vistas in Astronomy*, Vol. 2 (ed. A. Beer, London and New York, Pergamon Press), p. 1584; 1958, *A. J.*, **63**, 253.
2. Hubble, E. & Humason, M. 1931, *Ap. J.*, **74**, 43.
3. Shapley, H. 1934, *Harvard Annuals*, p. 761.
4. Shapley, H. 1961, in *Galaxies* (Cambridge, Harvard Univ. Press), p. 159.
5. Hubble, E. 1934, *Ap. J.*, **79**, 8.
6. Hubble, E. 1936, *Ap. J.*, **84**, 517.
7. Shane, C.D. & Wirtanen, C.A. 1967, *Pub. Lick Obs.*, Vol. 22, part I.
8. Peebles, P.J.E., Seldner, M. & Soneira, R.M. 1977, *Scientific American*, p. 76; Seldner, M., Siebers, B., Groth, E.J. & Peebles, P.J.E. 1977, *Ap. J.*, **82**, 249.

9. Abell, G.O. 1958, *Ap. J. S.*, **3**, 211.
10. Zwicky, F., Herzog, E., Wild, P., Karpowicz, M. & Kowal, C.T. 1961–68, *Catalog of Galaxies and Clusters of Galaxies*, Vols. 1–6 (Pasadena, California Institute of Technology).
11. Zwicky, F. 1938, *PASP*, **50**, 318.
12. Humason, M.L., Mayall, N.U. & Sandage, A.R. 1956, *A. J.*, **61**, 97.
13. de Vaucouleurs, G. & de Vaucouleurs, A. 1991, *Third Reference Catalogue of Bright Galaxies*, Vols. I–III (New York, Springer Verlag).
14. Palumbo, G., Tanzelli-Nitti, G. & Vettolani, G. 1983, *Catalogue of Radial Velocities of Galaxies* (New York, Gordon and Breach).
15. Huchra, J., Davis, M., Latham, D. & Tonry, J. 1983, *Ap. J. S.*, **52**, 89.
16. Huchra, J., Geller, M., de L'Apparent, B. & Corwin, H. 1990, *Ap. J. S.*, **72**, 433.
17. Huchra, J., Geller, M. & Corwin, H. 1995, *Ap. J. S.*, **99**, 391.
18. Geller, M. *et al.* 1997, *A. J.*, **114**, 2205.
19. Giovanelli, R. & Haynes, M. 1991, *ARA&A*, **29**, 499.
20. Da Costa, L.N. (& 8 co-authors) 1988, *Ap. J.*, **327**, 544.
21. Dressler, A. 1990, preprint.
22. Coleman, P.H. & Pietronero, M. 1992, *Phys. Reports*, **213**, 311.
23. Peebles, P.J.E. 1993, *Principles of Physical Cosmology* (Princeton Univ. Press, Princeton, N.J.).
24. Charlier, C.V.R. 1922, *Arkiv för Mat. Astron. Fys*, **16**, 1.
25. Broadhurst, T.J., Ellis, R.S., Koo, D.C. & Szalay, A. 1990, *Nature*, **343**, 726.
26. Einasto, J. & Gramman, M. 1993, *Ap. J.*, **407**, 443.
27. Abell, G.O., Corwin, H. & Olowin, R. 1989, *Ap. J. S.*, **70**, 1.
28. Einasto, J., Einasto, M., Gottlober, S., Muller, V., Sear, V., Starobinsky, A., Togo, E., Tucker, D., Audernach, H. & Frisch, P. 1997, *Nature*, **385**, 139.
29. Oort, J.H. 1932, *Bull. Astr. Neder.*, **6**, 249.
30. Oort, J.H. 1960, *Bull. Astr. Neder.*, **16**, 45.
31. Bahcall, J. & Soneira, R. 1984, *Ap. J. S.*, **55**, 67.
32. Gilmore, G., Wyse, R. & Kuijken, K. 1989, *ARA&A*, **27**, 62.
33. Slipher, V.M. 1914, *Lowell Observatory Bulletin*, **2**, p. 65.
34. Opik, E. 1922, *Ap. J.*, **55**, 406.
35. Babcock, H.W. 1939, *Bull. Lick Obs.*, **19**, 41.
36. Burbidge, G. & Burbidge, M. 1975, in *Stars and Stellar Systems*, Vol. 9 (eds. A. Sandage, M. Sandage & J. Kristian, Univ. Chicago Press), p. 81.
37. van Albada, T., Bahcall, J., Bergeman, K. & Sancisi, R. 1985, *Ap. J.*, **295**, 305.
38. Rubin, V., Burnstein, D., Ford, W.K. & Thonnard, N. 1985, *Ap. J.*, **289**, 81.
39. Milgrom, M. 1983, *Ap. J.*, **270**, 365 and 371.
40. Milgrom, M. 1994, *Ap. J.*, **429**, 540.
41. Begeman, K., Broeils, A. & Sanders, R. 1991, *M.N.R.A.S.*, **249**, 523.
42. Sanders, R. 1996, *Ap. J.*, **473**, 117.
43. Faber, S.M. & Gallagher, J.S. 1979, *ARA&A*, **17**, 135.
44. Ambartsumian, V.A. 1961, *A. J.*, **66**, 536; Ambartsumian, V.A. 1958, *Solvay Conference on Structure and Evolution of the Universe* (R. Stoops, Brussels), p. 241.
45. Zwicky, F. 1933, *Helvetica Phys. Acta*, **6**, 110.
46. Smith, S. 1936, *Ap. J.*, **83**, 23.
47. Einasto, J., Krasnik, A. & Saar, E. 1975, *Nature*, **250**, 309.
48. Ostriker, J., Yahil, A. & Peebles, P.J.E. 1974, *Ap. J. L.*, **193**, L1.
49. Burbidge, G. 1975, *Ap. J.*, **106**, L7.

50. Zwicky, F. 1937, *Phys. Rev.*, **51**, 290, 679.
51. Barnothy, J. & Barnothy, M.F. 1968, *Science*, **167**, 348; 1965, *A. J.*, **70**, 666.
52. Walsh, D., Carswell, R. & Weymann, R. 1979, *Nature*, **279**, 381.
53. Huchra, J., Gorenstein, M., Kent, S., Shapiro, I., Smith, G., Horine, E. & Perley, R. 1985, *A. J.*, **90**, 691.
54. Lynds, R. & Petrosian, V. 1986, *BAAS*, **18**, 1014; 1989, *Ap. J.*, **336**, 1.
55. Soucail, G., Fort, B., Mellier, Y. & Picat, J.P. 1987, *A&A*, **172**, L14.
56. Pacynski, B. 1986, *Ap. J.*, **304**, 1.
57. Alcock, C., Akenlof, C.W., Allsman, A. *et al.* 1993, *Nature*, **365**, 621; Alcock, C., Allsman, R.A., Axelrod, T.S. *et al.* 1996, *Ap. J.*, **461**, 84.
58. Aubourg, E., Bareyre, P., Bŕehin, S. *et al.* 1993, *Nature*, **365**, 623; Alard, C. & Guiberts, J. 1997, *A&A*, **326**, 1.
59. Udalski, A., Szymanski, M., Stanek, K.Z. *et al.* 1994, *Acta Astronomia*, **44**, 165.
60. Banerjee, S. and Narlikar, J.V. 1997, *Ap. J.*, **487**, 69.
61. Nayeri, A., Engineer, S., Narlikar, J.V. and Hoyle, F. 1999, *Ap. J.*, November 1.

21

A brief account of the radiation fields in the universe – the observations and their interpretation

As well as the large population of galaxies that we have just discussed, there are also diffuse fields of radiation which can be detected at all energies and wavelengths.

Measurements of this background radiation have been made at radio (meter, centimeter) wavelengths, in the microwave band, largely the cosmic microwave background (CMB), at infrared wavelengths, at optical and ultraviolet wavelengths, and at X-ray and γ-ray energies. Because together with the starlight there must be a corresponding flux of low energy neutrinos, it is assumed that a neutrino background also is present, but it has not been detected so far. The energy distribution of the background radiation as a function of wavelength (or photon energy) is shown in Fig. 21.1 (cf. ref. 1). It can be seen from this plot that the bulk of the energy is radiated in the microwave region of the spectrum.

From the time of the discovery of these background fluxes, it has been assumed that with the exceptions of the CMB and possibly the X-ray background, they are due to the integrated effect of large numbers of discrete sources. These are the galaxies, QSOs and radio sources that we have just been discussing, and any other populations of discrete sources that may be present but which have not yet been detected.

There are also diffuse background fluxes that can only come from our Galaxy – for example the soft X-ray background with photon energies $\lesssim 0.25$ keV which arises in the comparatively nearby interstellar gas. But we shall not discuss this here.

The diffuse optical light is the integrated radiation of stars from all of the galaxies. At radio wavelengths it has always been realized that the number counts must be reconciled with the radio background flux (cf. Longair[1]). For the general X-ray and γ-ray background there is an ongoing discussion concerning which kinds of active sources give rise to the background flux.

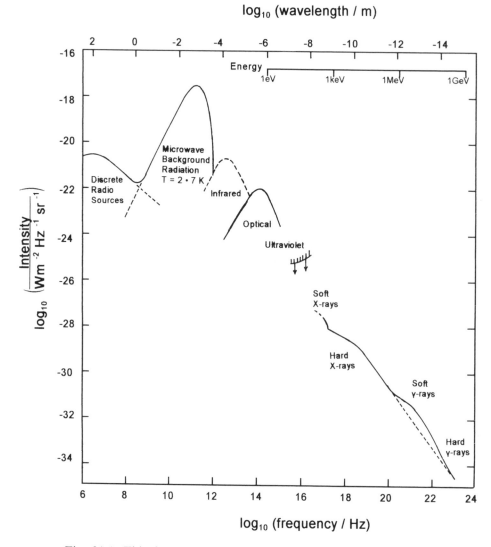

Fig. 21.1. This figure shows the energy fluxes per unit bandwidth of the radiation fields in the universe. It is adapted from a plot made by Longair[1].

Originally it was considered possible that the X-rays could be simply thermal bremsstrahlung from a very hot intergalactic gas. However, it now appears that the background is due to a population of active galactic nuclei including the known population of QSOs which may be able to explain the whole flux, though some argue that populations of lower luminosity objects so far undiscovered are required[2,3,4]. Burbidge and Hoyle[4] have shown that even if the QSOs have non-cosmological redshift components there is no problem in explaining the background. However, it is not yet clear whether in any of these models a good fit for the spectral shape of the background can be obtained.

But it is the microwave background radiation which stands out (a) because it dominates energetically, and (b) because its energy distribution follows a black-body form.

As we showed in Chapter 16, a case can be made within the framework of the QSSC for the view that the CMB, like all of the other diffuse radiation fields, comes from a large number of discrete sources. However, the argument that it is remnant radiation from a hot big bang has been so persuasive to most astronomers that discussion of the observations and their interpretation have largely been made in the context of the big-bang model. While we do not believe this to be correct, we shall in what follows give an account of this view, since contrary to most of the cosmological community, we believe that the reader is entitled to see arguments and evidence on both sides of important questions.

The microwave background – the popular view

In the standard big-bang model the radiation is thermalized within the first year after the beginning at a temperature greater than 10^7 K. Starting with the work of Penzias and Wilson in 1965, it became clear that the radiation is highly isotropic, with a black-body temperature close to 3 K. Early attempts to make measurements near the peak of the black-body curve suggested that there might be significant departures from a black-body energy distribution, but the instrument on the COBE satellite has shown that the curve has a very precise black-body form with $T = 2.728 \pm 0.01$ K (between 60 and 2880 Hz[5]). The measured CMB residual spectrum is shown in Fig. 21.2. To obtain this, a Planck black-body spectrum and a small galactic emission component have been subtracted from the measured spectrum in order to make the residuals visible.

There is a small dipole anisotropy (about 0.0033 K) associated with the motion of our Galaxy through the microwave background. Anisotropies on smaller scales than those associated with the motion of our Galaxy are extremely hard to detect, but it is believed that if they can be measured, it should be possible to differentiate between different versions of the so-called standard cosmological model. These anisotropies are all associated with the distortions of the radiation field which must have been caused by density fluctuations in the early universe, which are assumed to be present, since without them there would be no galaxies.

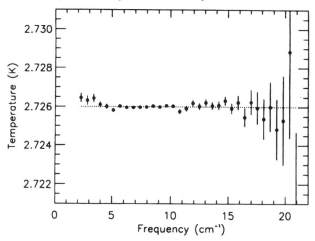

Fig. 21.2. The residual spectrum of the microwave background as a function of frequency (cm^{-1}). As was shown in Chapter 8, the black-body form is very precisely maintained at least out to radio frequencies.

It is worthwhile pointing out here that a fundamental distinction can be made between models in which the CMB originates in a big-bang universe and models in which the CMB arises in the QSSC. In the first case, the radiation and the density inhomogeneities are present initially and evolve together until the decoupling epoch. Thus the density fluctuations *must* leave an imprint on the CMB.

In the second case, it is the galaxies and the little big bangs in them which generate the energy in the CMB. The radiation is degraded into microwaves by interaction with matter which is ejected from the galaxies. Thus the radiation field is *generated by discrete objects* and becomes smooth through scattering and diffusion in space. The degree of smoothness depends completely on the details of this process.

Penzias and Wilson, in their original work, found that the radiation was isotropic to about 10%. The first theoretical ideas on galaxy formation in the big bang led to the idea that fluctuations of the order of 0.1–1% must be present due to simple gravitational effects: the Sachs–Wolfe effect. When it became clear that such anisotropies or temperature fluctuations (ΔT) were not present at this level, more theoretical modeling (often called more sophisticated modeling) led to a conclusion that smaller values of $\Delta T / T = 10^{-4}$ should be expected[6]. In the early 1980s, in the insistent drive to accept inflation and $\Omega_b = 1$, the proposal that most of the matter in the universe is non-baryonic was greeted with relief by those who were trying to find ways to push down the theoretical value of $\Delta T/T$ below 10^{-4}, since fluctuations at this level were not being seen.

If non-baryonic matter does dominate, then so-called 'refined' theoretical estimates showed that $\Delta T/T$ could be pushed down to $\sim 10^{-5}$. It was those two arguments – (a) to preserve the idea that the light elements were made in a big bang, and (b) to explain the very low level of the CMB fluctuations assumed to arise in a big bang – that led to non-baryonic matter being invented by the cosmologists, and it literally was *invented*. There was no other justification for it. The fact that it is being taken so seriously is a measure of the belief which many attach to a certain type of big-bang model.

The first real detection of anisotropies came from the COBE instruments in 1992[7] at the large angular scale $\sim 10°$ where it was found that $\Delta T/T \simeq 10^{-5} (\Delta T \sim 30 \pm 5 \ \mu K)$ (cf. Fig. 21.3). This was the result which led to the extravagant statements that were quoted in the preface to this book. These fluctuations indicate the presence of density variations on a scale of $\lesssim 100$ Mpc, as seen today.

COBE–DMR Map of CMB Anisotropy

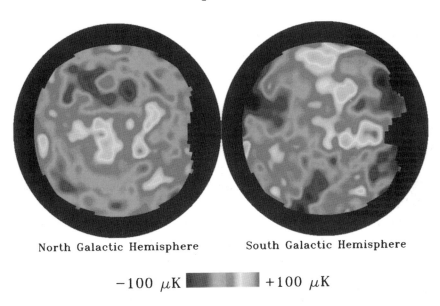

North Galactic Hemisphere   South Galactic Hemisphere

$-100 \ \mu K$ ▬▬▬▬ $+100 \ \mu K$

Fig. 21.3. This false color image shows tiny variations in the intensity of the cosmic microwave background measured in four years of observations by the Differential Microwave Radiometer on COBE.

By now variations of order $\Delta T/T \simeq 10^{-5}$ can be measured almost routinely, so that it is being claimed that by measuring such fluctuations and smaller ones on many angular scales, it will be possible to accurately measure cosmological parameters such as the baryon–photon ratio, the matter–photon ratio, H_0, the cosmological constant, and the optical depth to the last scattering surface. Currently there is extreme enthusiasm for this program, but of course no consideration of what can be discovered about other – non-big bang – cosmologies.

The physical causes of anisotropies with decreasing angular scale are[8]:

(a) the dipole anisotropy which measures our motion ($v_\odot$) relative to the standard of rest set by the CMB ($\Delta T/T = v_\odot/c$);

(b) the Sachs–Wolfe effect, or changes in the gravitational potential;

(c) adiabatic perturbations ($\Delta T/T = 1/3 \frac{\delta\rho}{\rho}$) (arising from inflation);

(d) isocurvature perturbations ($\Delta T/T = \frac{1}{3}.\delta S$) (fluctuations in the number density of particles which do not affect the energy density);

(e) Doppler shifts – a measure of when the radiation was last scattered ($\Delta T/T = v/c$);

(f) scattering by high-energy electrons in X-ray clusters (Zeldovich–Sunyaev effect) ($(\Delta T/T) = -2kT_e/(m_e c^2)$).

A summary recently published by Page[8] lists 28 ground-based experiments either completed, underway, or being planned, with only a small number completed, with resolution ranging from about $6°$ down to about $0.1°$ over a wide range of frequencies from about 10 GHz to 3000 GHz.

To analyze the temperature variations, it is customary to expand the temperature fluctuations in spherical harmonics and work in terms of multiple moments. It is also assumed that the initial (primordial) fluctuations can be represented by a power law in comoving wave numbers. We shall not discuss the details of the analysis, but the reader can consult White, Scott and Silk[9] and other references given there.

In this work the observed angular power spectrum is compared with the prediction for various models. The most popular model in recent years has been the cold dark matter (CDM) model in which the bulk of the mass is in the form of non-baryonic matter which is cold (meaning that its velocity dispersion is less than that of matter in a galaxy, i.e. $v \leq 300$ km s^{-1}). This model predicts a first peak in the plot of δT_l as l increases. We have already shown in Fig. 16.3 current data compared with calculated models.

Two new instruments which will make observations from space – the COBRAS–SAMBA (Planck Surveyor) satellite, and the MAP satellite –

are being planned, and it is from them and the results yet to be obtained from the ground that it is claimed that from the studies of the fluctuations on many scales, all of the cosmological parameters associated with the big-bang models will be obtained.

Despite the great enthusiasm, and the fact that large numbers of scientists and much money are being devoted to this work, it is clear that meaningful answers can come only if the microwave background originated in a big bang. We can confidently predict that, given the large number of parameters available in the big-bang scenario, *whatever anisotropy is discovered for large l* by these instruments, it will be argued that it confirms some theoretical prediction or other. If the past history of this field is any guide, it can be predicted that in the next few years there will be periodic news releases from NASA and some of the more vociferous big-bang believers making claims that yet another of their predictions has been fulfilled. Like earlier statements, these will tend to be timed to fit the annual requests from the groups to various Government Agencies for more money.

These are harsh judgments, but we believe that they are deserved as long as no attention whatever is paid to any alternative explanations. If our model is correct, small fluctuations will tell us very little. The reasons for this are as follows.

In QSSC the thermalization of the CMB is an ongoing process. Our approach to the uniformity of the background has been discussed in some detail already in Chapter 16. Uniformity comes, it will be recalled, from a mixing of radiation occurring dominantly at oscillatory maxima, on a time scale of ~ 100 billion years and on a distance scale $\sim 10^{29}$ cm. The uniformity so established is in the energy density of the radiation fields independent of their wavelength distributions. Subsequent absorptions and re-emissions by small particles, which we consider mostly to be thread-like whiskers of carbon and iron, do not change this uniformity. This is because the particles have little in the way of a heat capacity, and so they immediately re-radiate whatever energy they absorb, but at radically changed wavelengths. In the case of carbon whiskers, the re-radiation is essentially black-body at the temperature of the whiskers, which cannot be much different from that set by the CMB.

Because of the compression according to $S^{-3}(t)$ which occurs in the contracting phases of QSSC, thermalization by particles is most effective at and close to oscillatory minima. The thermalization can never be entirely perfect, however. There will be residual fluctuations from a strictly thermodynamic condition. The fluctuations will reflect the details of the sources of radiation and of the distribution of the thermalizing particles, just as there must still be very fine details in the deviations of the radiation inside a star from strict thermodynamics, reflecting the conditions under which the star formed from the interstellar gas. For stars, nobody has ever claimed such fluctuations have any deep-rooted significance, and according to our point of view so it is with small fluctuations in the CMB. These simply reflect the details of the way galaxies and clusters happen to be distributed. If we are interested in these, as well we might be, the sensible way to proceed is by observing galaxies and clusters directly, not by going through the extraordinarily convoluted process of examining very small fluctuations in the CMB.

References

Chapter 21

1. Longair, M.S. 1993, *The Physics of Background Radiation*, from The Deep Universe, Saas-Fee Advanced Course 23, (eds. A. R. Sandage, R. G. Kron & M. S. Longair), p. 323 (Swiss Society for Astrophysics and Astronomy, eds. B. Binggeli & R. Buser; Springer, Berlin 1995).
2. Hasinger, G., Burg, R., Giacconi, R., Hartner, G., Schmidt, M., Trümper, J. & Zamorani, G. 1993, *A&A*, **275**, 1.
3. Hasinger, G. 1995, *Wide Field Spectroscopy and the Distant Universe*, (eds. S. Maddox & A. Aragón-Salamanca, World Scientific Press), p. 321.
4. Burbidge, G. & Hoyle, F. 1996, *International Conference on X-ray Astronomy and Astrophysics – Roentgenstrahlung from the Universe* (eds. H. Ulrich Zimmermann, Joachim E. Trümper & Harold Yorke, MPE, Garching), p. 307.
5. Smoot, G. *et al.* 1991, *Ap. J.*, **371**, L1.
6. Wilson, M. L. & Silk, J., 1981, *Ap. J.*, **243**, 14.
7. Smoot, G. *et al.* 1992, *Ap. J.*, **396**, L1.
8. Page, Lyman, 1997, *Generation of Large-Scale Structure* (eds. D. Schramm & P. Galliotti; A. Wolfer), p. 75.
9. White, M., Scott, D. & Silk, J. 1994, *ARA&A*, **32**, 319.

22

A summary of the material contained in the previous chapters

History of conventional cosmology

In Chapters 1–8 we discussed the history of modern cosmology and the interplay between observation and theory. In Chapters 9 and 10 we gave a detailed account of the way that Gamow and his colleagues showed that the light elements could have been synthesized in the first few minutes of a big bang. But then we also discussed the history and background to that work and possible alternative ways of making the light isotopes.

Explosive cosmogony

In Chapters 11 and 12 we described at length the observational discoveries, starting in the 1960s, whose implications have largely been ignored by the general community. These results are of fundamental importance to our thesis that matter is created and ejected in the form of quasi-stellar objects and matter in other forms from the nuclei of galaxies. This is the cosmogony that is directly related to our cosmological model which we discussed in Chapters 15–18. As was described in Chapter 12, the original cosmogonical ideas are due to Ambartsumian. The most important results based on observation are as follows.

(1) That some kinds of extragalactic objects, mostly QSOs, are physically associated with apparently normal galaxies of stars, but they often have large redshifts compared with the galaxies. This is the most fundamental result since it implies:

    (a) that either the galaxies were ejected from the QSOs, or the QSOs were ejected from the galaxies. We consider it more likely that the latter hypothesis is correct. Once this premise is accepted it follows:

(b) that a component of the observed redshift of the QSO is intrinsic. In general, if z_Q and z_G are the redshifts of QSOs and galaxies respectively, $z_i = (1 + z_Q)/(1 + z_G) - 1$. We have attempted in Chapter 17 to understand the origin of the intrinsic redshifts. It is important to point out that independent of other evidence, unless all of the observational evidence for the phenomenon described in Chapter 11 can be shown to be incorrect, the existence of intrinsic redshifts – sometimes called non-cosmological redshifts which may range from very small values to values of $z_i \lesssim 2$ – must be accepted and the consequences taken into account. This argument applies *whether or not* the reader believes that our attempts to explain the phenomenon are on the right track. The major consequence of accepting (b) is that it leads to the result that:

(c) cosmological investigations based on QSO redshifts cannot be taken seriously since luminosity distances cannot be determined from the observed QSO redshifts unless the magnitudes of the intrinsic redshift components are known. The acceptance of (c) negates a great deal of extragalactic observational research currently underway. Only a cynic would argue that this is perhaps why the large amount of evidence has been ignored.

(2) The second important point made in Chapter 12 is that we consider it unlikely that the generally popular black hole–accretion disk explanation put forward to explain activity in galactic nuclei is correct.

(3) The third point which was originally made by Ambartsumian is that there is no good reason to believe that in *all* cases groups and clusters of galaxies are stable and bound. In some cases this is clearly true, but it is not a general rule. Thus, a large part of the argument for the presence of much dark matter goes away, since it is mostly based on the use of the virial theorem.

(4) The final point, also due to Ambartsumian originally, is that when we detect interacting systems of galaxies there is no good reason to believe that they are merging. This is what all of the observers like to assume, but from an observational point of view they may well be coming apart. Presumably the prejudice towards merging is based on the idea that gravity is the only force acting. But if explosive events are clearly seen, even if we do not understand them, the possibility that objects are separating rather than merging must be considered. Only in the cases in which tidal effects can clearly be seen to be present, as in the classical case of NGC 4038-39, is it reasonable to accept the merging hypothesis.

The microwave background and its smoothness

In Chapter 8 we discussed the history involving the discovery of the microwave background. While it is often stated that the detection of a microwave background radiation with a black-body spectrum is evidence that a big bang occurred, this is not necessarily the case. Those who say that seem unaware of the true sequence of discoveries in the field. As we pointed out in Chapter 8, the first detection of the microwave background was made by McKellar in 1941[1] from the sharp interstellar absorption lines due to CN originally identified by Adams[2]. McKellar showed the rotational levels indicated that the temperature must be about 2.3 K. This was based on an estimate of 5:1 for the relative intensity of the $R(0)$ line and the λ 3874.00 $R(1)$ line of the λ 3874.62 N band. His upper and lower limits given by putting $R(0)/R(1) = 2.5$ and $R(0)/R(1) = 10$ respectively were $T = 3.4$ K and $T = 1.8$ K.

The significance of this fundamental result seems to have been overlooked. For example, both George Gamow and one of us (FH) were aware of the result in the 1950s, but did not accept that it was related to big-bang radiation.

Also in Chapter 8 we pointed out that the investigations in the 1950s of Bondi, Gold and Hoyle[3] and one of us (GB)[4] concerning the origin of helium showed that there must have been a large release of energy if the helium was produced by hydrogen burning in stars. None of us seriously considered the possibility that this energy must ultimately have been degraded into black-body radiation. If that had been done, then based on the density of visible matter, we would have found that the temperature must be about 2.7 K. This then could have been related to McKellar's result. Alternatively if this value had been known and publicized in the 1950s, the detection by Penzias and Wilson would have been seen in a different light.

Of course, none of this occurred. But what it does show is that the existence of the microwave radiation does not necessarily have anything to do with a big bang, and certainly there is nothing of significance in the universal value of the temperature as far as the big bang is concerned. For big-bang cosmologists the fact that the value obtained above for the temperature of the CMB is equal to the measured value has to be a coincidence! We have discussed the natural way to produce it through helium production in stars in Chapter 8, and in creation processes in Chapter 16.

It seems very reasonable to suppose that the microwave radiation might very well have arisen from hydrogen burning in stars. We summarize here our earlier discussion showing how it can have arisen within the framework of the QSSC. First we work out the equilibrium average energy density of radiation in QSSC and then consider its homogeneity and isotropy. And lastly we consider its thermalization to a thermodynamic form. Let us now take these steps in turn.

(1) In QSSC the exponential expansion of the universe is determined by a factor in $S(t)$ of the form $\exp t/P$. In each exponentiation there are a number of oscillations of period Q. In the model considered in Chapter 16 the ratio of the exponentiation time scale P to the oscillatory period Q was about 20. In a universe with continuing expansion, the energy density of the radiation background is that which accumulates in $\frac{1}{4}P/Q$ oscillations, or about five oscillations.

Taking the background radiation to be derived from starlight and using $Q \simeq 100$ billion years as in Chapter 16, the equilibrium background energy density is that produced by stars in ~ 500 billion years. Now in the current cycle we know that stars have produced $\sim 10^{-14}$ erg cm^{-3} in about 10 billion years. For an approximately constant output with respect to time, the accumulation of energy in 500 billion years would then be $\sim 5 \times 10^{-13}$ erg cm^{-3}, which is just that of the microwave background. Hence in QSSC we obtain the observed energy density of the microwave background, not by purely *ad hoc* assumption as in big-bang cosmology, but from a simple calculation.

(2) In ~ 500 billion years, radiation travels a very long way, $\sim 10^5$ Mpc. At any spacetime point the Olbers effect guarantees that the radiation energy density will be extremely uniform. That coming from $\sim 10^5$ Mpc will dominate the local contribution from the nearest 10 Mpc by a factor $\sim 10^4$. Although local variations will introduce fluctuations they will be small, their nature depending on the details of how the distribution of galaxies happens to be in our region of the universe.

(3) Once the energy density of radiation becomes uniform it remains uniform so far as particle absorptions and re-emissions are concerned. The latter change the frequency distribution within the radiation but not the energy density. The circumstance that absorbing particles may be irregularly distributed has no effect at all on the energy density. And once the effect of absorptions and re-emissions reduces the radiation to a black-body form, irregularities in the thermalizing agent have no effect on the eventual result.

The epochs of thermalization are at oscillatory minima when the universal mean matter density is some hundreds of times higher than it is at present. Starlight accumulated in the immediately preceding cycle is mostly shifted in wavelength into the ultraviolet, where it is exposed to strong absorption by long thread-like carbon particles – whiskers. Because the emissivity of such particles is constant through the far infrared, the microwave region and into the radio band, the emission by carbon whiskers is black-body at the temperature of the whiskers, which will be close to that of the microwave background. Thus already at its first absorption and re-emission, starlight goes near where we want it to be – the microwave background – and with an immediate approximation to black-body form. The final step is then executed by metallic thread-like particles dominated by iron, which have an extremely high absorptivity in exactly the microwave region of the spectrum. It is worthwhile stressing that to metallurgists, carbon and metallic whiskers have been well known for half a century. They are the way vapors condense when, by laboratory standards, cooling is slow. In the laboratory they do, in fact, exist. Usually the universe has a way of being able to duplicate whatever is done in the laboratory.

The new theory

Following a critical discussion of standard cosmology in Chapter 14, we have described our new theory and the ways in which it can account for the current observations in Chapters 15–18. The major new development is that matter creation occurs with energy conservation (Chapters 14 and 15). Compensating for the positive energies of the usual particles and fields there is a negative component. The formulation of the mathematical properties of the negative component dates from the steady-state C-field of 1948. The C-field has both negative energy and negative pressure. The φ field of inflationary cosmology has negative pressure and is similar to the C-field in that respect.

Because these ideas can be put in an action principle form there is no vagueness about their mathematical consequences. Or about the conservation condition being satisfied. Or about the form of the gravitational equations. Because the term in the gravitational equation for $\dot{S}^2 + k$ (cf. 15.24) arising from the C-field is of opposite sign to the matter term and because it varies more steeply with $S(t)$, once this field exists it is impossible for the universe to collapse into a singularity of the big-bang type. Running time backwards, the universe inevitably bounces at the value of $S(t)$ for which $\dot{S} = 0$. This in our view occurs for $S_0/S(t) \simeq 6$, where S_0 is the present scale-factor, leading inevitably to the oscillatory model of Chapter 15, and so to a resolution of the microwave problem along the lines indicated above.

'Once this field exists' is a critical phase. Currently we have not proved theoretically that it does, or that matter creation takes place. Nor would we expect to do so from a classically formulated theory. Our position is that, *if* matter creation takes place, then we can describe its consequences. Our grounds for thinking that matter creation really does take place are observational. The entire class of violent events ranging from cases like NGC 4258 (Chapter 11) at the mild end up to the great radio sources (Chapter 12) requiring the release of upwards of 10^{60} erg, perhaps by several orders of magnitude, encourages us to bridge this critical gap.

Violent events are explained by believers in the big bang in terms of a black hole–accretion disk model in which a massive black hole is supposed to be present at the centers of most galaxies with a rotating accretion disk moving around it. Material falls into the system, impacting the disk and releasing explosive amounts of energy in the process. The first of our doubts about this scenario is whether the energy released could ever be as large as it is observed to be for the radio sources. But the major weakness of this proposal is that in more than two decades no convincing explanation for the presence of black holes at the centers of galaxies has been given. Observational evidence shows that it is likely that they do exist[5], and this is the usual answer given to this difficulty, supposedly removing the need for an explanation of it. However, if the origin of the black holes can be seen to demand the occurrence of creation events, this line of argument collapses.

Observation does not show the presence of massive black holes in the *disks* of spiral galaxies. Molecular clouds with masses upwards of $10^5 M_\odot$ are found, but only with average matter densities of $\sim 10^{-21} \mathrm{g\ cm}^{-3}$. The reason why there is no continuing condensation of such clouds is usually put down to rotary forces. When a cloud condenses, the inward gravitational forces increase as the inverse square of the scale of the cloud. But the rotary forces inhibiting condensation increase as the cube. Since the two are not far from being in balance initially, not much condensation is permitted before rotation becomes inimical to any further rise of the internal density.

The internal density inside a black hole of mass $10^6 M_\odot$, say, is about $10^4 \mathrm{g\ cm}^{-3}$. So a molecular cloud of initial average density $10^{-21} \mathrm{g\ cm}^{-3}$ would have to condense by upwards of a factor 10^8 in scale to produce a black hole, with rotary forces increasing by the immense factor of $\sim 10^{25}$. No amount of optimistic thinking can cope with an increase of that order. So what now is special about the center of a galaxy, instead of the disk, to permit a rise of rotary forces by this same enormous factor? Nothing. The center of a galaxy is certainly a unique point geometrically. But rotary forces are not suspended there. The situation is just the same there as for the disks of spirals.

When creation occurs, large numbers of particles of positive energy are produced, compensated by large numbers of negative energy particles. The wave equation for the latter, obtained from the action principle as we gave it in Chapter 14, is of massless particles without interaction with the positive energy particles, permitting the negative component to move through the positives, escaping freely unless seriously inhibited by gravitation. This is exactly what will happen to the C-field particles as the positive component is left behind. This latter component will build a gravitational barrier that prevents effective further dispersion of the negatives, a gravitational barrier consequent on the formation of a near-black hole.

A near-black hole is characterized by a barrier factor γ given by the ratio of time outside to time inside. For a near-black hole of mass $10^8 M_\odot$ the time scale for events inside is the time for the passage of light through a distance of the order of the gravitational radius $\sim 3 \times 10^{13}$ cm, i.e. a time $\sim 10^3$ s. Outside, however, propagation of the events through a barrier γ would be seen by an observer to occupy a time $\sim 10^3 \gamma$ s, which for $\gamma = 10^{10}$, say, is about 3×10^5 years. A complete-black hole has $\gamma \to \infty$ and events from its interior take too long ever to be observed externally. It is not emphasized enough that we do not see complete-black holes but only near-black holes.

Thus in our model, the existence of near-black holes is an indication that creation events have taken place. How do we visualize such events in physical terms? In particular, where do new particles come from if not from some previously existing particle? Our answer conceptually is from the basic fabric of spacetime.

We are well-used to thinking of the presence of a particle in spacetime in terms of 'paths', to which we can attach probabilities. A baryon is a triplet of quark paths weaving their way in close association in spacetime. The paths have abstract properties with respect to certain groups, quarks, with respect to SU_3 for the usual baryons.

The disposition among theoreticians seems to be to add more coordinates to the usual four of spacetime. A 'wrapping-up' of coordinates along a path in spacetime can be viewed as establishing an identity for the path, with other coordinates perhaps defining the nature of the particle, all within a single all-embracing group. This would seem an interesting way to go, especially as from our point of view it would bring together physically the properties of spacetime with those of the particles that in conventional language are supposed to move 'in' spacetime. It would imply, as we see it, that particle creation occurs in association with a transition that involves spacetime in a physical way, not just as a passive medium in which particles are supposed to exist.

In more graphic language, particle creation would be associated with an 'opening-up' of spacetime. The natural energy for provoking such a transition is that which we associate with a Planck particle, $(3/4\pi G)^{1/2}$, an energy of about 6×10^{18} GeV. This is essentially the energy at which the Compton wavelength of a particle is equal to its gravitational radius, where it is no longer possible to regard the particles as moving in locally flat spacetime. It is the threshold energy that we regard as being necessary for provoking a creation event. What happens then takes us into decisively new physics of which we have only the vaguest understanding. Although it would of course be possible to invent mathematical models, we doubt that at present they would really carry us much further towards reality.

Ironically, the highest energy encountered in the usual version of inflation in big-bang cosmology was $\sigma \simeq 2 \times 10^{15}$ GeV, short of where interesting things are expected to happen by a factor 3000.

In the QSSC model the positive energy component is born as Planck particles with energies $\sim 6 \times 10^{18}$ GeV. These decay to lower positive energy forms. At the same time compensating C-field particles come into existence with negative energies of magnitude $\sim 6 \times 10^{18}$ GeV. As they escape from the near-black hole of their birth, the C-field particles lose energy in magnitude by the factor $1/\gamma$, emerging into the cosmological background with negative energies of magnitude $\sim 6 \times 10^{18}/\gamma$ GeV, with the barrier factor that is appropriate to the particular near-black hole of their birth. We expect this to be variable from one C-field particle in the cosmological background to another. For $\gamma \gg 1$ the C-particle energies in the cosmological background are always small in magnitude compared with the creation threshold.

However, some C-field particles in the cosmological background will be incident on every near-black hole. And if for the moment we consider all black holes to have the same barrier factor γ, on entering a near-black hole the original C-field threshold energy $\sim 6 \times 10^{18}$ GeV may be recovered. The re-entry of cosmological C-field particles into black holes varies through each cycle of QSSC, however, the process being weakest at oscillatory maxima and strongest at minima. Because the magnitude of the C-field energy density varies with the scale factor $S(t)$ as S^{-6}, the individual particle energies vary as S^{-3}.

Since the surface area of a near-black hole varies as the square of its mass, this re-entry of C-field particles is more important for black holes of large mass than for those of small mass. Indeed, with the original production of particles by a black hole being only linearly proportional to the mass M, there is a value of M at which an oscillatory minimum adds as many C-field particles from the cosmological background as the black hole itself originally produced. These negative energy particles balance out the energy density of the positive component and permit destruction of the gravitational barrier, enabling the positive component to expand apart. Calculation suggests that the order of mass at which this comparatively gentle process, occurring through oscillatory minima, takes place is 10^{10}–$10^{12} M_\odot$, or the mass range of many galaxies. What happens is that the positive mass interior to near-black holes of galactic mass is disgorged very gently over time scales of the order of some hundreds of millions of years, as seen by an outside observer. This process, described in some detail in Chapter 18, is that which in our view supplies the material going to form the stars in galaxies. The process is a gentle one because it is contingent on the rise of the energy density of the cosmological C-field and this, again as seen externally, takes place over billions of years.

Our picture of violent events depends on the barrier factor γ being variable from one near-black hole to another. When a C-field particle falls into a black hole with a γ-factor higher than that of the black hole from which it originally escaped, the energy is lifted above threshold, above $\sim 6 \times 10^{18}$ GeV, in some cases by a considerable amount. It is then, in our view, that the spacetime structure becomes exposed, with new creation taking place on an appreciable scale – i.e. with high probability. Given a sufficient energy rise occurring to a sufficient number of incident C-field particles from the cosmological background, creation becomes essentially explosive on the internal time scale of the black hole.

Should the negative pressure associated with the newly created C-field particles become irregularly distributed, and should the negative pressure be large in magnitude, there will be a tendency for the black hole in question to breakup into fragments, which is emphatically what we see observationally, on a large scale in the case of the big radio sources, and on a smaller scale in cases like NGC 4258. In the latter case it seems as if the original black hole ejected a number of smaller ones that we see as X-ray sources distributed around the original larger mass. The picture is of black holes sometimes ejecting large quantities of positive energy particles and sometimes of them breaking up violently into two or more pieces. The first of these processes leads in our view to the formation of galaxies around black holes, while the second produces a cluster of developing galaxies around an initial black hole of large mass. This is a situation we associate observationally with a cluster of smaller galaxies around a highly massive central galaxy.

The picture is basically centrifugal, not the normal centripetal picture in which clusters of galaxies are supposed to have evolved from a more distended state. Although star formation in clouds of gas exuded from a central object could still proceed centripetally. These would be diffuse galaxies without a central black hole, such as quite likely occur in considerable numbers towards the lower end of the galaxy luminosity function.

Because the basic physics of these processes lies in uncharted waters we have necessarily been limited here only to a general overall sketch. We have given such an outline because we believe that this scenario seems to be what the observations described in Chapters 11 and 12 indicate.

The current observational picture

In the first part of the book we described the interplay between theory and observation from the earliest days up to the 1960s and beyond in a few areas – for example the determination of H_0. In order to complete the discussion, we have given in Chapters 19, 20 and 21, an abbreviated account of the current observations as they relate to the numbers and distribution of related objects in the cosmos, the large-scale distribution of matter and radiation fields as they are currently being investigated. Here we have attempted to concentrate on observed properties alone and keep the interpretation to a minimum. Unfortunately, modern extragalactic observers very rarely describe their data without a great deal of interpretation, and literally all of this is done almost without exception by assuming a Friedmann model for the universe, and innumerable simple assumptions concerning galactic evolution, and almost always if active galactic nuclei are involved, the black hole–accretion disk paradigm prevails. Thus, it is difficult to sort out what is actually observed from what is claimed to have been seen. But we have tried.

This completes our summary. We have tried in the first 21 chapters to give an account of how modern cosmology has developed, and why we find the conventional model unsatisfactory. In developing a new model, we feel quite strongly that this is a move in the right direction.

At the same time, we continue to believe that observations are the real driver in the field of extragalactic astronomy. Thus in the next and final chapter of the book we describe a number of unsolved problems that should serve as a challenge for the future.

References

Chapter 22

1. McKellar, A. 1941, *Pub. Dom. Astrophys. Observatory, Victoria*, Vol. 7, No. 5.
2. Adams, W.S. 1941, *Ap. J.*, **93**, 11.
3. Bondi, H., Gold, T. & Hoyle, F. 1955, *Obser. Mag.*, **75**, 80.
4. Burbidge, G. 1958, *PASP*, **70**, 83.
5. Kormendy, J. & Richstone, D. 1995, *ARA&A*, **33**, 581.

23

Some unsolved problems

By now the reader will have got a good idea of our view of the way that the observations should be driving cosmological ideas. While we have concentrated in this book on evidence that appears to bear directly on current cosmology or cosmogony we discuss below what we believe are some of the most difficult problems which are worthy of study.

We have shown that some of the basic phenomena originally interpreted as evidence of a big-bang model can also be interpreted differently in the quasi-steady-state model, and that in our view the cosmogony is pointing in this latter direction. But there are still many problems well grounded in observed phenomena which have not been understood quantitatively in any cosmological model and it is worthwhile describing some of them.

The origin of angular momentum

As far as we are aware, all astronomical bodies are rotating. Their angular momentum must be related to their origin and interactions. The origin of the angular momentum of galaxies if they were formed from initial fluctuations in a Friedmann universe was suggested long ago by one of us (FH[1]) to be due to tidal interactions between condensing systems. This idea has been further taken up by Peebles[2], but it is still not clear whether or not the mechanism gives an adequate solution.

If galaxies and other condensed objects are ejected from dense configurations as we have proposed, the angular momentum must arise from the relative motions of different parts of an ejected object, but we understand little of how this has taken place.

Related to this general problem, we also have the difficulty of understanding how objects which are formed by gravitational collapse and condensation, i.e. normal stars, manage to get rid of large amounts of angular momentum which they clearly must do. A significant fraction of the initial angular momentum of a collapsing molecular cloud must be removed so that the remainder can form a rotating disk and/or ultimately a binary or multiple star system, or a single star with a planetary system. In recent years there has been a great deal of work done on the star-formation problem and it is clear observationally that much matter is ejected in the formation process.

However, we know very much less about the way that these processes occur on the galactic scale. In our view the angular momentum problem on the cosmological scale is clearly related to the problem of the origin of magnetic fields in the universe, but it is not yet understood.

Magnetic fields

Magnetic fields with strengths of 1–10 μG are widespread in the cosmos – in galaxies, extended radio sources, and in clusters of galaxies. Within the framework of the big-bang cosmology the magnetic fields must either have originated from some exceedingly weak seed field generated in the inflationary epoch, or in the plasma epoch[3,4,5], and then amplified by dynamo action in galaxies and the like, or in the framework of the QSSC the flux may have been generated in the creation process in regions of very strong gravitational fields in the centers of galaxies. Zeldovich and Novikov[6] noted that the Friedmann model allows a weak magnetic field as an initial condition.

A great deal of attention has been given to the dynamo amplification of the seed field in galactic evolution. The problem is that the field needs to be amplified by very large factors, from 10^{-17}–10^{-18} G (assumed to be the initial seed field) to $\sim 10^{-5}$–10^{-6} G, in times $\sim 10^9$–10^{10} years.

Most of the models have been devoted to the amplification of an initial seed field by dynamo action in spiral galaxies. It is argued that the weak random magnetic fields in the interstellar gas will reach equipartition with the turbulent gas[7,8], i.e. $B_{eq} \approx (4\pi\rho v^2)^{1/2}$ where ρ is the density and v is the r.m.s. turbulent velocity. The values of B_{eq} in our Galaxy are ~ 5 μG. A different argument concerning equipartition is made when it is supposed that the relativistic electrons and the accompanying cosmic-ray protons in the galactic disks and halos are also close to equipartition with the magnetic fields. Here the measurements come from the observations of the non-thermal (synchrotron) radio emission which tells us directly the magnitude of the electron flux which we assume has together with the proton flux reached rough equipartition with the magnetic field energy. Radio observations of many spiral galaxies then give equipartition fields ranging from ~ 4 μG in M33 to ~ 12 μG in NGC 6946 and NGC 1566. Values up to 20 μG can be found in spiral arms, and in M82 in which a large-scale outburst is occurring, $B \simeq 50$ µG.

The regular field strengths $\bar{B}_\perp$ as obtained from the intensity of polarized emission are approximately a few microgauss. They are consistent with the regular field strengths $\bar{B}_\parallel$ which can be obtained from the Faraday rotation data if we assume typical electron densities of $\sim 3 \times 10^{-2}$ cm^{-3}. Since $\bar{B}_\perp \approx \bar{B}_\parallel$, the filling factor must be small, perhaps 0.5.

Since spiral galaxies do show large-scale magnetic fields of this order, the main challenge to the theorists is to show how these arise. It is argued that large-scale flows involving shear and compression can amplify weak seed fields, thereby converting small-scale fields into large-scale fields. This is the so-called α effect[9].

However, the feedback of the magnetic field on the α effect and on turbulent diffusion may be a problem[10] though Parker[11], one of the few experts in the field, disagrees. Recent work[12] has also suggested that the α effect may be drastically quenched. Clearly the question remains unresolved. Kulsrud and Anderson[13] have also raised the difficulty that the growth of large-scale fields (which are certainly present) will be supported by ambi-polar diffusion at small scales.

Our understanding of magnetic fields in spiral galaxies is very limited. An observation which we believe is highly significant is that for many nearby spiral galaxies the measured interstellar magnetic fields falls off much more slowly with radius than the matter density, i.e. in the outer parts of such galaxies the magnetic stresses become comparable to gravity. Not only is this hard to understand in theoretical modelling, but it is taking place at distances where the rotation curve is flat (Chapter 20) and the flatness is usually interpreted as evidence for the presence of a halo of dark matter. Thus the neglect of this effect may have a bearing on the conventional dark matter arguments. For example, if the dark matter were non-baryonic it is hard to see how it could interact with a magnetic field.

What about magnetic fields in the central regions of galaxies where activity is present?

The existence of jets due to synchrotron radiation is a well-known phenomenon in radio galaxies, and the model of Blandford and Rees[14] of a relativistic jet fed from gravitational energy released at the center is a popular theoretical model which has been assumed in many investigations. It is clear from the observations that the magnetic fields in small jets near the centers are $\sim 100\text{--}1400$ µG, and one of the main questions is how effective such jets are in carrying energy and magnetic flux outward into much larger volumes.

The nearest galaxy in which we can see nuclear activity that, in this case, is probably due to stars, is M82, and it is clear that not only is the magnetic field generated in the starburst region very strong[15], ~ 50 µG, but that magnetic flux along with plasma is being ejected.

While we have no difficulty in believing that viable models are possible for this type of activity involving starbursts and also perhaps in non-thermal conditions in the centers of radio galaxies, the most baffling problem appears to be to understand how magnetic fields as strong as 1–5 µG can be built up by such activity to fill regions with sizes 0.1–1 Mpc in the lobes of radio galaxies, and in the intergalactic medium in the central parts of clusters of galaxies. The observational evidence for this is now quite strong[16]. For more details about the magnetic fluids in general the reader should consult the recent reviews of Kronberg[16] and of Beck *et al.*[17].

If we look at the observations in terms of the energy density in various forms, we have in different environments the following properties.

The Milky Way (probably typical of spiral galaxies)

$$\epsilon_m \approx \epsilon_t \approx \epsilon_{cr} \approx \epsilon_{cmb},$$

where the magnetic energy $\epsilon_m = \bar{B}^2/8\pi$ and we have put $B \simeq 4$ µG, the turbulent energy $\epsilon_t = \rho v^2$ where we have put $\rho \approx 10^{-24}$ gm cm^{-3}, $\bar{v} \approx 10$ km s^{-1}; ϵ_{cr} is the measured cosmic-ray energy; and ϵ_{cmb} is the measured microwave background energy.

A typical large radio source

$$\epsilon_m \approx \epsilon_{cmb} \approx \epsilon_{cr} \gg \epsilon_t.$$

Intergalactic gas in clusters

$$\epsilon_m \approx \epsilon_{cmb} \approx \epsilon_{cr} \gg \epsilon_t.$$

In the latter two cases, we know that the gas density $\rho \le 10^{-27}$ gm cm^{-3}. For the intergalactic gas we have put $v \approx 500$ km s^{-1}.

These relations are hard to explain, particularly for radio sources and clusters. In our own Galaxy, the fact that $\epsilon_m = \epsilon_t = \epsilon_{cr}$ was originally used by Fermi as evidence that a Galactic acceleration mechanism for cosmic rays is at work. It has always been assumed that the fact that ϵ_{cmb} has the same numerical value is simply a coincidence! In the framework of the QSSC, on the other hand, we see a direct link between ϵ_{cmb} and the energy density of starlight.

For radio sources and clusters the dominant role of the magnetic energy shows how hard it is to understand its origin. It must be argued that after ejection of magnetic flux from the centers of active galaxies, instead of getting weaker, it is amplified by turbulent action involving the much lower density, but much faster moving, hot inter-galactic gas, but no quantitative theory demonstrating this has been developed, and we find it extremely difficult to understand.

Peculiarities in the redshift distribution

Among all of the observational discoveries of the past 30 years, it has been those which involve the measured redshifts which cause the most problems. They are so difficult to understand and so unexpected, that discussion of them has been almost completely left out of other books on cosmology.

We have described some of the data in Chapter 11 and have tried to take these into account in our theoretical discussion. But there are some phenomena that we have not so far described since we also have not been able to understand or explain them. Nevertheless we have concluded that they cannot be ignored since we believe that the data are good and will ultimately affect our view of the universe.

We divide the discussion into two categories. In the first, we discuss redshift anomalies in the spectra of normal galaxies. In the second we discuss redshift peaks and periodicities in the spectra of QSOs and related objects.

Redshifts of normal galaxies

Starting more than 20 years ago, Tifft[18] showed that the differential redshifts of galaxies in the Coma cluster showed a distinct periodicity with a value of $c \triangle z \sim 72 \, \mathrm{km \, s^{-1}}$. One of us (GB) was the third referee of his paper. The first referee (now deceased) reached the conclusion that the observations had been made correctly, but that there must be something the matter with the statistical analysis. So this paper was sent to a second referee, a distinguished statistician who worked on astronomical problems, who is also now deceased. That referee reached the conclusion that the statistical analysis had been done correctly, but that there must be something wrong with the observations. The third referee then got the paper and recommended that it be published. Tifft went on to find the same effect in members of nearby groups. These results were confirmed by Weedman, and over the years the phenomenon has been found in pairs of galaxies[19], and in the redshift differences between satellite galaxies and the central galaxies in small groups[20]. Tifft[21,22] extended the work to the global scale, arguing that the redshifts of galaxies occurring anywhere in the sky are periodic when a suitable correction is made for the solar motion. Thus there is both a global periodicity for galaxies anywhere in the sky as well as a periodicity in the differential redshifts of adjacent galaxies. Most recently, Guthrie and Napier[23] have done a comprehensive study confirming that these quantized effects are present in the redshifts of normal galaxies within the local supercluster ($cz < 2000$ km s^{-1}). Their work is based on extremely accurate redshifts measured using the 21 cm line. They have done an exceedingly careful study using two independent samples, each containing about 100 spirals with redshifts cz measured to an accuracy of ≤ 4 km s^{-1} and they have shown that the redshifts are strongly periodic with $P \simeq 37.6$ km s$^{-1} = c\Delta z$, about half the value first obtained by Tifft.

We show a power spectrum of one of the samples of redshifts in Fig. 23.1. This most remarkable discovery due to Tifft has been totally ignored by the

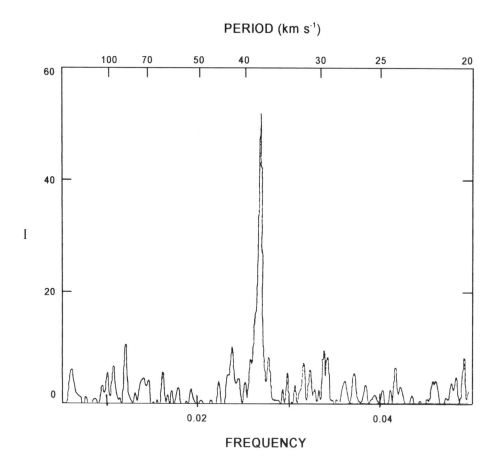

Fig. 23.1. Power spectrum associated with the redshifts of 97 spiral galaxies corrected for the solar motion taking the solar vector $(v_u, \ell_u, b_u) = (217$ km s^{-1}, $95°$, $-12°)$ (ref. 23).

cosmological community. In more than 20 years no one has ever been invited to talk about these data at any of the innumerable cosmological conferences devoted to conventional cosmology. These results have not been anticipated in any theory of which we are aware, nor is anyone except Tifft and Cocke making any attempt to do theoretical work in this field. The fact that the basic value of $c\triangle z$ is very small makes it hard to understand how a Doppler component can easily contribute to the observed value of z_0.

The other evidence for periodicities in the redshifts of normal galaxies comes from the pencil beam surveys of faint galaxies[24], and from studies of the redshift distribution of comparatively nearby clusters of galaxies[25] with $z < 0.12$. As was mentioned in Chapter 20, these apparent periodicities are interpreted as periodicities in the sizes of structures with scales of ~ 120 Mpc (ref. 24) and 240 Mpc (ref. 25). For $H_0 = 60$ km s^{-1} Mpc^{-1} these correspond to values of $c\triangle z \simeq 12\,800$ km s^{-1} and 25 600 km s^{-1}, immensely larger than the values of $c\triangle z$ in the Tifft effect. Of course, for the faint galaxies and the clusters the redshifts are not known with anything like the precision which is available for the bright galaxies used to discover the Tifft effect.

Apart from periodicities, Arp has found examples of apparently normal galaxies connected by optical features to other galaxies with very different redshifts[26,27]. The best case is NGC 7603 and its companion which is clearly physically connected to it[26]. The two galaxies have redshifts $c\triangle z = 8100$ km s^{-1} and 16 400 km s^{-1} respectively. We show this pair in Fig. 23.2. Arp has found other examples of pairs of galaxies with discrepant redshifts[27], but none as spectacular as NGC 7603. However, no one else has done any serious work on this since Arp was removed from the telescopes. If such pairs are found accidentally, the mind-set of the observers is to immediately conclude that a foreground galaxy is confused with a background galaxy (*because* two different redshifts are involved). After all, look what happened to Arp (cf. Chapter 11).

We know that the effect must be rare, otherwise a good Hubble relation would not be found for normal galaxies. In other words, if anomalous redshifts were widespread in galaxies as they apparently are in QSOs, there would be a large scatter in the Hubble diagram. Even so, it must be telling us something.

Redshifts of QSOs and related objects

We discussed in Chapter 11 the extensive evidence which shows that many QSOs have large redshift components that are not of cosmological origin.

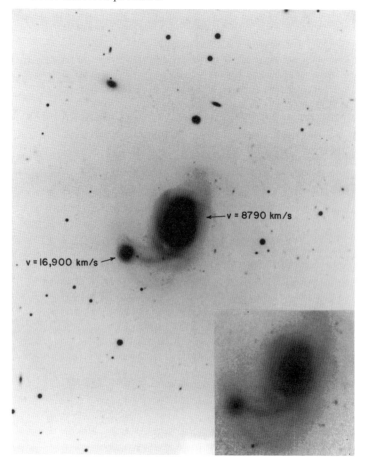

Fig. 23.2. This shows NGC 7603 and its companion galaxy. The luminous bridges between the two are clearly visible. With the physical connection between the two galaxies well established by the existence of the bridges, the very different redshifts are best understood as either being due to the smaller galaxy being ejected from the brighter one with a line-of-sight velocity component of 8100 km s^{-1}, or else a large part of that redshift has some other (intrinsic) origin. The inset image is printed at high contrast showing that there is a very strong compact nucleus in the companion galaxy. We are indebted to Dr Nigel Sharp who obtained this picture using the 4-meter telescope at KPNO.

The tendency for different QSOs to have very similar redshifts was seen in some of the first data. The first value of z that stood out[28] was $z = 1.955$.

Soon after that it became clear that if all of the objects which were either low-redshift QSOs or related objects with very similar emission line spectra, were examined as a class – there were only 72 of them in 1968 – the redshift distribution appeared to be quantized[29] with $\triangle z = 0.061$. Peaks could easily be seen at $n\triangle z$ where n ranged from 1 to 10. We show in Fig. 23.3 the original figure from the 1968 paper. With time the number of objects in this category has vastly increased and by 1990 more than 700 objects were known with $z < 0.2$. In this larger set the peaks have become even more prominent particularly at $z = 0.061$. Ninety-four objects out of the sample have redshifts[30] in the interval $0.056 \leq z \leq 0.065$. A histogram of this distribution is shown in Fig. 23.4. Burbidge and O'Dell[31] and later Duari *et al.*[32] carried out statistical tests of the samples and showed that the strong periodicity is real, the exact value of $\triangle z_0$ is 0.0565, and its significance is increased when the redshifts are transformed to the galactocentric frame[32]. A second period $\triangle z = 0.0128$, is also found at a high significance level.

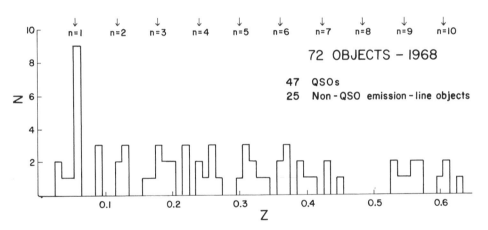

Fig. 23.3. Histogram of the number of objects n as a function of z based on all of the objects available in 1968 (ref. 29).

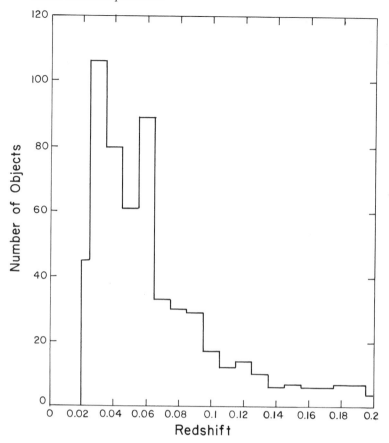

Fig. 23.4. This is an updated version of Fig. 23.3 made in 1990 (ref. 30). It is made from the restricted range of redshift $z \leq 0.2$. More than 500 objects are included and the huge peak at $z = 0.06$ is due to the fact that there are 94 objects with redshifts between $z = 0.056$ and $z = 0.065$.

As the number of QSO redshifts increased, it became apparent that there are major peaks in the QSO redshift distribution at $z = 0.30, 0.60, 0.96, 1.41$ and 1.96. We show in Fig. 23.6 the observed distribution as it appeared in 1977[33].

Analysis of these peaks were carried out by several groups including Depaquit, Pecker and Vigier[34] who confirmed the effect, and Box and Roeder[35] who did not. There were also claims that the peaks were due to optical selection effects. It was with this in mind that Fig. 23.5 was originally prepared with the emission lines found in the spectra of QSOs and the red-shift ranges over which they can be used, marked on the figure. But the best demonstration that the redshift peaks are real is obtained when the QSOs are chosen using criteria which have nothing to do with their spectral energy distribution. This is the case for the sample plotted in Fig. 23.6 which is all of the QSOs originally identified as 3C or 3CR radio sources, i.e. they were chosen from their radio flux characteristics and optical morphology. This is also the case for the QSOs in Table 23.1 where they have been chosen because of their proximity to and/or physical connection to bright comparatively nearby galaxies.

Karlsson[36] first showed that these redshift peaks could be fitted to a periodic sequence such that $\triangle \log(1 + z) = 0.089$, or $(1 + z_{n+1})/(1 + z_n) = 1.227$. According to this formula, the next peaks beyond $z = 1.96$ should occur at $z = 2.62$, 3.44 and 4.45. However, the redshift distribution starts to fall off quite rapidly near $z = 2.5$.

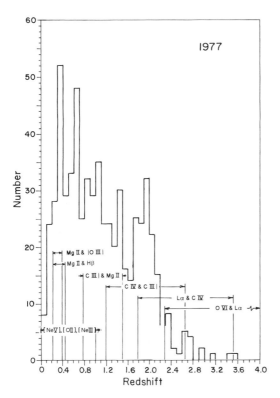

Fig. 23.5. This histogram was made in 1977 when only about 600 QSOs had their redshifts measured. They were predominantly QSOs identified originally as radio sources, were comparatively bright $< 18^m5$, and were distributed all over the sky. Here peaks at 0.30, 0.60, 0.96, 1.41 and 1.96 are obviously present. We now believe that many of these QSOs are quite close by so that the cosmological redshift terms z_c, are very small. The plot is taken from ref. 38.

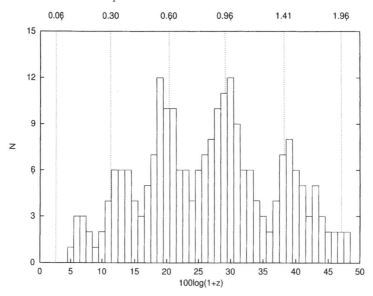

Fig. 23.6. This histogram has been made by W. Napier and G. Burbidge. The QSOs are all of those identified as radio sources in the 3C and 3CR catalogues. Thus, they were all identified from their radio flux properties and optical morphology. They make up a complete set and no selection effects associated with their optical energy distribution – lines or continuum – have entered into their identification. The data are taken from the catalog of Hewitt and Burbidge[39].

The most remarkable aspect of these peaks and periodicities is that they can be seen at all. The argument goes as follows. We all accept the basic premise that we are observing extragalactic objects, and that the universe is expanding. Consequently, the observed redshift z_0 must contain a component due to the expansion, z_c. In addition we always assume that at least all galaxies have some motions in addition to the cosmological expansion. Thus there must always be a Doppler component z_d. If for QSOs the major redshift component of intrinsic origin is z_i, then

$$(1 + z_0) = (1 + z_c)(1 + z_d)(1 + z_i).$$

But for a unique value of z at a peak to stand out, $z_0 \approx z_i$. Thus $z_c \ll 1$, and $z_d \ll 1$. For galaxies the random motions indicate that $z_d \leq 0.001$. Thus z_d is very much less than 1. However, a range of values of z_c from 0.001 to 0.05, say, which corresponds to a modest cosmological distance, would completely smear out the peaks shown in Figs. 23.5 and 23.6. Thus the appearance of sharp peaks forces us to the conclusion that the QSOs with these redshifts must be comparatively local objects.

As the number of redshifts of QSOs increased, the peaks just described became less prominent (cf. Fig. 23.7), though the peaks at $z_0 = 0.3$, 1.41 and 1.96 are still noticeable. In our view this is partly because many of the large numbers of QSOs identified by optical methods have cosmological redshift terms with a wide range of non-negligible values of z_c, and in part because there *are* optical selection effects present, because many of the radio quiet QSOs in the catalogues are initially identified from their colors.

That the cosmological redshift term z_c is responsible for some of the smearing is clear. For example, the statistical analysis showing the association of QSOs with the galaxies in the Lick catalog and other samples[37] shows that these QSOs must have values of $z_c \approx 0.1$. On the other hand, the peaks are most prominent in QSOs which can be clearly associated with nearby galaxies with very small values of z_c.

This is clear from the data in Table 23.1. This table contains an updated list of QSOs which are clearly physically associated with nearby active galaxies. Many of them were discussed earlier in Chapter 11. In each case the QSO redshift is very close to one of the peaks in the distribution discovered earlier. It appears that in these cases we can argue that $(1 + z_0) = (1 + z_i)(1 + z_c)(1 + z_d)$ where we have now added a term z_d which measures the true Doppler velocity in the line of sight of the QSOs ejected from the parent galaxy. Since z_0, z_i and z_c are known, it is a simple matter to calculate z_d. Both redshifted and blueshifted velocities cz_d are seen with magnitude $\leq 0.1c$.

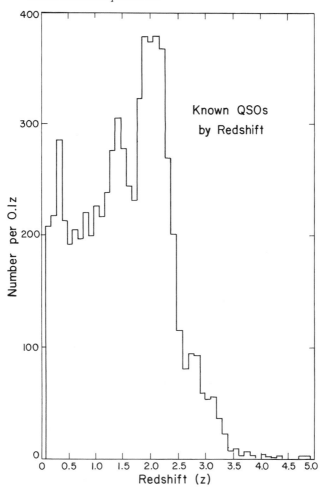

Fig. 23.7. This histogram is taken from Hewitt and Burbidge (ref. 39); 7300 QSOs are plotted. The peaks at $z \simeq 0.3$, 1.4, and 1.9–2.0 are very obvious.

Table 23.1 *Nearby galaxies with associated QSOs*

Galaxy	z_c	Evidence for association	z_Q	Appropriate z_i
NGC 470	0.0085	QSOs very close	1.875	1.96
			1.533	1.41
NGC 622	0.0174	QSOs very close	0.91	0.96
			1.46	1.41
NGC 1068 (Classical Seyfert)	0.004	Alignment of X-ray QSOs	0.655	0.60
			0.385	0.30
NGC 1073	0.0040	QSOs very close	0.599	0.60
			1.411	1.41
			1.945	1.96
NGC 2639	0.011	Alignment of X-ray QSOs	0.305	0.30
			0.332	0.30
NGC 3067	0.0049	QSO very close	0.533 (3CR 232)	0.60
NGC 3516 (Classical Seyfert)	0.0087	Alignment of X-ray QSOs	0.328	0.30
			0.690	0.60
			0.930	0.96
			1.399	1.41
			2.1	1.96
NGC 3842	0.020	X-ray QSOs very close	0.335	0.30
			0.946	0.96
			2.200	1.96
NGC 4138	0.0034	QSO very close	1.400 (3CR 268.4)	1.41
NGC 4258	0.0016	Alignment of X-ray QSOs	0.398	0.30
			0.653	0.60
NGC 4651	0.0026	QSO very close	0.577 (3CR 275.1)	0.60
NGC 5273	0.004	QSO very close	0.33	0.30
NGC 5548	0.017	Alignment of X-ray QSOs	0.560	0.60
			0.674	0.60
			0.89	0.96
			0.18	0.30
NGC 5689	0.0076	Alignment of X-ray QSOs	1.358	1.41
			1.94	1.96
			2.397	(2.62)
NGC 5832	0.0014	QSO very close	0.905 (3CR 309.1)	0.96
NGC 6212	0.03	QSO very close +23 other close QSOs	0.595 (3C 345)	0.60
NGC 7413	0.032	QSO very close	0.543 (3C 455)	0.60

Note that z_c is the redshift of the galaxy while z_Q is the redshift of the QSO, denoted in the text by z_0. We may take z_c to be the cosmological component of the QSO redshift, on the assumption that the QSO and the galaxy are physically associated.

Concluding remarks

In this last chapter we have described in outline a number of observed phenomena whose origins we do not understand either within the framework of the big-bang cosmology or within the framework of the QSSC. The universe is an immensely complicated place. There is good reason to start with simple models, but there is no excuse for ignoring observations which do not apparently fit into a picture which is largely based on some well accepted results, but also a number of preconceived ideas.

If nothing else, we hope that we have made both theorists and observers aware that observations remain primary in this field.

References

Chapter 23

1. Hoyle, F. 1949, *Problems of Cosmical Aerodynamics* (eds. J. Burgers & H.C. van de Hulst, Central Air Documents Office, Dayton Ohio).
2. Peebles, P.J.E. 1969, *Ap. J.* **155**, 393.
3. Zweibel, E. 1988, *Ap. J. Lett.* **329**, L1.
4. Harrison, E.R. 1970, *M.N.R.A.S.* **147**, 279.
5. Turner, M.S. & Widrow, L.M. 1988, *Phys. Rev. D.*, **37**, 2743.
6. Zeldovich, Ya.B. & Novikov, I.D. 1982, *Structure and Evolution of the Universe* (University of Chicago Press).
7. Kraichnan, R.H. 1965, *Phys. Fluids* **8**, 1385.
8. Zweibel, E. & McKee, C.F. 1995. *Ap. J.* **439**, 778.
9. Parker, E.N. 1955, *Ap. J.* **122**, 293.
10. Piddington, J.H. 1970, *Australian J. Physics*, **23**, 731.
11. Parker, E.N. 1973, *Astrophys. & Space Science*, **22**, 279.
12. Vainshtein, S. & Cattaneo, F. 1992, *Ap. J.*, **393**, 165.
13. Kulsrud, R. & Anderson, S. 1992, *Ap. J.*, **396**, 606.
14. Blandford, R.D. & Rees, M.J. 1974, *M.N.R.A.S.*, **169**, 395.
15. Reuter, H.P., Klein, V., Lesch, H., Weilibinski, R. & Kronberg, P. 1994, *Astron. and Ap.*, **256**, 10.
16. Kronberg, P. 1994, *Rep. Prog. Physics*, **57**, 325.
17. Beck, R., Brandenberg, A., Moss, D., Shukurov, A. & Sokoloff, D. 1996, *ARAA*, **34**, 155.
18. Tifft, W. 1976, *Ap. J.*, **206**, 38.
19. Tifft, W. & Cocke, W.J. 1989, *Ap. J.*, **336**, 128.
20. Arp, H.C. & Sulentic, J. 1985, *Ap. J.*, **291**, 88.
21. Tifft, W. 1995, *Astrophys. & Space Science*, **227**, 25.
22. Tifft, W. 1996, *Ap. J.*, **468**, 491.
23. Guthrie, B.N. & Napier, W.M. 1996, *Astron. & Ap.*, **310**, 353.
24. Broadhurst, T.J., Ellis, R., Koo, D. & Szalay, A. 1990, *Nature*, **343**, 726.
25. Einasto, J., Einasto, M., Gottlober, S., Muller, V., Sear, V., Starbinsky, A., Togo, E., Tucker, D., Andernach, H. & Frisch, P. 1997, *Nature*, **385**, 139.
26. Arp, H.C. 1970, *Astrophys. Letters*, **7**, 221.
27. Arp H.C. 1982, *Ap. J.*, **263**, 54.

28. Burbidge, G. & Burbidge, E.M. 1967, *Quasi-Stellar Objects* (San Francisco and London, W.H. Freeman and Co).
29. Burbidge, G. 1968, *Ap. J. Lett.*, **154**, L41.
30. Burbidge, G. & Hewitt, A. 1990, *Ap. J.*, **359**, L33.
31. Burbidge, G. & O'Dell, S. 1972, *Ap. J.*, **178**, 583.
32. Duari, D., Das Gupta, P. & Narlikar, J.V. 1992, *Ap. J.*, **384**, 35.
33. Hewitt, A. & Burbidge, G. 1980, *Ap. J. S.*, **43**, 57.
34. Depaquit, S., Pecker, J.C. & Vigier, J.-P. 1984, *Astron. Nach.*, **305**, 339.
35. Box, T.C. & Roeder, R.C. 1984, *Astron. & Ap.*, **134**, 234.
36. Karlsson, K.G. 1977, *Astron. & Ap.*, **58**, 237.
37. Chu, Y., Xhu, X., Burbidge, G. & Hewitt, A. 1984, *Astron. & Ap.*, **138**, 408
38. Burbidge, G. 1980, *Texas Symposium Proc. Ann. New York Academy of Sciences.*
39. Hewitt, A. & Burbidge G. 1993, *Ap. J. S.*, **87**, 451.

References

Abell, G.O. 1958, *Ap. J. S.*, **3**, 211. (Ch. 6, Ref. 1; Ch. 20, Ref. 9)

Abell, G.O., Corwin, H. & Olowin, R. 1989, *Ap. J. S.*, **70**, 1. (Ch. 20, Ref. 27)

Abraham, R.G., Tanvir, N.R., Santiago, B.X., Ellis, R.S., Glazebrook, K. & van den Bergh, S. 1996, *M.N.R.A.S.*, **279**, L47. (Ch. 19, Ref. 25)

Adams, W.S. 1941, *Ap. J.*, **93**, 11. (Ch. 22, Ref. 2)

Adams, W.S., Joy, A.H. and Humason, M.L. 1929, *Pub. Astron. Soc. Pacific*, **41**, 195. (Ch. 3, Ref. 11)

Alcock, C., Akenlof, C.W., Allsman, A. *et al.* 1993, *Nature*, **365**, 621; Alcock, C., Allsman, R.A., Axelrod, T.S. *et al.* 1996, *Ap. J.*, **461**, 84. (Ch. 20, Ref. 57)

Alfvén, H. & Herlofson, N. 1950, *Phys. Rev.*, **78**, 616. (Ch. 5, Ref. 3)

Alpher, R.A. & Herman, R. 1948, *Nature*, **162**, 774. (Ch. 8, Ref. 10)

Alpher, R.A. & Herman, R. 1949, *Phys. Rev.*, **75**, 1084. (Ch. 9, Ref. 4)

Alpher, R.A. & Herman, R. 1950, *Rev. Mod. Phys.*, **22**, 153. (Ch. 8, Ref. 11; Ch. 9, Ref. 5)

Alpher, R.A., Fellin, J.W., Jr., & Herman, R. 1953, *Phys. Rev.*, **92**, 1347. (Ch. 8, Ref. 12)

Ambartsumian, V.A. 1958, *Solvay Conference on Structure and Evolution of the Universe* (R. Stoops, Brussels), p. 241. (Ch. 12, Ref. 4; Ch. 20, Ref. 44)

Ambartsumian, V.A. 1961, *A. J.*, **66**, 536. (Ch. 12, Ref. 3; Ch. 20, Ref. 44)

Ambartsumian, V.A. 1965, *Structure and Evolution of Galaxies, Proc. 13th Conf. on Physics, University of Brussels* (New York, Wiley Interscience). (Ch. 12, Ref. 5)

Archer, S., Baldwin, J., Edge, D., Elsmore, B., Scheuer, P. & Shakeshaft, J. 1959, *Proc. Paris Symposium on Radio Astronomy* (Stanford Press), p. 487. (Ch. 7, Ref. 10)

Arp, H. 1966, *Atlas of Peculiar Galaxies* (California Institute of Technology). (Ch. 11, Ref. 12)

Arp, H. 1967, *Ap. J.*, **148**, 321. (Ch. 11, Ref. 11)

Arp, H.C. 1968, *PASP*, **80**, 129. (Ch. 11, Ref. 10)

Arp, H.C. 1970 *Astrophys. Letters*, **7**, 221. (Ch. 23, Ref. 26)

Arp, H.C. 1982, *Ap. J.*, **263**, 54. (Ch. 23, Ref. 27)

Arp, H.C. 1983, *Ap. J. L.*, **271**, L41. (Ch. 11, Ref. 21)

Arp, H.C 1987, *Quasars, Redshifts and Controversies* (Berkeley, Interstellar Media). (Ch. 11, Ref. 33)

Arp, H.C. 1990, *A&A*, **229**, 93. (Ch. 11, Ref. 35)

Arp, H.C. 1997, *A&A*, **319**, 33. (Ch. 11, Ref. 38)

Arp, H., Burbidge, E.M., Mackay, C. & Strittmatter, P. 1972, *Ap. J. L.*, **171**, L41. (Ch. 11, Ref. 15)

Arp, H.C. & Crane, P. 1992, *Phys. Lett. A.*, **168**, 6. (Ch. 11, Ref. 61)

Arp, H.C. & Sulentic, J. 1985, *Ap. J.*, **291**, 88. (Ch. 23, Ref. 20)

Aston, F. 1929, *Nature*, **123**, 313. (Ch. 4, Ref. 34)

Aubourg, E., Bareyre, P., Bréhin, S. *et al.* 1993, *Nature*, **365**, 623; Alard, C. & Guiberts, J. 1997, *A&A*, **326**, 1. (Ch. 20, Ref. 58)

Baade, W. 1944, *Ap. J.*, **100**, 137. (Ch. 4, Ref. 11)

Baade, W. 1953, *Trans. IAU*, **8** (Cambridge U. Press), p. 387. (Ch. 4, Ref. 13)

Baade, W. & Minkowski, R. 1954, *Ap. J.*, **119**, 206. (Ch. 5, Ref. 11)

Baade, W. & Zwicky, F. 1934, *Proc. Natl. Acad. Sci.*, **20**, 254. (Ch. 3, Ref. 8)

Babcock, H.W. 1939, *Bull. Lick Obs.*, **19**, 41. (Ch. 20, Ref. 35)

Bagla, J.S., Padmanabhan, T. & Narlikar, J.V. 1996, *Com. Astrophys.*, **18**, 275. (Ch. 4, Ref. 41)

Bahcall, J. & Soneira, R. 1984, *Ap. J. S.*, **55**, 67. (Ch. 20, Ref. 31)

Bahcall, J.N., Jannuzi, B.T., Schneider, D.P., Hartig, G.F., Bohlin, R. & Junkkarinen, V. 1991, *Ap. J.*, **377**, L5. (Ch. 11, Ref. 79)

Bahcall, J.N., Kirhakos, S. & Saxe, D.H. 1997, *Ap. J.*, **479**, 642. (Ch. 11, Ref. 49)

Bahcall, J.N., McKee, C.F. & Bahcall, N. 1971, *Astroph. Letters*, **10**, 147. (Ch. 11, Ref. 18)

Banerjee, S. and Narlikar, J.V. 1997, *Ap. J.*, **487**, 69. (Ch. 20, Ref. 60)

Barlow, T. 1997, private communication. (Ch. 11, Ref. 77)

Barnes, E.W., Bishop of Birmingham, 1931, *British Assoc. for the Adv. of Sci., Centenary Mtg.* (London, Spottiswoode, Ballantyne), p. 598. (Ch. 2, Ref. 9)

Barnothy, J. & Barnothy, M.F. 1968, *Science*, **167**, 348; 1965, *A. J.*, **70**, 666. (Ch. 20, Ref. 51)

Bartelmann, M. & Schneider, P. 1993, *A&A*, **271**, 421. (Ch. 11, Ref. 26)

Bartelmann, M. & Schneider, P. 1994, *A&A*, **284**, 1. (Ch. 11, Ref. 27)

Beck, R., Brandenberg, A., Moss, D., Shukurov, A. & Sokoloff, D. 1996, *ARAA*, **34**, 155. (Ch. 23, Ref. 17)

Begeman, K., Broeils, A. & Sanders, R. 1991, *M.N.R.A.S.*, **249**, 523. (Ch. 20, Ref. 41)

Bennett, A.S. *et al.* 1962, *Mem. Roy. Astr. Soc.*, **68**, 163. (Ch. 19, Ref. 2)

Bennett, T.L., Turner, M.S. & White, M. 1997, *Phys. Today*, **50**, 32. (Ch. 16, Ref. 10)

Bergeron, J. & Boissé, J. 1991, *A&A*, **243**, 344. (Ch. 11, Ref. 67)

Bergeron, J., D'Odorico, S. & Kunth, D. 1987, *A&A*, **180**, 1. (Ch. 11, Ref. 81)

Blaauw, A. & Morgan, H.-R. 1954, *Bull. A. Ned.*, **12**, 95. (Ch. 4, Ref. 14)

Blandford, R.D. & Rees, M.J. 1974, *M.N.R.A.S.*, **169**, 395. (Ch. 23, Ref. 14)

Boesgaard, A. M. & Steigman, G. 1985, *Ann. Rev. A & A*, **23**, 319. (Ch. 8, Ref. 5)

Bolton, J. & Wall, J. 1970, *Aust. J. Physics*, **23**, 789. (Ch. 11, Ref. 17)

Bolton, J.G. 1948, *Nature*, **162**, 141. (Ch. 5, Ref. 8)

Bolton, J.G. & Stanley, G. 1948, *Nature*, **161**, 312. (Ch. 5, Ref. 6)

Bondi, H. & Gold, T. 1948, *M.N.R.A.S.*, **108**, 252. (Ch. 7, Ref. 1)

Bondi, H., Gold, T. & Hoyle, F. 1955, *Observatory*, **75**, 80. (Ch. 8, Ref. 13; Ch. 22, Ref. 3)

Bonifacio, P. & Molaro, P. 1997, *M.N.R.A.S.*, **285**, 847. (Ch. 9, Ref. 12)

Borgeest, V. & Melhert, D. 1993, *A&A*, **275**, L21. (Ch. 11, Ref. 69)

Borgeest, V., Dietrich, M., Hopp, V., Kollatschny, W. & Schramm, K.-J. 1991, *A&A*, **243**, 93. (Ch. 11, Ref. 83)

Boroson, T. & Oke, J.B. 1982, *Nature*, **296**, 397. (Ch. 11, Ref. 52)

Boroson, T. & Oke, J.B. 1984, *Ap. J.*, **281**, 535. (Ch. 11, Ref. 54)

Bowen, E.G. 1987, *Radar Days* (Bristol, Institute of Physics Publishing Co. Ltd., Adam Hilger). (Ch. 5, Ref. 16)

Box, T.C. & Roeder, R.C. 1984, *Astron. & Ap.*, **134**, 234. (Ch. 23, Ref. 35)

Boyle, B.J., Shanks, T. & Peterson, B.A. 1990, *M.N.R.A.S.*, **243**, 1. (Ch. 19, Ref. 4)

Branch, D. & Tammann, G.A. 1992, *ARA&A*, **30**, 359. (Ch. 4, Ref. 27)

Broadhurst, T.J., Ellis, R., Koo, D. & Szalay, A. 1990, *Nature*, **343**, 726. (Ch. 23, Ref. 24)

Broadhurst, T.J., Ellis, R.S. & Shanks, T. 1988, *M.N.R.A.S.*, **235**, 827. (Ch. 19, Ref. 16)

Broadhurst, T.J., Ellis, R.S., Koo, D.C. & Szalay, A. 1990, *Nature*, **343**, 726. (Ch. 20, Ref. 25)

Browne, I.A.W. 1982, *Ap. J.*, **263**, L7. (Ch. 11, Ref. 20)

Burbidge, E.M. 1995, *A&A*, **298**, L1. (Ch. 11, Ref. 42; Ch. 19, Ref. 8)

Burbidge, E.M. 1997, *Ap. J.*, **484**, L99. (Ch. 11, Ref. 43; Ch. 19, Ref. 9)

Burbidge, E.M., private communication. (Ch. 19, Ref. 10)

Burbidge, E.M. & Burbidge, G.R. 1975, *Ap. J.*, **202**, 287. (Ch. 11, Ref. 74)

Burbidge, E.M. & Burbidge, G. 1997, *Ap. J. L.*, **477**, L13. (Ch. 11, Ref. 40)

Burbidge, E.M., Burbidge, G., Fowler, W.A. & Hoyle, F. 1957, *Rev. Mod. Phys.*, **29**, 547. (Ch. 4, Ref. 37; Ch. 9, Ref. 6)

Burbidge, E.M., Burbidge, G., Solomon, P.M. & Strittmatter, P.A. 1971, *Ap. J.*, **170**, 223. (Ch. 11, Ref. 13)

Burbidge, G. 1958, *Publ. Astr. Soc. Pacific*, **70**, 83. (Ch. 8, Ref. 14; Ch. 22, Ref. 8)

Burbidge, G. 1959, *Ap. J.*, **129**, 849. (Ch. 12, Ref. 6)

Burbidge, G. 1968, *Ap. J. Lett.*, **154**, L41. (Ch. 23, Ref. 29)

Burbidge, G. 1975, *Ap. J.*, **106**, L7. (Ch. 20, Ref. 49)

Burbidge, G. 1980, *Texas Symposium Proc. Ann. New York Academy of Sciences.* (Ch. 23, Ref. 38)

Burbidge, G. 1985, *A. J.*, **90**, 1399. (Ch. 11, Ref. 60)

Burbidge, G. 1996, *A&A*, **309**, 9. (Ch. 11, Ref. 32)

Burbidge, G. & Burbidge, E.M. 1964, *The Structure and Evolution of Galaxies* (ed. Professor Prigogine) *Proc. 13th Solvay Conf. on Physics*, Sept. 1964, p. 137. (Ch. 12, Ref. 7)

Burbidge, G. & Burbidge, M. 1967, *Quasi-Stellar Objects* (San Francisco and London, W. H. Freeman and Co.). (Ch. 11, Ref. 1; Ch. 23, Ref. 28)

Burbidge, G. & Burbidge, M. 1975, in *Stars and Stellar Systems*, Vol. 9 (eds. A. Sandage, M. Sandage & J. Kristian, Univ. Chicago Press), p. 81. (Ch. 20, Ref. 36)

Burbidge, G. & Hewitt, A. 1987, *A. J.*, **93**, 1. (Ch. 11, Ref. 84; Ch. 19, Ref. 5)

Burbidge, G. & Hewitt, A. 1990, *Ap. J.*, **359**, L33. (Ch. 23, Ref. 30)

Burbidge, G. & Hoyle, F. 1996, *International Conference on X-ray Astronomy and Astrophysics – Roentgenstrahlung from the Universe* (eds. H. Ulrich Zimmermann, Joachim E. Trümper & Harold Yorke, MPE, Garching), p. 307. (Ch. 21, Ref. 4)

Burbidge, G. & Hoyle, F. 1998, *Ap. J.*, **509**, L1. (Ch. 9, Ref. 18; Ch. 10, Ref. 1)

Burbidge, G. & O'Dell, S. 1972, *Ap. J.*, **178**, 583. (Ch. 23, Ref. 31)

Burbidge, G., Hewitt, A., Narlikar, J.V. & Das Gupta, P. 1990, *Ap. J. S.*, **74**, 675. (Ch. 11, Ref. 24)

Burbidge, G., Hoyle, F. & Schneider, P. 1997, *A&A*, **320**, 8. (Ch. 11, Ref. 58)

Burbidge, G., O'Dell, S.L., Strittmatter, P.A. 1972, *Ap. J.*, **175**, 601. (Ch. 11, Ref. 16)

Butcher, H. & Oemler, A. 1978, *Ap. J.*, **219**, 18. (Ch. 6, Ref. 9; Ch. 19, Ref. 18)

Butcher, H. & Oemler, A. 1984, *Ap. J.*, **226**, 559. (Ch. 19, Ref. 19)

Cameron, A.G.W. & Fowler, W.A. 1971, *Ap. J.*, **104**, 111. (Ch. 9, Ref. 15)

Canizares, C.L. 1981, *Nature*, **291**, 620. (Ch. 11, Ref. 34)

Chambers, K.C. & Charlot, S. 1990, *Ap. J.*, **348**, L1. (Ch. 6, Ref. 13)

Chambers, K.C., Miley, G. & van Breugel, W. 1987, *Nature*, **329**, 604. (Ch. 6, Ref. 10)

Charlier, C.V.R. 1922, *Arkiv för Mat. Astron. Fys*, **16**, 1. (Ch. 20, Ref. 24)

Christianson, G.E. 1995, *Edwin Hubble* (New York, Farrar, Strauss and Giroux). (Ch. 3, Ref. 20)

Chu, Y., Wei, J., Hu, J., Zhu, X. & Arp, H. 1998, *Ap. J.*, **500**, 596. (Ch. 11, Ref. 44)

Chu, Y., Zhu, X., Burbidge, G. & Hewitt, A. 1984, *A&A*, **138**, 408. (Ch. 11, Ref. 25; Ch. 23, Ref. 37)

Clerke, Agnes M. 1905, *The System of the Stars* (London, Adam & Charles Black), p. 349. (Ch. 1, Ref. 1)

Code, A. D. 1959, *Pub. Astr. Soc. Pacific*, **71**, 118. (Ch. 7, Ref. 5)

Coleman, P.H. & Pietronero, M. 1992, *Phys. Reports*, **213**, 311. (Ch. 20, Ref. 22)

Coleman, S. & Weinberg, E. 1973, *Phys. Rev. D.*, **7**, 1888. (Ch. 14, Ref. 4)

Colless, M., Ellis, R., Taylor, K. & Hook, R.N. 1990, *M.N.R.A.S.*, **244**, 408. (Ch. 19, Ref. 15)

Courant, R. & Hilbert, D. 1962, *Methods of Mathematical Physics* (New York, Interscience). (Ch. 17, Ref. 2)

Cowan, J.J., Thielemann, F.-K. & Truran, J.W. 1991, *ARA&A*, **29**, 447. (Ch. 4, Ref. 40)

Curtis, H.D. 1917, *Lick Obs. Bull.*, **9**, 108; 1919, *Proc. Natl. Acad. Sci.*, **9**, 218. (Ch. 3, Ref. 5)

Curtis, H.D. 1933, *The Nebulae*, Hdb V/2 (Springer), p. 861. (Ch. 3, Ref. 10)

Da Costa, L.N. (& 8 co-authors) 1988, *Ap. J.*, **327**, 544. (Ch. 20, Ref. 20)

Dar, A. 1991, *Ap. J. L.*, **382**, L1. (Ch. 11, Ref. 63)

Das Gupta, P., Narlikar, J.V. & Burbidge, G.R. 1988, *A. J.*, **95**, 5. See also *Non-evolving Luminosity Functions for Radio Galaxies*, P. Das Gupta 1988, Ph.D. Thesis, Bombay University. (Ch. 7, Ref. 18)

Dent, W.A. 1965, *Nature*, **205**, 487. (Ch. 11, Ref. 5)

Depaquit, S., Pecker, J.C. & Vigier, J.-P. 1984, *Astron. Nach.* **305**, 339. (Ch. 23, Ref. 34)

de Sitter, W. 1917, *Proc. Akad. Weteusch Amsterdam*, **19**, 1217. (Ch. 2, Ref. 4)

de Vaucouleurs, G. 1948, *Comptes Rendus*, **237**, 466. (Ch. 7, Ref. 4)

de Vaucouleurs, G. 1953, *A. J.*, **58**, 30; 1956, in *Vistas in Astronomy*, Vol. 2 (ed. A. Beer, London and New York, Pergamon Press), p. 1584; 1958, *A. J.*, **63**, 253. (Ch. 20, Ref. 1)

de Vaucouleurs, G. *see* Abell, G. 1975, in *Galaxies and the Universe* (eds. A. and M. Sandage, Chicago, University Chicago Press), Ch. 15. (Ch. 4, Ref. 23)

de Vaucouleurs, G. 1982, *Lectures, Austr. Natl. Univ. (Canberra)*. (Ch. 4, Ref. 20)

de Vaucouleurs, G. & de Vaucouleurs, A. 1964, *Bright Galaxy Catalogue* (Austin, University of Texas Press). (Ch. 11, Ref. 84)

de Vaucouleurs, G. & de Vaucouleurs, A. 1991, *Third Reference Catalogue of Bright Galaxies*, Vols. I–III (New York, Springer Verlag). (Ch. 20, Ref. 13)

Dingle, H. 1953, *Observatory*, **73**, 42. (Ch. 7, Ref. 21)

Dose, A. 1927, *Astron. Nacht.*, **229**, 157. (Ch. 3, Ref. 16)

Douglas, A.E. & Herzberg, G. 1941, *Ap. J.*, **94**, 381. (Ch. 8, Ref. 3)

Dressler, A. 1990, preprint. (Ch. 20, Ref. 21)

Dressler, A. & Gunn, J.E. 1992, *Ap. J. S.*, **78**, 1. (Ch. 6, Ref. 8)

Drinkwater, M., Webster, R.L. & Thomas, P. 1993, *A. J.*, **106**, 848. (Ch. 11, Ref. 72)

Duari, D. & Narlikar, J.V. 1995, *Int. J. Phys. D.*, **4**, 367. (Ch. 11, Ref. 70)

Duari, D., Das Gupta, P. & Narlikar, J.V. 1992, *Ap. J.* **384**, 35. (Ch. 23, Ref. 32)

Duc, P.-A. & Mirabel, I.F. 1991, *A&A*, **289**, 83. (Ch. 12, Ref. 2)

Eales, S.A. & Rawlings, S. 1996, *Ap. J.*, **460**, 68. (Ch. 6, Ref. 15)

Einasto, J. & Gramman, M. 1993, *Ap. J.*, **407**, 443. (Ch. 20, Ref. 26)

Einasto, J., Einasto, M., Gottlober, S., Muller, V., Sear, V., Starobinsky, A., Togo, E., Tucker, D., Audernach, H. & Frisch, P. 1997, *Nature*, **385**, 139. (Ch. 20, Ref. 28; Ch. 23, Ref. 25)

Einasto, J., Krasnik, A. & Saar, E. 1975, *Nature*, **250**, 309. (Ch. 20, Ref. 47)

Einstein, A. 1915, *Preuss. Akad. Wiss. Berlin, Sitzber*, 778, 799, 844. (Ch. 1, Ref. 2; Ch. 2, Ref. 1)

Einstein, A. 1917, *Preuss. Akad. Wiss. Berlin, Sitzber*, 142. (Ch. 2, Ref. 3)

Einstein, A. & de Sitter, W. 1932, *Proc. Natl. Acad. Sci.* **18**, 213. (Ch. 2, Ref. 10)

Ellis, R. 1987, in *Observational Cosmology* IAU Symp. 124, eds. A. Hewitt, G. Burbidge & L.Z. Fang (Dordrecht, Reidel), p. 367. (Ch. 19, Ref. 14)

Ellis, R.S. 1997, *ARAA*, **35**, 389. (Ch. 19, Ref. 17)

Elsmore, B., Ryle, M. & Leslie, P. 1959, *Mem. R.A.S.*, **68**, 62. (Ch. 5, Ref. 19)

Faber, S.M. & Gallagher, J.S. 1979, *ARA&A*, **17**, 135. (Ch. 20, Ref. 43)

Fabian, A. 1994, *ARAA*, **32**, 277. (Ch. 19, Ref. 7)

Fabian, A. & Barcons, X. 1992, *ARAA*, **30**, 429. (Ch. 19, Ref. 6)

Fowler, W.A. 1987, *Q. J. R. Astron. Soc.*, **28**, 87. (Ch. 4, Ref. 39)

Fowler, W.A., Caughlan, G. & Zimmerman, B. 1975, *ARA&A*, **13**, 69. (Ch. 10, Ref. 3)

Freedman, W. L. *et al.* 1994, *Nature*, **371**, 757. (Ch. 4, Ref. 24)

Friedmann, A., 1922, *Z. Phys.*, **10**, 377. (Ch. 2, Ref. 5)

Friedmann, A. 1924, *Z. Phys.*, **21**, 326. (Ch. 2, Ref. 6)

Gamow, G. 1946, *Phys. Rev.*, **70**, 572. (Ch. 8, Ref. 6; Ch. 9, Ref. 2)

Gamow, G. 1948, *Nature*, **162**, 680. (Ch. 8, Ref. 9)

Gamow, G. 1948, *Phys. Rev.*, **74**, 505. (Ch. 9, Ref. 3)

Geller, M. *et al.* 1997, *A. J.*, **114**, 2205. (Ch. 20. Ref. 18)

Giavalisco, M. 1997, *The Universe at Low and High Redshifts* (Woodbury, AIP Press), in press. (Ch. 19, Ref. 22)

Giavalisco, M., Livio, M., Bohlin, R.C., Macchetto, F.D. & Stecher, T.P. 1996, *A. J.*, **112**, 369. (Ch. 19, Ref. 23)

Giavalisco, M., Steidel, C.C. & Macchetto, F.D. 1996, *Ap. J.*, **470**, 189. (Ch. 19, Ref. 26)

Gilmore, G., Wyse, R. & Kuijken, K. 1989, *ARA&A*, **27**, 62. (Ch. 20, Ref. 32)

Ginzburg, V.L. 1951, *Dokl. Akad. Nauk SSSR*, **76**, 377; see also 1984 *In the Early Years of Radio Astronomy* (ed. W. Sullivan, Cambridge University Press), p. 289. (Ch. 5, Ref. 4)

Giovanelli, R. & Haynes, M. 1991, *ARA&A*, **29**, 499. (Ch. 20, Ref. 19)

Glanz, J. 1998, *Science*, **282**, 1249. (Ch. 16, Ref, 71)

Gold, T. & Hoyle, F. 1959, *Proc. Paris Symposium No. 9 on Radio Astronomy*, (Stanford Univ. Press), p. 583. (Ch. 7, Ref. 14)

Greenstein, J.L. & Matthews, T.A. 1963, *Nature*, **197**, 1037. (Ch. 5, Ref. 26)

Gunn, J.E. & Oke, J.B. 1975, *Ap. J.*, **195**, 255. (Ch. 6, Ref. 2)

Guth, A. 1981, *Phys. Rev. D.*, **23**, 347. (Ch. 14, Ref. 1)

Guthrie, B.N. & Napier, W.M. 1996, *Astron. & Ap.*, **310**, 353. (Ch. 23, Ref. 23)

Hamuy, M., Phillips, M.M., Maza, J., Suntzeff, N.B., Schommer, R.A. & Avilez, R. 1995, *A. J.*, **109**, 1. (Ch. 4, Ref. 29; Ch. 16, Ref. 8)

Hanbury-Brown, R. 1991, *Boffin* (Bristol, Institute of Physics Publishing Co. Ltd., Adam Hilger). (Ch. 5, Ref. 15)

Harrison, E.R. 1970, *M.N.R.A.S.* **147**, 279. (Ch. 23, Ref. 4)

Hasinger, G. 1995, *Wide Field Spectroscopy and the Distant Universe*, eds. S. Maddox & A. Aragón-Salamanca (World Scientific Press), p. 321. (Ch. 21, Ref. 3)

Hasinger, G., Burg, R., Giacconi, R., Hartner, G., Schmidt, M., Trümper, J. & Zamorani, G. 1993, *A&A*, **275**, 1. (Ch. 21, Ref. 2)

Hazard, C., Mackey, M.B. & Shimmins, A.J. 1963, *Nature*, **197**, 1037. (Ch. 5, Ref. 24)

Heckman, T. & Lehnert, M. 1997, *The Hubble Space Telescope and the High Redshift Universe* (eds. N.R. Tanvir, A. Aragon-Salamanca and J.V. Wall, World Scientific Pub. Singapore), p. 377. (Ch. 11, Ref.50)

Hewitt, A. & Burbidge, G. 1980, *Ap. J. S.* **43**, 57. (Ch. 23, Ref. 33)

Hewitt, A. & Burbidge, G. 1987, *Ap. J. S.*, **63**, 1. (Ch. 11, Ref. 22)

Hewitt, A. & Burbidge, G. 1987, *Ap. J. S.*, **69**, 1. (Ch. 11, Ref. 3)

Hewitt, A. & Burbidge, G. 1993, *Ap. J. S.*, **87**, 451. (Ch. 11, Ref. 4; Ch. 19, Ref. 33; Ch. 23, Ref. 39)

Hey, J.S., Parsons, S.J. & Phillips, J.W. 1946, *Nature*, **158**, 234. (Ch. 5, Ref. 5)

Holmes, A. & Lawson, R. 1927, *Amer. J. Sci.*, **13**, 327. (Ch. 4, Ref. 32)

Hoyle, F. 1946, *M.N.R.A.S.*, **106**, 343. (Ch. 5, Ref. 1; Ch. 8, Ref. 7; Ch. 9, Ref. 1)

Hoyle, F. 1948, *M.N.R.A.S.*, **108**, 372. (Ch. 7, Ref. 2; Ch. 13, Ref. 1)

Hoyle, F. 1949, *Problems of Cosmical Aerodynamics* (eds J. Burgers & H.C. van de Hulst, Central Air Documents Office, Dayton Ohio). (Ch. 23, Ref. 1)

Hoyle, F. 1954, *Ap. J. S.*, **1**, 121. (Ch. 5, Ref. 2)

Hoyle, F. 1959, *M.N.R.A.S.*, **119**, 124. (Ch. 4, Ref. 36)

Hoyle, F. 1969, *Proc. R. Soc. Lond.* A, **308**, 1. (Ch. 7, Ref. 8)

Hoyle, F. 1987, in *Highlights in Gravitation and Cosmology*, Proc. International Conference on Gravitation and Cosmology at Goa, India (eds. B.R. Iver, A. Kembavi, J.V. Narlikar, Cambridge), p. 236. (Ch. 8, Ref. 20)

Hoyle, F. 1992, *Astroph. & Space Sci.*, **198**, 177. (Ch. 10, Ref. 2)

Hoyle, F. & Burbidge, G.R. 1966, *Nature*, **210**, 1346. (Ch. 11, Ref. 2)

Hoyle, F. & Fowler, W.A. 1960, *Ann. Phys.*, **10**, 280. (Ch. 4, Ref. 38)

Hoyle, F. & Narlikar, J.V. 1962, *M.N.R.A.S.*, **123**, 133. (Ch. 7, Ref. 15)

Hoyle, F. & Narlikar, J.V. 1963, *M.N.R.A.S.*, **125**, 13. (Ch. 7, Ref. 16)

Hoyle, F. & Narlikar, J.V. 1974 *Action at a Distance in Physics and Cosmology* (San Francisco, W.H. Freeman). (Ch. 17, Ref. 1)

Hoyle, F. & Schwarzschild, M. 1955, *Ap. J. S.*, **2**, 1; Schwarzschild, M. 1958, *Structure and Evolution of the Stars* (Princeton University Press). (Ch. 4, Ref. 35)

Hoyle, F. & Tayler, R. 1964, *Nature*, **203**, 1108. (Ch. 9, Ref. 8)

Hoyle, F. & Wickramasinghe, C. 1988, *Astrophys. & Space Sci.*, **147**, 248. (Ch. 8, Ref. 18)

Hoyle, F., Burbidge, G. & Narlikar, J.V. 1993, *Ap. J.*, **410**, 437. (Ch. 16, Ref. 1)

Hoyle, F., Burbidge, G. & Narlikar, J.V. 1994, *A&A*, **289**, 729. (Ch. 16, Ref. 3)

Hoyle, F., Burbidge, G. & Narlikar, J.V. 1994, *M.N.R.A.S.*, **267**, 1007. (Ch. 16, Ref. 2)

Hoyle, F., Burbidge, G. & Sargent, W.L.W. 1966, *Nature*, **209**, 751. (Ch. 11, Ref. 7)

Hubble, E. 1926, *Ap. J.*, **63**, 236. (Ch. 3, Ref. 9)

Hubble, E. 1929, *Proc. Natl. Acad. Sci.*, **15**, 168. (Ch. 3, Ref. 17)

Hubble, E. 1934, *Ap. J.*, **79**, 8. (Ch. 20, Ref. 5)

Hubble, E. 1936, *The Realm of the Nebulae* (Oxford University Press). (Ch. 4, Ref. 1)

Hubble, E. 1936, *Ap. J.*, **84**, 517. (Ch. 20, Ref. 6)

Hubble, E. & Humason, M.L. 1931, *Ap. J.*, **74**, 43. (Ch. 3, Ref. 19; Ch. 20, Ref. 2)

Huchra, J., Davis, M., Latham, D. & Tonry, J. 1983, *Ap. J. S.*, **52**, 89. (Ch. 20, Ref. 15)

Huchra, J., Geller, M. & Corwin, H. 1995, *Ap. J. S.*, **99**, 391. (Ch. 20, Ref. 17)

Huchra, J., Geller, M., de L'Apparent, B. & Corwin, H. 1990, *Ap. J. S.*, **72**, 433. (Ch. 20, Ref. 16)

Huchra, J., Gorenstein, M., Kent, S., Shapiro, I., Smith, G., Horine, E. & Perley, R. 1985, *A. J.*, **90**, 691. (Ch. 11, Ref. 59; Ch. 20, Ref. 53)

Humason, M.L. 1929, *Proc. Natl. Acad. Sci.*, **15**, 167. (Ch. 3, Ref. 18)

Humason, M.L. 1931, *Ap. J.*, **74**, 35. (Ch. 3, Ref. 3)

Humason, M.L. 1932, quoted by Curtis, H.D. 1933, *The Nebulae* (Berlin, Springer-Verlag), Chap. 6, Table 14. (Ch. 3, Ref. 4)

Humason, M.L. 1936, *Ap. J.*, **83**, 10. (Ch. 4, Ref. 3)

Humason, M.L. & Wahlquist, H.D. 1955, *A. J.*, **60**, 254. (Ch. 4, Ref. 8)

Humason, M.L., Mayall, N.U. & Sandage, A.R. 1956, *A. J.*, **61**, 97. (Ch. 4, Ref. 4; Ch. 20, Ref. 12)

Jacoby, G.H. *et al.* 1992, *Pub. Astron. Soc. Pacific*, **104**, 599. (Ch. 4, Ref. 22)

Karlsson, K.G. 1977, *Astron. & Ap.*, **58**, 237. (Ch. 23, Ref. 36)

Kazanas, D. 1980, *Ap. J.*, **241**, L59. (Ch. 14, Ref. 2)

Kellermann, K. & Wall, J.V. 1987, *IAU Symposium No. 124* (eds. A. Hewitt, G. Burbidge, & L.-Z. Fang, Dordrecht, D. Reidel), p. 545. (Ch. 7, Ref. 19)

Kennicutt, R.C., Freedman, W.L. & Mould, J.R. 1995, *A. J.*, **110**, 1476. (Ch. 4, Ref. 26)

Khembavi, A.K. 1978, *M.N.R.A.S.*, **185**, 807. (Ch. 17, Ref. 3)

Kilkenny, D., O'Donoghue, D., Koen, C., Stobie, R.S. & Chen, A. 1997, *M.N.R.A.S.*, **287**, 867. (Ch. 11, Ref. 82)

Kinman, T. 1965, *Ap. J.*, **142**, 1241. (Ch. 5, Ref. 29)

Kippenhahn, R. & de Vries, H.L. 1974, *Astroph. Space Sci.*, **26**, 131. (Ch. 11, Ref. 14)

Kipper, A. 1931, *Astron. Nacht.*, **241**, 249. (Ch. 3, Ref. 12)

Koo, D.C. & Kron, R.G. 1992, *ARAA*, **30**, 613. (Ch. 19, Ref. 12)

Kormendy, J. & Richstone, D. 1995, *ARA&A*, **33**, 581. (Ch. 22, Ref. 5)

Kraichnan, R.H. 1965, *Phys. Fluids*, **8**, 1385. (Ch. 23, Ref. 7)

Kristian, J. 1973, *Ap. J. L.*, **179**, L11. (Ch. 11, Ref. 48)

Kristian, J., Sandage, A. & Westphal, J.A. 1978, *Ap. J.*, **221**, 383. (Ch. 6, Ref. 3)

Kronberg, P. 1994, *Rep. Prog. Physics*, **57**, 325. (Ch. 23, Ref. 16)

Kulsrud, R. & Anderson, S. 1992, *Ap. J.*, **396**, 606. (Ch. 23, Ref. 13)

Kundic, T. *et al.* 1997, *Ap. J.*, **482**, 75. (Ch. 11, Ref. 62)

Lanzetta, K.M., Wolfe, A.M., Turnshek, D.A. 1995, *Ap. J.*, **440**, 435. (Ch. 11, Ref. 73)

Le Brun, V., Bergeron, J., Boissé, P. & Christian, C. 1993, *A&A*, **279**, 23. (Ch. 11, Ref. 68)

Lemaitre, G. 1927, *Ann. de la Societe Scientifique de Bruxelles*, **47**, 49. (Ch. 2, Ref. 7)

Letter quoted by Woodruff T. Sullivan, III, 1988, in *Modern Cosmology in Retrospect* (eds. R. Balbinot *et al.*, Cambridge University Press). (Ch. 7, Ref. 13)

Lilly, S.J. & Longair, M.S. 1984, *M.N.R.A.S.*, **211**, 833. (Ch. 6, Ref. 12)

Longair, M.S. 1993, *The Physics of Background Radiation*, from The Deep Universe, Saas-Fee Advanced Course 23, (eds. A.R. Sandage, R.G. Kron & M.S. Longair), p. 323 (Swiss Society for Astrophysics and Astronomy, eds. B. Binggeli & R. Buser; Springer, Berlin 1995). (Ch. 21, Ref. 1)

Lovell, B. 1991, *Echoes of War, the Story of H₂S Radar* (Bristol, Institute of Physics Publishing Co. Ltd., Adam Hilger). (Ch. 5, Ref. 14)

Lu, L., Sargent, W.L.W. & Barlow, T. 1997, *Ap. J.*, **484**, 131. (Ch. 19, Ref. 31)

Lundmark, K. 1924, *M.N.R.A.S.*, **84**, 747. (Ch. 3, Ref. 7; Ch. 3, Ref. 15)

Lynds, R. 1971, *Ap. J. L.*, **164**, L73. (Ch. 11, Ref. 65)

Lynds, R. & Millikan, A.G. 1972, *Ap. J.*, **176**, L5. (Ch. 11, Ref. 31)

Lynds, R. & Petrosian, V. 1986, *BAAS*, **18**, 1014; 1989, *Ap. J.*, **336**, 1. (Ch. 20, Ref. 54)

Madau, P., Ferguson, H.C., Dickinson, M.E., Giavalisco, M., Steidel, C.C. & Fruchter, A. 1996, *M.N.R.A.S.*, **283**, 1388. (Ch. 19, Ref. 27)

Maddox, S.J., Sutherland, W.J., Efstathiou, G., Loveday, J. & Peterson, B.A. 1990, *M.N.R.A.S.*, **247**, 1P. (Ch. 19, Ref. 13)

Mather, J.C. *et al.* 1990, *Ap. J.*, **354**, L37. (Ch. 8, Ref. 23)

Mather, J.C. *et al.* 1994, *Ap. J.*, **420**, 439. (Ch. 8, Ref. 24)

Matthews, T.A. & Sandage, A.R. 1963, *Ap. J.*, **138**, 30. (Ch. 5, Ref. 23)

Matthews, T.A., Morgan, W.W. & Schmidt, M. 1964, *Ap. J.*, **140**, 35. (Ch. 5, Ref. 17)

Mattig, W. 1958, *Astron. Nacht.*, **284**, 109. (Ch. 4, Ref. 2)

Mayall, N.U. 1935, *Pub. Astron. Soc. Pacific*, **47**, 319. (Ch. 4, Ref. 5)

McCarthy, P. 1993, *Ann. Rev. A.&A.*, **31**, 639. (Ch. 6, Ref. 14)

McCarthy, P.J., Spinrad, H. & van Breugel, W. 1995, *Ap. J. S.*, **99**, 27. (Ch. 6, Ref. 11)

McKellar, A. 1940, *Pub. Astr. Soc. Pacific*, **52**, 307. (Ch. 8, Ref. 2)

McKellar, A. 1940, *Pub. Astr. Soc. Pacific*, **52**, 407. (Ch. 9, Ref. 14)

McKellar, A. 1941, *Pub. Dom. Astrophys. Observatory, Victoria, B.C.*, **7**, 251. (Ch. 8, Ref. 4; Ch. 22, Ref. 1)

Milgrom, M. 1983, *Ap. J.*, **270**, 365 and 371. (Ch. 20, Ref. 39)

Milgrom, M. 1994, *Ap. J.*, **429**, 540. (Ch. 20, Ref. 40)

Miller, J. 1998, private communication. (Ch. 11, Ref. 55)

Mills, B.Y. 1953, *Austr. J. Phys.*, **6**, 452. (Ch. 5, Ref. 12)

Mills, B.Y. & Slee, O.B. 1957, *Austr. J. Phys.*, **10**, 162. (Ch. 7, Ref. 12)

Mineur, H. 1952, *C.R. Acad. Paris*, **220**, 495. (Ch. 4, Ref. 12)

Minkowski, R. 1960, *Ap. J.*, **132**, 908; *Pub. Astron. Soc. Pacific*, **72**, 354. (Ch. 5, Ref. 20)

Minkowski, R. 1962, *I.A.U. Symp. No. 15* (New York, MacMillan), p. 379. (Ch. 5, Ref. 21)

Morris, S.L., Weymann, R.J., Savage, B.D. & Gilliland, B.D. 1993, *Ap. J.*, **419**, 524. (Ch. 11, Ref. 80)

Nabarrow, F.R.N. & Jackson, P.J. 1958, in *Growth and Perfection in Crystals* (eds. R.H. Doremus, P.W. Roberts & D. Turnbull, New York, J. Wiley), 65, 388. Gomez, R.J. 1957, *Chem. Phys.* **26**, 1333; 1958, **28**, 457. Dittmar, W. & Neumann, K. 1958, in *Growth and Perfection in Crystals* (eds. R. H. Doremus, P.W. Roberts & D. Turnbull, New York, J. Wiley). (Ch. 8, Ref. 17)

Narlikar, J.V. 1973, *Nature*, **242**, 35. (Ch. 15, Ref. 1)

Narlikar, J.V. 1993, *Introduction to Cosmology*, 2nd edn. (Cambridge University Press) (Ch. 2, Ref. 2)

Narlikar, J.V., Edmunds, M. & Wickramasinghe, C. 1976, in *Far Infrared Astronomy*, Proc. 1975 Conf. of Roy. Astron. Soc. (ed. M. Rowan Robinson, Pergamon), p. 131. (Ch. 8, Ref. 16; Ch. 16, Ref. 4)

Nayeri, A., Engineer, S., Narlikar, J.V. and Hoyle, F. 1999, *Ap. J.*, November 1. (Ch. 20, Ref. 61)

Oke, J.B. 1963, *Nature*, **197**, 1040. (Ch. 5, Ref. 25)

Oke, J.B. & Sandage, A. R. 1968, *Ap. J.*, **154**, 21. (Ch. 7, Ref. 7)

Oort, J.H. 1932, *Bull. Astr. Neder.*, **6**, 249. (Ch. 20, Ref. 29)

Oort, J.H. 1960, *Bull. Astr. Neder.*, **16**, 45. (Ch. 20, Ref. 30)

Opik, E. 1922, *Ap. J.*, **55**, 406. (Ch. 20, Ref. 34)

Osterbrock, D.E. 1983, *PASP*, **95**, 12. (Ch. 19, Ref. 3)

Ostriker, J. 1989, in *BL Lac Objects: Lecture Notes* (ed. L. Maraschi, Berlin, Springer Verlag), Vol. 334. (Ch. 11, Ref. 36)

Ostriker, J., Yahil, A. & Peebles, P.J.E. 1974, *Ap. J. L.*, **193**, L1. (Ch. 20, Ref. 48)

Pacynski, B. 1986, *Ap. J.*, **304**, 1. (Ch. 20, Ref. 56)

Page, Lyman, 1997, *Generation of Large-Scale Structure* (eds. D. Schramm & P. Galliotti; A. Wolfer), p. 75 (Ch. 21, Ref. 8)

Palumbo, G., Tanzelli-Nitti, G. & Vettolani, G. 1983, *Catalogue of Radial Velocities of Galaxies* (Gordon and Breach, New York). (Ch. 20, Ref. 14)

Parker, E.N. 1955, *Ap. J.*, **122**, 293. (Ch. 23, Ref. 9)

Parker, E.N. 1973, *Astrophy. & Space Science*, **22**, 279. (Ch. 23, Ref. 11)

Pease, F.G. 1915, *Publ. Astron. Soc. Pacific*, **27**, 133. (Ch. 3, Ref. 2)

Peebles, P.J.E. 1966, *Ap. J.*, **146**, 542; *Phys. Rev. Lett.*, **10**, 410. (Ch. 9, Ref. 7)

Peebles, P.J.E. 1969, *Ap. J.*, **155**, 393. (Ch. 23, Ref. 2)

Peebles, P.J.E. 1993, *Principles of Physical Cosmology* (Princeton University Press, Princeton, N.J.). (Ch. 20, Ref. 23)

Peebles, P.J.E., Schramm, D., Turner, E. & Kron, R. 1991, *Nature*, **352**, 769. (Ch. 11, Ref. 45)

Peebles, P.J.E., Seldner, M. & Soneira, R.M. 1977, *Scientific American*, p. 76; Seldner, M., Siebers, B., Groth, E.J. & Peebles, P.J.E. 1977, *A. J.*, **82**, 249. (Ch. 20, Ref. 8)

Penzias, A.A. & Wilson, R.W. 1965, *Ap. J.*, **142**, 419. (Ch. 7, Ref. 20; Ch. 8, Ref. 15)

Perlmutter, S. *et al.* 1999, *Ap. J.*, **517**, 565. (Ch. 16, Ref. 5)

Peterson, B. 1997, *An Introduction to "Active Galactic Nuclei"* (Cambridge University Press). (Ch. 11, Ref. 46)

Pettini, M., Smith, L., King, D. & Hunstead, R. 1997, *Ap. J.*, **486**, 665. (Ch. 19, Ref. 32)

Pettini, M., Steidel, C.C., Dickinson, M., Kellogg, M., Giavalisco, M. & Adelberger, K. 1997, *The Universe at Low and High Redshifts* (Woodbury, AIP Press), in press. (Ch. 19, Ref. 24)

Pettit, E. 1954, *Ap. J.*, **120**, 413. (Ch. 4, Ref. 6)

Piddington, J.H. 1970, *Australian J. Physics*, **23**, 731. (Ch. 23, Ref. 10)

Pietsch, W., Vogler, A., Kahabka, P., Jain, A. & Klein, V. 1994, *A&A*, **284**, 386. (Ch. 11, Ref. 41)

Prochaska, J. & Wolfe, A.M. 1997, *Ap. J.*, **487**, 73. (Ch. 19, Ref. 30)

Radecke, H.D. 1997, *A&A*, **319**, 18. (Ch. 11, Ref. 39)

Rees, M. 1967, *M.N.R.A.S.*, **135**, 345. (Ch. 11, Ref. 9)

Rees, M.J. 1984, *ARA&A*, **22**, 471; Blandford, R. & Rees, M.J. 1974, *M.N.R.A.S.*, **169**, 395. (Ch. 11, Ref. 6; Ch. 12, Ref. 8)

Reeves, H., Fowler, W.A. & Hoyle, F. 1970, *Nature*, **226**, 727. (Ch. 9, Ref. 13)

Reuter, H.P., Klein, V., Lesch, H., Weilibinski, R. & Kronberg, P. 1994, *Astron. and Ap.*, **256**, 10. (Ch. 23, Ref. 15)

Richards, G., York, D., Yanny, B., Kollgaard, R., Laurent-Muehleisen, S. & Vander Berk, D. 1999, *Ap. J.*, **513**, 576. (Ch. 11, Ref. 71)

Richstone, D. & Oke, J.B. 1977, *Ap. J.*, **213**, 8. (Ch. 11, Ref. 53)

Riess, A. *et al.* 1998, *A. J.*, **116**, 1009. (Ch. 16, Ref. 6)

Riess, A.G., Press, W.H. & Kirshner, R.P. 1995, *Ap. J.*, **438**, L17. (Ch. 4, Ref. 30)

Robertson, H.P. 1928, *Phil. Mag.*, **5**, 835. (Ch. 2, Ref. 8)

Robertson, H.P. 1935, *Ap. J.*, **82**, 248. (Ch. 2, Ref. 11)

Rubin, V., Burnstein, D., Ford, W.K. & Thonnard, N. 1985, *Ap. J.*, **289**, 81. (Ch. 20, Ref. 38)

Rutherford, E. 1929, *Nature*, **123**, 314. (Ch. 4, Ref. 33)

Ryle, M. 1955, *Observatory*, **75**, 137. (Ch. 7, Ref. 9)

Ryle, M. & Smith, F.G. 1948, *Nature*, **162**, 462. (Ch. 5, Ref. 7)

Sachs, R., Narkilar, J. & Hoyle, F. 1996, *A&A*, **313**, 703. (Ch. 16, Ref. 9)

Sandage, A. 1958, *Ap. J.*, **127**, 513. (Ch. 4, Ref. 15)

Sandage, A. 1961, *Sky & Telescope*, **21**, 148. (Ch. 5, Ref. 22)

Sandage, A. 1962, in *Problems of Extragalactic Research*, Proc. IAU Symp. 15 (Kluwer, Dordrecht), p. 359. (Ch. 4, Ref. 16)

Sandage, A. 1972, *Ap. J.*, **178**, 1. (Ch. 6, Ref. 16)

Sandage, A. 1995, in *The Deep Universe* (Saas-Fee Advanced Course 23, eds. A.R. Sandage, R.G. Kron, M.S. Longair, Berlin, Heidelberg, Springer), p. 77, 117, Fig 4.2. (Ch. 4, Ref. 19; Ch. 6, Ref. 6)

Sandage, A. & Perelmutter, J.-M., 1990, *Ap. J.*, **350**, 481. (Ch. 3, Ref. 21)

Sandage, A. & Tammann 1974a, *Ap. J.*, **190**, 525. (Ch. 4, Ref. 17)

Sandage, A. & Tammann 1974b, *Ap. J.*, **194**, 559. (Ch. 4, Ref. 18)

Sandage, A. & Tammann, G. 1990, *Ap. J.*, **365**, 1. (Ch. 4, Ref. 25)

Sandage, A. & Tammann, G.A. 1983, in *Large Scale Structure of the Universe, Cosmology, and Fundamental Physics* (eds. S. Setti & L. Van Hove, Geneva, ESO/CERN), p. 127. (Ch. 6, Ref. 5)

Sandage, A. & Tammann, G.A. 1993, *Ap. J.*, **415**, 1. (Ch. 4, Ref. 28)

Sandage, A. & Tammann, G.A. 1998, *M.N.R.A.S.*, **293**, L23 (Ch. 4, Ref. 42)

Sandage, A., Saha, A., Tammann, G.A., Labhardt, L., Panagia, N. & Macchetto, F.D. 1996, *Ap. J.*, **460**, L15; *Ap. J.*, **466**, 55; *Ap. J. S.*, **107**, 693. (Ch. 4, Ref. 31)

Sandage, A.R. 1965, *Ap. J..*, **141**, 1560. (Ch. 5, Ref. 28)

Sanders, R. 1996, *Ap. J.*, **473**, 117. (Ch. 20, Ref. 42)

Sargent, W. & Boroson, T. 1977, *Ap. J.*, **212**, 382. (Ch. 11, Ref. 75)

Sargent, W.L.W., Young, P., Boksenberg, A. & Tytler, D. 1980, *Ap. J. S.*, **42**, 41. (Ch. 11, Ref. 66)

Sato, K. 1981, *M.N.R.A.S.*, **195**, 467. (Ch. 14, Ref. 3)

Schmidt, M. 1963, *Nature*, **197**, 1040. (Ch. 5, Ref. 25)

Schmidt, M. 1965, *Ap. J.*, **141**, 1. (Ch. 5, Ref. 18)

Schmidt, M. 1965, *Ap. J.*, **141**, 1295. (Ch. 5, Ref. 27)

Schneider, D.P., Gunn, J.E. & Hoessel, J.G. 1983, *Ap. J.*, **264**, 337. (Ch. 6, Ref. 7)

Seyfert, C.K. 1943, *Ap. J.*, **97**, 28. (Ch. 11, Ref. 47)

Shain, C.A. 1959, *Proc. IAU Symposium No. 9* (ed. R.N. Bracewell, Stanford University Press), p. 451. (Ch. 5, Ref. 13)

Shakeshaft, J., Ryle, M., Baldwin, J., Elsmore, B. & Thomson, D. 1955, *Mem. R.A.S.*, **67**, 97. (Ch. 7, Ref. 11)

Shane, C.D. & Wirtanen, C.A. 1967, *Pub. Lick Obs.*, Vol. 22, part I. (Ch. 20, Ref. 7)

Shapley, H. 1918, *Ap. J.*, **48**, 89. (Ch. 4, Ref. 9)

Shapley, H. 1934, *Harvard Annuals*, p. 761. (Ch. 20, Ref. 3)

Shapley, H. 1961, in *Galaxies* (Cambridge, Harvard University Press), p. 159. (Ch. 20, Ref. 4)

Shapley, H. & Curtis, H.D. 1921, *Bull. Natl. Research Council*, Vol. **2**, part 3, No. 11. (Ch. 3, Ref. 6)

Slipher, V.M. 1914, *Lowell Observatory Bulletin*, **2**, p. 65. (Ch. 20, Ref. 33)

Slipher, V.M. 1914, *Lowell Bull.*, **2**, 56; 1917, *Astron. Nacht.*, **213**, 47; 1917, *Proc. Amer. Phil. Soc.*, No. 5. (Ch. 3, Ref. 1)

Smith, F.G. 1951, *Nature*, **168**, 555. (Ch. 5, Ref. 10)

Smith, S. 1936, *Ap. J.*, **83**, 23. (Ch. 20, Ref. 46)

Smoot, G. *et al.* 1991, *Ap. J.*, **371**, L1. (Ch. 21, Ref. 5)

Smoot, G. *et al.* 1992, *Ap. J.*, **396**, L1. (Ch. 8, Ref. 22; Ch. 21, Ref. 7)

Soifer, B.T., Houck, J. & Neugebauer, G. 1987, *ARAA*, **25**, 187. (Ch. 19, Ref. 1)

Soucail, G., Fort, B., Mellier, Y. & Picat, J.P. 1987, *A&A*, **172**, L14. (Ch. 20, Ref. 55)

Spinrad, H., Djorgovski, S., Marr, J. & Aguilar, L. 1985, *Pub. Astron. Soc. Pacific*, **97**, 932. (Ch. 7, Ref. 17)

Spite, M. & Spite, F. 1985, *Ann. Rev. A&A*, **23**, 225. (Ch. 9, Ref. 10)

Stebbins, J. & Whitford, A.E. 1948, *Ap. J.*, **108**, 413. (Ch. 7, Ref. 3)

Stebbins, J. & Whitford, A.E. 1952, *Ap. J.*, **115**, 284. (Ch. 4, Ref. 7)

Steidel, C.C. & Dickinson, M. 1992, *Ap. J.*, **394**, 81. (Ch. 11, Ref. 83)

Steidel, C.C., Pettini, M. & Hamilton, D. 1995, *A. J.*, **110**, 2519. (Ch. 19, Ref. 21)

Stockton, A. 1978, *Ap. J.*, **223**, 747. (Ch. 11, Ref. 56)

Sulentic, J. & Arp, H.C. 1987, *Ap. J.*, **319**, 687. (Ch. 11, Ref. 30)

Sulentic, J.W. & Tifft, W.G. 1973, *The Revised New General Catalog of Nonstellar Astronomical Objects* (The University of Arizona Press, Tucson, Arizona). (Ch. 11, Ref. 23)

Sullivan, W.T. 1988, in *Modern Cosmology in Retrospect* (eds. R. Balbinot *et al.*, Cambridge U. Press). (Ch. 5, Ref. 9)

Swings, P. & Rosenfeld, L. 1937, *Ap. J.*, **86**, 483. (Ch. 8, Ref. 1)

Thompson, D.J. *et al.* 1995, *Ap. J. S.*, **101**, 259; 1996, *Ap. J. S.*, **107**, 227. (Ch. 19, Ref. 11)

Thorburn, J. A. 1994, *A. J.*, **421**, 318. (Ch. 9, Ref. 11)

Tifft, W. 1976, *Ap. J.*, **206**, 38. (Ch. 23, Ref. 18)

Tifft, W. 1995, *Astrophys. & Space Science*, **227**, 25. (Ch. 23, Ref. 21)

Tifft, W. 1996, *Ap. J.*, **468**, 491. (Ch. 23, Ref. 22)

Tifft, W. & Cocke, W.J. 1989, *Ap. J.*, **336**, 128. (Ch. 23, Ref. 19)

Tinsley, B.M. & Gunn, J.E. 1976, *Ap. J.*, **203**, 52. (Ch. 6, Ref. 4)

Tolman, R.C. 1934, *Relativity, Thermodynamics and Cosmology* (Oxford, Clarendon Press), p. 455. (Ch. 3, Ref. 22)

Toomre, A. & Toomre, J. 1972, *Ap. J.*, **178**, 623. (Ch. 12, Ref. 1)

Turner, M.S. & Widrow, L.M. 1988, *Phys. Rev. D.*, **37**, 2743. (Ch. 23, Ref. 5)

Tytler, D., Fan, X.-M. & Burles, S. 1996, *Nature*, **381**, 207. (Ch. 9, Ref. 17; Ch. 10, Ref. 4)

Udalski, A., Szymanski, M., Stanek, K.Z. *et al.* 1994, *Acta Astronomia*, **44**, 165. (Ch. 20, Ref. 59)

Vainshtein, S. & Cattaneo, F. 1992, *Ap. J.*, **393**, 165. (Ch. 23, Ref. 12)

van Albada, T., Bahcall, J., Bergeman, K. & Sancisi, R. 1985, *Ap. J.*, **295**, 305. (Ch. 20, Ref. 37)

van den Bergh, S. 1992, *Pub. Astron. Soc. Pacific*, **104**, 861. (Ch. 4, Ref. 21)

Vangioni-Flam, E., Cassé, M., Fields, B. & Olive, K. 1996, *Ap. J.*, **468**, 199. (Ch. 9, Ref. 16)

van Maanen, A. 1916–30, *Mt. Wilson Contr. Nos.* 111, 136, 158, 182, 204, 237, 270, 290, 321, 356, 391, 405–8. (Ch. 3, Ref. 13)

Wade, C. M. 1960, *Observatory*, **30**, 235. (Ch. 11, Ref. 37)

Wagoner, R.V., Fowler, W.A. & Hoyle, F. 1967, *Ap. J.*, **148**, 3. (Ch. 8, Ref. 8; Ch. 9, Ref. 9)

Walker, A.G. 1936, *Proc. London Math. Soc.*, **42**, 90. (Ch. 2, Ref. 12)

Walsh, D., Carswell, R. & Weymann, R. 1979, *Nature*, **279**, 381. (Ch. 11, Ref. 57; Ch. 20, Ref. 52)

Wampler, E.J., Chugai, N.N. & Petitjean, P. 1995, *Ap. J.*, **443**, 586. (Ch. 11, Ref. 76)

Wampler, E.J., Robinson, L., Burbidge, E.M. & Baldwin, J. 1975, *Ap. J. L.*, **198**, 49. (Ch. 11, Ref. 51)

Webster, A. 1982, *M.N.R.A.S.*, **200**, 47. (Ch. 11, Ref. 19)

Weedman, D. 1970, *Ap. J. L.*, **161**, L113. (Ch. 11, Ref. 29)

White, G.E. & Woods, S.B. 1959, *Phil. Trans. R. Soc. Lond.* A**251**, 273. (Ch. 8, Ref. 21)

White, M., Scott, D. & Silk, J. 1994, *ARA&A*, **32**, 319. (Ch. 21, Ref. 9)

Whitford, A.E. 1952, *Ap. J.*, **120**, 599. (Ch. 7, Ref. 6)

Williams, R.E. (& 14 co-authors) 1996, *A. J.*, **112**, 1335. (Ch. 19, Ref. 20)

Wilson, M.L. & Silk, J., 1981, *Ap. J.*, **243**, 14. (Ch. 21, Ref. 6)

Wilson, R.E. 1923, *A. J.*, **35**, 35; *Ap. J.*, **89**, 218. (Ch. 4, Ref. 10)

Wirtz, C. 1924, *Astron. Nacht.*, **222**, 21. (Ch. 3, Ref. 14)

Wolfe, A.M., Lanzetta, K.M. & Bowen, D.V. 1992, *Ap. J.*, **391**, 48. (Ch. 19, Ref. 29)

Wolfe, A.M., Turnshek, D., Lanzetta, K.M. & Lu, L. 1993, *Ap. J.*, **404**, 480. (Ch. 11, Ref. 78)

Wolfe, A.M., Turnshek, D., Smith, H.E. & Cohen, R.D. 1986, *Ap. J. S.*, **61**, 249. (Ch. 11, Ref. 64; Ch. 19, Ref. 28)

Woltjer, L. 1966, *Ap. J.*, **146**, 597. (Ch. 11, Ref. 8)

Wright, E.L. 1982, *Ap. J.*, **255**, 401. (Ch. 8, Ref. 19)

Zeldovich, Ya.B. & Novikov, I.D. 1982, *Structure and Evolution of the Universe* (University of Chicago Press). (Ch. 23, Ref. 6)

Zhu, X.-F. & Chu, Y.-Q. 1995, *A&A*, **297**, 300. (Ch. 11, Ref. 28)

Zweibel, E. 1988, *Ap. J. Lett.*, **329**, L1. (Ch. 23, Ref. 3)

Zweibel, E. & McKee, C.F. 1995. *Ap. J.*, **439**, 778. (Ch. 23, Ref. 8)

Zwicky, F. 1933, *Helvetica Phys. Acta*, **6**, 110. (Ch. 20, Ref. 45)

Zwicky, F. 1937, *Phys. Rev.*, **51**, 290, 679. (Ch. 20, Ref. 50)

Zwicky, F. 1938, *PASP*, **50**, 318. (Ch. 20, Ref. 11)

Zwicky, F., Herzog, E., Wild, P., Karpowicz, M. & Kowal, C.T. 1961–68, *Catalog of Galaxies and Clusters of Galaxies*, Vols. 1–6 (Pasadena, California Institute of Technology). (Ch. 20, Ref. 10)

Index